Applied Mathematics

by

R. Jesse Phagan

Woodstock Academy
Woodstock, Connecticut

Publisher
THE GOODHEART-WILLCOX COMPANY, INC.
Tinley Park, Illinois

Copyright 1997

by

THE GOODHEART-WILLCOX COMPANY, INC.

Previous edition copyright 1992

Library of Congress Catalog Card Number 94-30267
International Standard Book Number 1-56637-117-1

1 2 3 4 5 6 7 8 9 10   97   00 99 98 97 96

Library of Congress Cataloging in Publication Data

Phagan, R. Jesse.
    Applied mathematics / by R. Jesse Phagan.
Rev. ed.

    p. cm.
  Includes index.
  ISBN 1-56637-117-1
  1. Mathematics. I. Title.
QA39.2.P55 1997
513'.14--dc20   94-30267
                            CIP

R. Jesse Phagan has an extensive teaching and writing
background in areas of math and electronics. He is the author
of two electronics math books and an electronics book. He
holds degrees in Industrial Arts/Technology Education,
Vocational Education, and Electronics Technology. In addition
to his teaching experience, he has worked in manufacturing
and for the federal government.

Introduction

Mathematics is considered a basic life skill—a skill as important as the ability to read and write. **Applied Mathematics** has been developed as a complete math course. It is particularly relevant to the needs of students in vocational and technical programs. Through the use of realistic applications, an understanding of the importance of math is developed. Skills in problem solving are also developed. This textbook answers the question, "Where will I use this math concept?"

Applied Mathematics contains an abundance of Sample Problems, Practice Problems, and Problem-Solving Activities. With this combination, there should be no difficulty mastering objectives and developing sound math skills. Other features of this textbook include:

- Clear, readable text in a well-organized, easy-to-follow format, allowing students with diverse abilities to understand basic math.
- Coverage of the basics that builds understanding of math principles presented in later chapters.
- Exercises and activities that develop problem-solving abilities and provide examples of applications.
- Principles tied to real-life applications and current technologies.
- Key terms presented in boldface type and defined in context when introduced.

In addition to this textbook, a workbook is available. The workbook consists of Practice Problems and Problem-Solving Activities that follow the format of the text to ensure continuity.

Skill in mathematics greatly simplifies the learning of vocational and technical subject matter. This, in turn, builds enthusiasm for learning and promotes successful achievement in technical areas. Sound math skills can foster confidence, which will contribute to success both in the classroom and on the job, as well as in meeting the challenges of everyday life.

Contents

Part I

Basic Math

Chapter 1

Solving Word Problems with Whole Numbers

OBJECTIVES

After studying this chapter, you will be able to:
- Identify operations needed to solve word problems.
- Set up word problems and perform the required arithmetic.
- By checking, state whether or not answers to word problems are correct.

When mathematics is applied to everyday life, seldom do the problems come simply in number form. Usually math problems come in forms of either worded questions or statements. The first chapter in this textbook of APPLIED MATHEMATICS is logically, then, an explanation in the techniques used to solve word problems.

To best serve the purpose of this chapter, the given problems deal only with *whole numbers*. These are numbers that do not contain decimals or fractions. The numbers in the problems are all positive. No negative numbers are used. All of the problems are **arithmetic operations**. That is, they deal with basic addition, subtraction, multiplication, and division.

PLACE VALUES OF WHOLE NUMBERS

The location of a digit in a number establishes the value of the digit. The location of a digit is called its **place value.** Refer to Fig. 1-1 for the place value of each location in a six-digit number. A digit is multiplied by its place value to obtain its value in the number. The complete number is formed by adding the products obtained by multiplying each digit by its place value.

Example	Place value	Multiply digit by	Value
3 1 4 , 5 6 2			
└──── ones	1	$2 \times 1 = 2$
└────── tens	10	$6 \times 10 = 60$
└──────── hundreds	100	$5 \times 100 = 500$
└────────── thousands	1,000	$4 \times 1,000 = 4,000$
└──────────── ten-thousands	10,000	$1 \times 10,000 = 10,000$
└────────────── hundred-thousands	100,000	$3 \times 100,00 = 300,000$
		Add:	314,562

Fig. 1-1. Place value chart.

Addition and place values

When performing addition, align the numbers so that digits with the same place value are located in the same column. Addition cannot be performed when digits with the same place value are located in different place value columns. When the addition of the numbers in any column results in a number greater than 9, a *carry* is brought to the next higher place value column.

Example: add 475 + 2,120 + 10,021

```
        1  ← carried from 10s column.
      475
    2,120
 + 10,021
   12,616
```

Subtraction and place values

Subtraction requires place value columns to be aligned, as in addition. When subtracting, it is often necessary to *borrow* from the digit in the next higher place value column. The borrow reduces the digit it is borrowed from by 1 and adds 10 to the digit needing the borrow.

Example: subtract 3,517 − 243

```
                  1  ← borrowed from 100s column. The 1 in the 10s column
  3,517     3,417     becomes 11.
 −  243   −   243
            3,274
```

GENERAL PROCEDURES FOR SOLVING WORD PROBLEMS

Every word problem will have different techniques and procedures to follow in order to arrive at the final solution. It is possible, however, to establish a set of basic steps to follow. These steps will apply to every problem encountered. (You may not always see each of these steps included in the presentation of material in chapters that follow, but you can be sure the steps were performed.)

The first step is to identify the operation. Addition, subtraction, multiplication, and division are the operations of arithmetic. These operations will be used to solve the problem. At times, you can determine which operation to perform by first skimming the problem for key words. Of course, it is not enough merely to look for key words. You must think about what the question is asking. Also, problems do not *always* have key words, but the required operation can be determined by giving extra thought to the question. Making a sketch of the problem can sometimes be helpful.

The next step is to perform the arithmetic, always including the units. Each type of problem requires a standard format by which the numbers should be arranged. This is laid out in the sample problems of this chapter. Writing the numbers in the standard format is part of performing the arithmetic operation. Using this format will make the task much easier and will reduce mistakes. Units must be included with the final answer. They help to make it clear if the question asked in the word problem has been answered.

The final step is to check the answer. The answer should be checked in reference to the problem to see if it seems correct. This usually requires some knowledge of the subject of the problem, which comes with experience. Also, you can quickly check your arithmetic by using approximate figures.

> ## RULES FOR SOLVING WORD PROBLEMS
> 1. Identify the operation.
> 2. Perform the arithmetic. Include the units.
> 3. Check the answer.

WORD PROBLEMS USING ADDITION

Although it may seem obvious, keep in mind there will be an increase in a quantity whenever addition of positive numbers is involved. Use the following key words to help identify word problems of addition.

- **Sum** — the result of adding together two or more numbers.
  Example: The *sum* of the numbers 1, 2, and 3 is 6. (1 + 2 + 3 = 6)
- **Total** — suggests the complete amount or the result of adding amounts together.
  Example: A crew of 2 welders, 2 pipefitters, and 1 laborer makes a *total* of 5 workers. (2 + 2 + 1 = 5)
- **In addition to** — says that something has been added.
  Example: A quantity of 5, *in addition to* an existing quantity of 2, makes a quantity of 7. (5 + 2 = 7)
- **Plus** — similar to the word "bonus," meaning an additional amount is added.
  Example: Most students know 2 *plus* 3 is 5. (2 + 3 = 5)
- **Increase** — signifies that the original amount has been made larger.
  Example: An *increase* of 5 students to the class of 20 brought the count to 25. (20 + 5 = 25)
- **More than** — shows a larger number than the original amount.
  Example: If there were 3 before, and now there are 12 *more than* that, then there are now 15. (3 + 12 = 15)
- **And** — states that two numbers are combined.
  Example: Most students know 2 *and* 2 are 4. (2 + 2 = 4)

Sample Problem 1-1.

A supply of lumber costs a customer $15, and nails $9. What is the sum of the purchase?

Step 1. Identify the operation.
The key word is *sum*, which means addition.

Step 2. Perform the arithmetic. Include the units. In addition, the best format is for the numbers to be in columns, one above the other.

$$\begin{array}{r} 15 \\ +\ 9 \\ \hline \$24 \end{array}$$

Step 3. Check the answer.
The figure of $24 seems to be the correct answer because it is higher than both individual numbers. Double-check the arithmetic.

Sample Problem 1-2.

When the lights in an automobile are on, the two headlights each use 120 watts of power. Each of the two taillights uses 9 watts of power. What is the total power used for the lights?

Step 1. Identify the operation.
The key word is *total*, which means addition.

Step 2. Perform the arithmetic. Include the units.

$$\begin{array}{r} 120 \\ 120 \\ 9 \\ +\ 9 \\ \hline 258\ W \end{array}$$

Step 3. Check the answer.

It is sometimes helpful to estimate the answer using numbers that are easy to add. In this case, each 9 can be changed to a 10. Doing so, the approximate answer would be 260. Notice that the numbers appearing in the problem are used twice. They are doubled because the problem states the wattage for "each" light.

Sample Problem 1-3.

A carpenter laid 375 roofing shingles one day, 520 the next day, and 460 the third day to finish the job. How many shingles were used for the job, not including waste?

Step 1. Identify the operation.

The problem does not have a specific key word. However, it should make sense that each day, more shingles were *added* to the roof.

Step 2. Perform the arithmetic. Include the units.

$$\begin{array}{r} 375 \\ 520 \\ +\ 460 \\ \hline 1355 \ \textit{shingles} \end{array}$$

Step 3. Check the answer.

The easiest way to check this answer is by using numbers that are easy to add. Round the numbers to the nearest multiple of 50 as follows:

$$375 \rightarrow 400$$
$$520 \rightarrow 500$$
$$460 \rightarrow 450$$

Now, add the numbers together.

$$\begin{array}{r} 400 \\ 500 \\ +\ 450 \\ \hline 1350 \ \text{(estimated)} \end{array}$$

The estimated figure is approximately equal to the actual answer.

Practice Problems

Complete the following problems on a separate sheet of paper. Include units of measure (inches, feet, minutes, dollars, etc.).

1. During a snowstorm, a snow plow operator cleared roads with lengths of 3 miles, 5 miles, 1 mile, 2 miles, and 6 miles. How many miles of road were cleared by this operator?

2. For a particular construction job, the contractor totaled his costs as follows: $42 for 2 × 4s, $122 for drywall, $38 for paint, $95 for miscellaneous supplies, and $680 for labor. What were the contractor's total costs?

3. In a given week, an automotive technician worked 12 hours on Monday, 8 hours on Tuesday, 9 hours on Wednesday, 6 hours on Thursday, 11 hours on Friday, and 4 hours on Saturday. How many total hours did the technician work during the week?

4. To determine the needs for a circuit, an electrician estimated the current requirements for the following loads: computer = 5 amps, television = 4 amps, light bulb = 1 amp, and stereo = 6 amps. If all these loads were operated at the same time on one circuit, what is the total estimated current?

5. What is the full length of the wall illustrated in Fig. 1-2?

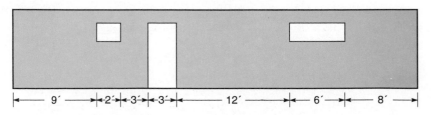

Fig. 1-2. Calculate the total length of the wall shown above.

6. The total resistance of a series electrical circuit is found by adding the individual resistances. Find the total resistance of the series circuit illustrated in Fig. 1-3.

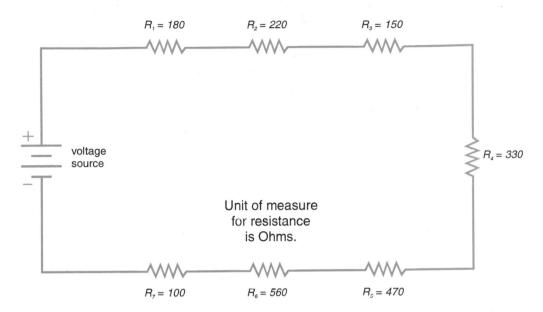

Fig. 1-3. *Calculate the total resistance in this series circuit.*

7. A quality-control technician performs destructive testing on a support beam by placing a weight of 1500 pounds on the beam. The weight is increased as follows until the beam breaks: 500 pounds, 875 pounds, 350 pounds, and 240 pounds. How many pounds made the beam break?
8. During a month-long blood drive, Red Cross volunteers visited area high schools and community colleges. How many pints of blood were donated during the drive if the first week brought in 150 pints, the second week brought in 238 pints, the third week brought in 484 pints, and the fourth week brought in 329 pints?
9. What length of fence is needed to enclose a play area with sides of 55 meters, 30 meters, 60 meters, and 28 meters?
10. A flooring contractor placed 45 red tiles, 68 green tiles, 71 yellow tiles, and 150 white tiles. How many tiles were placed?

WORD PROBLEMS USING SUBTRACTION

With subtraction, numbers are deducted. Since all numbers in this chapter will be positive, the larger number will be decreased by the smaller number. Only two numbers can be subtracted at a time, unlike addition, where many numbers can be added together in one operation. The number that is being deducted from is called the **minuend**. The number that is deducted is called the **subtrahend**. The answer is called the **difference**.

Example: 5 (minuend)
            − 3 (subtrahend)
              2 (difference)

Quite often, word problems state the total or end result of having done something. The question is to find the "missing step" of what was originally an addition problem. (This will become clearer as you continue to read.)

Use the following key words to help identify word problems of subtraction:
• **Difference**—the amount between two numbers.
  Example: The *difference* between 12 and 5 is 7. (12 − 5 = 7)
• **Fewer**—means the total amount has been lowered.
  Example: With 5 *fewer* bolts in a box that originally had 12, there are 7 left. (12 − 5 = 7)

- **Less than** — suggests that something has been taken away from the original amount.
  Example: If there were 8 before, and now there are 6 *less than* that, then there are now 2. (8 − 6 = 2)
- **Reduce** — the process of making something smaller.
  Example: The class began with 20 students but was *reduced* by 5, leaving 15 students. (20 − 5 = 15)
- **Decrease** — also makes the amount smaller.
  Example: After a sale, a supply of 50 tires *decreased* by 10, so that 40 tires were left. (50 − 10 = 40)
- **Minus** — something has been removed.
  Example: A carton once had 6 cans but now has 4 because it is *minus* 2. (6 − 2 = 4)

Sample Problem 1-4.

The electricity used in a house on the coast of Maine was 1538 kilowatts (kW) during the month of September. During that month, a windmill supplied 925 kilowatts to the house. How many kilowatts were supplied by the power company?

Step 1. Identify the operation.
This is an example of the "missing step" in an addition problem. The power company *plus* the windmill supplied the total electricity. The *difference* between the total and the windmill is what the power company supplied.

Step 2. Perform the arithmetic. Include the units.

$$\begin{array}{r} 1538 \\ -\ 925 \\ \hline 613 \end{array}\ kW \text{ from the power company}$$

Step 3. Check the answer.
The answer can be checked two ways: first, by comparing to what makes sense and, second, by addition. First, when 613 is compared to the question, it makes sense that the amount should be less than the total kilowatts. Second, in subtraction problems, add the difference (answer) to the subtrahend. The answer you get should be the same number as the minuend.

$$\begin{array}{r} 613\ \text{(difference)} \\ +\ 925\ \text{(subtrahend)} \\ \hline 1538\ \text{(minuend)} \end{array}$$

Note: The addition method is an exact check to see if the subtraction has been performed correctly. It is always a good idea, however, to also see if the answer makes sense in answering the question.

Sample Problem 1-5.

A lumber yard delivers 144 — 2 x 4s. After a day of framing walls, the contractor counts the 2 x 4s and finds 29 remaining. How many boards were used?

Step 1. Identify the operation.
This problem suggests subtraction because the total amount was *reduced* by what was used.

Step 2. Perform the arithmetic. Include the units.

$$\begin{array}{r} 144 \\ -\ 29 \\ \hline 115 \end{array}\ \textit{boards} \text{ used}$$

Step 3. Check the answer.
Add the difference to the subtrahend.

$$\begin{array}{r} 115\ \text{(difference)} \\ +\ 29\ \text{(subtrahend)} \\ \hline 144\ \text{(minuend)} \end{array}$$

Sample Problem 1-6.

A 55-gallon drum of paint thinner is in an auto body shop. The shop used the following amounts during the month: 3 gallons, 9 gallons, 4 gallons, 14 gallons, 2 gallons. How much is remaining in the drum?

Step 1. Identify the operation.

There are two methods of solving this type of problem. One is to start with the total and subtract each amount as it is taken. The other is to add all the amounts taken, then subtract from the original total. The end result with either method will be the amount remaining in the drum at the end of the month.

Step 2. Perform the arithmetic. Include the units.

First method:

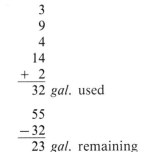

$$\begin{array}{r} 55 \\ -\ 3 \\ \hline 52 \end{array} \qquad \begin{array}{r} 52 \\ -\ 9 \\ \hline 43 \end{array} \qquad \begin{array}{r} 43 \\ -\ 4 \\ \hline 39 \end{array}$$

$$\begin{array}{r} 39 \\ -14 \\ \hline 25 \end{array} \qquad \begin{array}{r} 25 \\ -\ 2 \\ \hline 23 \end{array} \textit{gal.}\ \text{remaining}$$

Second method:

$$\begin{array}{r} 3 \\ 9 \\ 4 \\ 14 \\ +\ 2 \\ \hline 32 \end{array} \textit{gal.}\ \text{used}$$

$$\begin{array}{r} 55 \\ -32 \\ \hline 23 \end{array} \textit{gal.}\ \text{remaining}$$

Step 3. Check the answer.

Using two methods to solve the same problem is a good way to check your work. Also, the answer makes sense because if gallons are removed from the drum, the amount remaining should be less than the starting amount.

Practice Problems

Complete the following problems on a separate sheet of paper. Include units of measure (inches, feet, minutes, dollars, etc.).

1. An electric meter read 14,082 kilowatt hours at the end of January. At the end of February, it read 15,287 kilowatt hours. How many kilowatt-hours of electricity were used during the month of February?

2. An appliance warehouse delivered 35 refrigerators in one week. How many refrigerators remain in the warehouse if there were 87 in stock at the beginning of the week?

3. A tree farm had 150 blue spruce trees ready for its annual three-day sale. On the first day of the sale, 27 trees were sold. On the second day, 18 trees were sold. On the third day, a wholesaler offered to buy all the remaining trees. How many trees were left for the wholesaler to buy?

4. Determine the thickness of the material used for the base in Fig. 1-4.

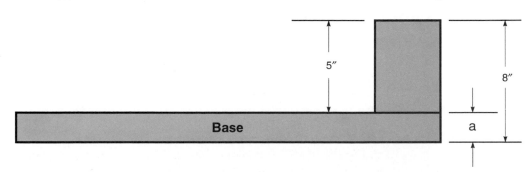

Fig. 1-4. What is the thickness of the base material shown above?

5. A crop duster carries 1500 pounds of chemical fertilizer. The pilot releases 750 pounds of fertilizer on a field. How much fertilizer remains on the plane?

6. A plumber was paid $2250 for a job that took one week. His expenses were as follows: $350 for pipe, $105 for fittings, $780 for fixtures, $53 for miscellaneous supplies, and $285 for his helper's wages. How much money was left for the plumber?

7. A printing company employee opened a 55-gallon drum of cleaning fluid and removed 12 gallons. A month later, the employee removed another 12 gallons. Surprisingly, this emptied the drum—the balance of the fluid was lost through evaporation. How much fluid evaporated?

8. The total resistance in a series circuit is the sum of the resistances of the individual resistors. What is the value of resistor R_2 in Fig. 1-5?

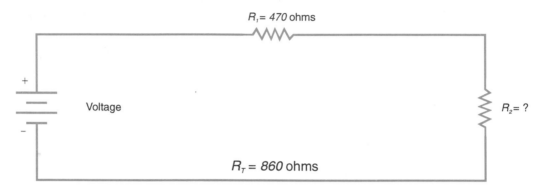

Fig. 1-5. Calculate the value of R_2.

9. To determine the weight of a puppy, a person steps on a scale and determines her weight to be 150 pounds. Then, while holding the puppy, she steps on the scale again. This time, the scale reads 178 pounds. How much does the puppy weigh?

10. A sawmill is monitoring its supply of 6×6 timbers. The mill ships out the timbers as follows: 40 timbers on the first truck, 28 timbers on the second truck, 56 timbers on the third truck, 18 timbers on the fourth truck, and 29 timbers on the fifth truck. If there were 250 timbers in stock before the first shipment, how many timbers remain after the fifth truck leaves the mill?

WORD PROBLEMS USING MULTIPLICATION

Multiplication is a form of addition. It is like adding the same number many times. To recognize multiplication in a word problem, look for words that suggest a number is repeated. Keep in mind that multiplication (with whole numbers) will always result in an increase in the original number. The numbers being multiplied are **factors**; the answer is the **product.**

A multiplication problem can be set up in several formats. Each of the following formats are read "6 times 2 equals 12."

$6 \times 2 = 12$

$6 * 2 = 12$

$6 \cdot 2 = 12$

$(6)(2) = 12$ (adjacent quantities enclosed in parentheses)

Sometimes, as you will learn in upcoming chapters, letters will be used in equations and formulas to represent different numbers. When they are, you may also see a multiplication problem in another format. In this format, letters or numbers and letters are written side by side. For example:

$ab = c$ is read "a times b equals c" or "a-b equals c".

$2b = 12$ is read "2 times b equals 12" or "2-b equals 12"

Use the following key words to help identify word problems of multiplication:

• **Product**—the result of multiplication.

Example: The *product* of 2 and 3 is 6. ($2 \times 3 = 6$)

- **At** — often used with pricing items and states the cost of the item.
  Example: A case of oil (24 cans) *at* $1 each costs $24. (1 × 24 = 24)
- **Times** — describes how often a number is repeated.
  Example: If you swim 1 mile, 3 *times* today, you will swim a total of 3 miles. (1 × 3 = 3)
- **By** — often used with dimensions to represent surface area, which is calculated with multiplication.
  Example: The garage is 20 feet *by* 30 feet. The area is 600 square feet. (20 × 30 = 600)
- **Rate** — an amount that is repeated.
  Example: A worker, paid at the *rate* of $6 for each hour of work, makes $48 in an 8-hour day. (6 × 8 = 48)
- **Per** — indicates how often an event is repeated.
  Example: A car traveling 60 miles *per* hour for 3 hours will cover 180 miles. (60 × 3 = 180)

  Note: The word *per* is also used with division. Use multiplication when the individual numbers are given, and the question asks for the total.

Sample Problem 1-7.

A worker earns $7 per hour. He works 40 hours each week. What is the weekly gross pay (pay before payroll deductions)?

Step 1. Identify the operation.
Each hour, $7 is earned, which is repeated for 40 hours. Another way to interpret this problem is to recognize $7 as a *rate* of pay. Rate becomes the key word for multiplication.

Step 2. Perform the arithmetic. Include the units.

$$\begin{array}{r} 40 \\ \times\ 7 \\ \hline \$280\ \text{weekly pay} \end{array}$$

Step 3. Check the answer.
The best way to check this type of problem is to see if the answer makes sense in the problem. This answer does make sense based on an approximation of how much money should be earned.

Sample Problem 1-8.

A stack of lumber is 32 boards wide by 16 boards high. How many boards are in the stack?

Step 1. Identify the operation.
Multiplication is the indicated operation because the stack is 16 layers high, with 32 boards in each layer. Therefore, 32 is repeated 16 times. Another way to look at this problem is to recognize the key word *by* as an indication of multiplication.

Step 2. Perform the arithmetic. Include the units.

$$\begin{array}{r} 32 \\ \times\ 16 \\ \hline 512\ \textit{boards}\ \text{in the stack} \end{array}$$

Step 3. Check the answer.
With any multiplication problem, it is possible to repeatedly add the basic number enough times. In this problem, 32 would be repeated by addition, 16 times. A quicker check is to estimate the answer by changing to numbers that are easy to multiply. Round to the nearest multiple of 5 as follows:

$$32 \rightarrow 30$$
$$16 \rightarrow 15$$

Multiply the estimated numbers together.

$$\begin{array}{r} 30 \\ \times\ 15 \\ \hline 450\ \text{estimated} \end{array}$$

Note: The estimated answer is not quite near the exact answer. However, it is close enough to indicate the answer is likely correct.

Sample Problem 1-9.

An 8-foot-long, pressure-treated, 6 x 8 post costs $9. If 50 posts will be needed for a fence, what will be the total cost of the fence posts?

Step 1. Identify the operation.

Recognize the fact that a price indicates *rate*. This tells you to multiply. Also, notice that $9 is to be repeated 50 *times*.

Step 2. Perform the arithmetic. Include the units.

$$\begin{array}{r} 50 \\ \times\ \ 9 \\ \hline \$450 \end{array}$$ total cost for fence posts

Step 3. Check the answer.

The estimate method could be done by rounding 9 to 10 and multiplying to get an estimated 500.

Practice Problems

Complete the following problems on a separate sheet of paper. Include units of measure (inches, feet, minutes, dollars, etc.).

1. If a car is traveling at an average of 55 miles per hour, how many miles will it travel in 6 hours?
2. A certain gasoline-powered electric generator burns fuel at the rate of 4 gallons per hour. A contractor plans to use this generator for five 8-hour days. How much gasoline should the contractor plan to use?
3. A room is 14 feet wide and 16 feet long. If one floor tile measures 1 foot × 1 foot, how many tiles would be needed to cover this floor?
4. A drafting job is available for $8 per hour. What is the gross pay (before deductions) for a 40 hour week?
5. The veterinarian's instructions on a certain canine medicine read "5 mg per pound of body weight." What is the dosage for a 62 pound dog?
6. The formula for torque is as follows:

   Torque = Length × Force

   How much torque is exerted if 30 pounds of force is applied to a 2-foot-long lever? (The unit of measure for torque is ft-lb.)
7. The formula for power in an electrical circuit is as follows:

   Power = Current × Voltage

   Power is measured in watts, current is measured in amps, and voltage is measured in volts. How much power is used by an electric drill drawing 8 amps of current when connected to 120 volts?
8. Using the screw placement shown in Fig. 1-6, determine the number of screws estimated for 15 sheets of plywood.

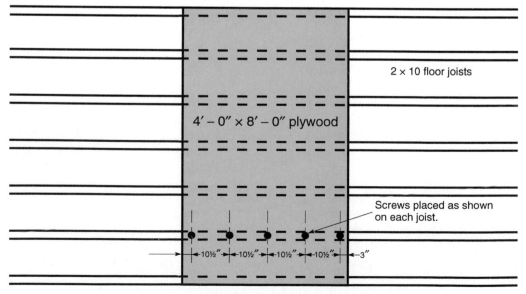

Fig. 1-6. Screw placement for a plywood subfloor. Note that placement for only one joist is shown. Screws must be placed in each joist.

9. Fig. 1-7 shows a 6-foot-long piece of plastic tubing divided into 2-foot-long pieces. How many 2-foot-long pieces can be obtained from 56 6-foot-long pieces of the tubing?

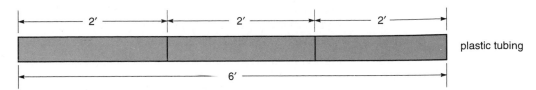

Fig. 1-7. How many 2-foot-long pieces can be cut from 56 tubes that are 6-feet long each?

10. An inventory control specialist determines that 50 plastic brackets weigh 1 pound. The total number of brackets in inventory weighs 7 pounds. How many brackets are in stock?

WORD PROBLEMS USING DIVISION

Division is the mathematical process that is opposite multiplication. It is usually stated in a way that suggests the number can be divided into many pieces of equal size. The number being divided is the **dividend.** The number by which the dividend is divided is the **divisor.** The answer is the **quotient.** A division problem can be set up in several formats. Each of the following formats are read "12 divided by 6 equals 2."

$$12 \div 6 = 2$$

$$6 \overline{\smash{)}12}^{\,2}$$

$$\frac{12}{6} = 2 \qquad \text{(or } 12/6 = 2\text{)}$$

Note: Be sure to read the division symbol as "divided by" to make certain the correct order of division is followed.

Use the following key words to help identify word problems of division:
- **Quotient** – the result of having performed division.
  Example: The *quotient* of 12 divided by 6 is 2. (12 ÷ 6 = 2)
- **Divided into** – states that an amount has been split into several pieces of equal size.
  Example: If $50 in ones is *divided into* 5 bags evenly, each bag will have $10. (50 ÷ 5 = 10)
- **Per** – states the amount for each item.
  Example: The cost for 5 boxes of bolts is $10. They are priced $2 *per* box. (10 ÷ 5 = 2)
  Note: The word *per* is also used with multiplication. Use division when the total is given, and the individual numbers are needed.

Sample Problem 1-10.

How many 6-inch-long pieces can be cut from a 48-inch-long board? (Ignore waste.)

Step 1. Identify the operation.

Division is used for this problem. This is because the board is 48 inches total and will be divided into several pieces, all 6 inches long.

Step 2. Perform the arithmetic. Include the units.

$$6 \overline{\smash{)}48}^{\,8 \; pieces}$$

Step 3. Check the answer.

With division problems, it is extremely important to examine the answer in the original problem. This is to make certain that it makes sense. If the result is incorrect, it may be that the division operation was performed backwards. To check a division problem, multiply the quotient by the *divisor*.

$$
\begin{array}{r}
8 \\
\times\ 6 \\
\hline
48
\end{array}
$$

Sample Problem 1-11.

A car traveled 273 miles on 13 gallons of gas. How many miles per gallon did this car get?

Step 1. Identify the method.

Per is the key word, making this a division problem. Any division problem using the word per can be written as "divided by." This problem, in effect, would read "miles divided by gallons."

Step 2. Perform the arithmetic. Include the units.

$$13\overline{)273}\ \ 21\ mpg$$

Step 3. Check the answer.

The answer, 21 mpg, makes sense in this problem. Also, check by multiplication.

$$
\begin{array}{r}
13 \\
\times\ 21 \\
\hline
273
\end{array}
$$

Sample Problem 1-12.

An electrician installs 40 outlet boxes in a house for a cost of $240, including wire. What is the installed cost on a per-box basis?

Step 1. Identify the method.

The total installed cost is equally divided among the boxes.

Step 2. Perform the arithmetic. Include the units.

$$40\overline{)240}\ \ \$6\ per\ box$$

Step 3. Check the answer.

The best way to check this problem is by using multiplication.

$$
\begin{array}{r}
40 \\
\times\ 6 \\
\hline
240
\end{array}
$$

Practice Problems

Complete the following problems on a separate sheet of paper. Include units of measure (inches, feet, minutes, dollars, etc.).

1. A case of engine oil (12 bottles) sells for $24. What is the price per bottle?
2. How many 8-inch-long pieces can be cut from a 96-inch length of copper pipe?
3. An aircraft technician receives a paycheck of $520 for 40 hours of work. What is the technician's hourly rate of pay?
4. To determine current in an electrical circuit when given power, the formula is Current (amps) = Power (watts) / Voltage (volts). How many amps of current are required for a toaster having a power rating of 1200 watts and a voltage of 120 volts?
5. What is the gas mileage (miles per gallon) for a car that travels 270 miles and uses 15 gallons of gas?

6. Determine the number of studs required to frame the section of wall shown in Fig. 1-8. Include the first stud by adding 1 to the total.

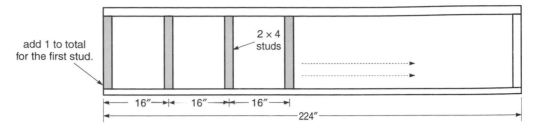

add 1 to total for the first stud.

2 × 4 studs

16" 16" 16" 224"

Fig. 1-8. The framing studs in this wall are spaced at 16 inches on center.

7. How many rafters are needed for the roof of a garage if the rafters are spaced 2 feet on center and the garage is 32 feet long? Add 1 for the first rafter.
8. What is the distance between the equally spaced holes in the steel plate shown in Fig. 1-9?

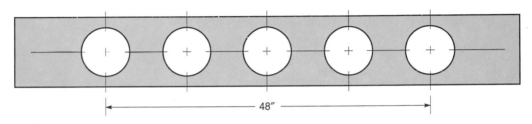

48"

Fig. 1-9. Bolt holes in a section of steel plate.

9. A cosmetologist bought 560 ounces of shampoo in a bulk container. How many 16-ounce bottles can be filled with shampoo?
10. A machine can produce 3,840 plastic bottles in a 24-hour period. At this rate, how many bottles are produced each hour?

WORD PROBLEMS USING COMBINATION ARITHMETIC

Many word problems contain a situation where it is necessary to perform more than one operation to arrive at the answer. For example, some numbers might first be added together. Their sum, then, might be divided into another number. Any combination of required operations may be performed. There are no shortcuts to such problems. Every operation must be performed.

Sample Problem 1-13.

On a trip touring the country, a motor home had a starting odometer reading of 62,960 miles and an ending reading of 65,424 miles. If the vehicle burned 176 gallons of gasoline, what was the miles per gallon during this trip?

Step 1. Identify the operation.
Two are involved:
a. Determine the miles traveled by subtracting the ending odometer reading from the starting odometer reading.
b. Find the miles per gallon by dividing the miles by the gallons.

Step 2. Perform the arithmetic. Include the units.
Miles traveled:

$$\begin{array}{r} 65{,}424 \\ -62{,}960 \\ \hline 2464 \ \textit{miles traveled} \end{array}$$

Miles per gallon:

$$176\overline{)2464} \quad \dfrac{14 \ \textit{mpg}}{}$$

Step 3. Check the answer.

With combination problems, you need to check each step. Be sure the final result answers the question. This question asks for miles per gallon, not miles traveled. This is a good reason to include units with the answer.

Sample Problem 1-14.

Using the following list of materials, determine the total cost:

150—2 x 4s at $2 each
35—2 x 6s at $3 each
15 sheets of 1/2-inch plywood at $20 each
5 pounds of nails at $2 per pound

Step 1. Identify the operation.

Since the materials are priced per each, multiply the quantity times the price. This will give the cost for each type of material. Then, add the individual costs for a total.

Step 2. Perform the arithmetic. Include the units.

2 x 4s:

$$\begin{array}{r} 150 \\ \times\ \ 2 \\ \hline \$300 \end{array}$$

2 x 6s:

$$\begin{array}{r} 35 \\ \times\ \ 3 \\ \hline \$105 \end{array}$$

Plywood:

$$\begin{array}{r} 15 \\ \times\ 20 \\ \hline \$300 \end{array}$$

Nails:

$$\begin{array}{r} 5 \\ \times\ 2 \\ \hline \$10 \end{array}$$

Total cost:

$$\begin{array}{r} 300 \\ 105 \\ 300 \\ +\ \ 10 \\ \hline \$715 \end{array}$$

Step 3. Check the answer.

Each answer makes sense. The final result is reasonable and appears to answer the question asked.

Sample Problem 1-15.

A bridge painter works a different number of hours every week. What is the average weekly pay if weekly gross income for five weeks work is: $685, $470, $520, $375, $530?

Step 1. Identify the operation.

"Average" problems require adding to find the total and dividing by the amount of numbers added. In this problem, add the dollar amounts and divide by 5.

Step 2. Perform the arithmetic. Include the units.

Total income:

$$\begin{array}{r} 685 \\ 470 \\ 520 \\ 375 \\ +\ \ 530 \\ \hline \$2580 \end{array}$$

Average income:

$$\frac{\$516}{5\overline{)2580}} \textit{ per week average}$$

Step 3. Check the answer.

Check the answer by comparing to the original numbers to see that it falls near the middle of the range of numbers. The average in this problem does. It is likely correct.

Practice Problems

Complete the following problems on a separate sheet of paper. Include units of measure (inches, feet, minutes, dollars, etc.).

1. An automotive technician must choose between two different job offers. Dealer ABC is offering 35 hours per week at $10 per hour. Dealer XYZ is offering 45 hours per week at $7 per hour. Which of these jobs offers the highest weekly pay? How much more?

2. A roofing contractor estimates 2 bundles of shingles for one section of a roof, 5 bundles for another section, and 15 bundles for the final section. If the bundles cost $15 each, nails cost $5, and other materials cost $50, how much is the estimated cost of labor if the final estimate is $600?

3. A pilot is considering taking a load of freight that weighs 750 pounds. The maximum load of the plane is 1200 pounds. The pilot weighs 150 pounds, and the estimated weight of fuel is 7 pounds per gallon. It is estimated that the trip will require 30 gallons of fuel. Is the total load under the maximum? If so, is it safe to bring along a 120-pound passenger?

4. To build a set of chairs, sixteen 18-inch-long pieces of wood are needed for the legs. The only stock available is 8 feet long (96 inches). How many 8-foot boards will be needed?

5. A truck driver does the end-of-the-month reports. Mileage at the start of the month was 35,481. At the end of the month it was 36,681. What is the truck's gas mileage (mpg) if 150 gallons were used during the month?

6. A cook opened a new bag of flour and removed 2 cups of flour for cookies, 8 cups for bread, and 12 cups for pancakes, which emptied the bag. If a bag 5 times as large were used, how many cups of flour could be expected?

7. What is dimension A in Fig. 1-10?

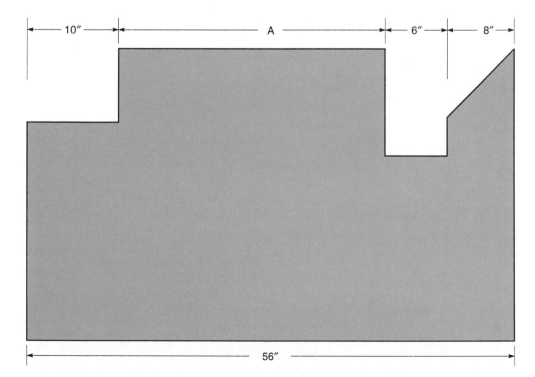

Fig. 1-10. Find dimension A in the above drawing.

8. If the temperature on one October day reached a high of 65°F and a low of 35°F, what was the average for the day?
9. How many shelves can be placed in the unit shown in Fig. 1-11?

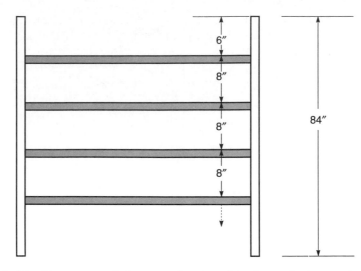

Fig. 1-11. How many shelves can be placed in the rack shown above?

10. Five members of the wrestling team weighed in as follows: 235 pounds, 278 pounds, 262 pounds, 250 pounds, and 180 pounds. What was the average weight of the wrestlers?

TEST YOUR SKILLS

Do *not* write in this book. Use a separate sheet of paper to complete the following problems. Show your work and your final answer. Include units of measure (for example, inches, feet, minutes, etc.).

1. A planing machine can plane pine lumber at a rate of 14 linear feet per minute. How many linear feet can be planed in 90 minutes?
2. Electrical appliances operate in a kitchen. All are connected to the same circuit. The appliances draw current as follows: toaster, 8 amps; blender, 2 amps; deep-fryer, 3 amps; light bulb, 1 amp. What is the total amount of electrical current being drawn on this circuit?
3. A delivery truck weighs 6870 pounds with a load of machine parts. What does the load weigh if the truck weighs 2350 pounds when empty?
4. In Fig. 1-12, what is the width of the window opening?

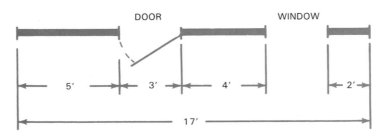

Fig. 1-12. Dimensions given for a wall.

5. A shopper finds the price of a case of canned vegetables to be $48. What is the price of each can with 24 in a case?
6. A plumber connects a water heater in a home. The following materials are used: 18 feet of copper pipe costing $3 per foot; 4 elbows at $1 each; 2 shut-off valves at $6 each; miscellaneous supplies worth $15. What is the total cost?

7. A farmer separated a 9000-pound supply of grain into three equal-size bins. If each bin feeds the livestock for 4 days, how much grain is fed out of each bin per day?
8. An air conditioning unit supplies 150,000 Btu per hour to 6 rooms of equal size. What is the average Btu per room?
9. In a math class, a student receives the following test scores: 75, 90, 87, 63, and 85 percent. What is the average of the test scores?
10. A painting contractor determines the outside walls of a house to have 18,065 square feet of surface, including 3921 square feet that will not be painted. How much area will the contractor paint?

PROBLEM-SOLVING ACTIVITIES

Activity 1-1.
Length of electrical wiring

Objective: To estimate the length of wire needed to go from a switch near a door to a duplex outlet box on the opposite side of the room.

Instructions:
1. Locate a switch near the door of your classroom and an outlet box on the opposite side of the room.
2. Assume that the wire between the switch and the box is housed in metal conduit and is routed above the ceiling.
3. Measure the distance from the switch to the outlet box, rounding all dimensions to the nearest foot.
4. Connections to the power source should not be included in your calculations.

Activity 1-2.
Nails by the pound

Objective: To use weight to estimate quantity.

Instructions:
1. Weigh a container that is large enough to hold 1 pound of nails.
2. Add nails of the same type to the container until they weigh one pound (be sure to subtract the weight of the container).
3. Count the nails in the container.
4. Use the quantity in one pound to estimate the number of nails in 5 pounds, 10 pounds, and 50 pounds.

Activity 1-3.
Estimating the number of wall studs

Objective: To estimate the number of 2 × 4 studs needed to frame a wall.

Instructions:
1. Assuming that a wall will be built to divide your classroom into two sections, measure the length of the room. Round your measurement to the nearest foot.
2. Convert feet to inches by multiplying by 12.
3. If the 2 × 4 studs are placed 16 inches on center, determine the number of 2 × 4 studs needed.
4. When calculating the number of studs, be sure to add 1 for the starter and add 1 if your calculations result in a fraction or decimal.

Activity 1-4.
Estimating the cost of electrical installation

Objective: To estimate the cost of an electrical installation based on a standard price.

Instructions:
1. Count the total number of duplex outlets, switches, light fixtures, etc., in your classroom.

2. Using the figures given in the following chart, estimate the cost of installing electrical fixtures in your classroom.

Electrical Fixture	Cost of Installation
Duplex outlet box	$4
Each single pole (on/off) light switch	$5
Light fixture (installation only)	$3
8-foot fluorescent fixture with 2 bulbs	$20
8-foot fluorescent fixture with 4 bulbs	$35
4-foot fluorescent fixture with 4 bulbs	$15

Activity 1-5.
Chair parts inventory
   Objective: To determine the inventory of parts needed to build the chair/desk assembly in your classroom.
   Instructions:
   1. Closely examine how one of the chair/desk assemblies in your classroom is made.
   2. Make a list of the parts. Include screws, bolts, tabletop, seat, legs, etc. If two or more parts are welded together, count each piece before it is welded.
   3. Use the list to make a bill of materials for one chair/desk assembly.
   4. Make a complete bill of materials to build all the chair/desk assemblies in the classroom.

Chapter 2

Fractions

OBJECTIVES

After studying this chapter, you will be able to:
- Explain basic terminology and functions of fractions.
- Reduce fractions to lowest terms.
- Find lowest common denominators.
- Perform operations of arithmetic on fractions.

Fractions are used throughout mathematics. Indeed, they are widely used in everyday life. For example, when an apple pie is cut into pieces, fractions are represented. Another example of the use of fractions is a ruler. It has inches that are divided into fractional parts. An inch is a fractional part of a foot. A **fraction** is a portion of a whole amount. It describes into how many parts a whole is divided, and how many parts are used. (Some fractions have more parts used than the whole has part, for example, 5/3, 6/5, or 3/2. Such fractions are discussed later on in this chapter.)

READING FRACTIONS

When reading (and saying) a fraction, read the top number first, then the bottom number. As examples:

- $\frac{1}{2}$ —"one-half" or "one over two."

- $\frac{3}{5}$ —"three-fifths" or "three over five."

When reading mixed numbers, read the whole number first and connect it to the fraction with the word "and." As examples:

- $4\frac{2}{3}$ —"four and two-thirds" or "four and two over three."

- $2\frac{1}{8}$ —"two and one-eighth" or "two and one over eight."

Practice Problems

Write out (in words) each of the following fractions on a separate sheet of paper.

1. $\frac{7}{16}$

2. $\frac{6}{8}$

3. $\frac{5}{10}$

4. $\frac{3}{4}$

5. $3\frac{5}{6}$

6. $5\frac{2}{7}$

7. $1\frac{1}{9}$ 8. $10\frac{1}{3}$

9. $4\frac{1}{4}$ 10. $9\frac{2}{5}$

VALUE OF FRACTIONS

A fraction may be an amount between two whole numbers. For example, 2 1/2 is between the whole numbers 2 and 3. A whole unit can be divided into any number of parts, or fractions of the unit. In Fig. 2-1, circles and squares are divided into fractional parts.

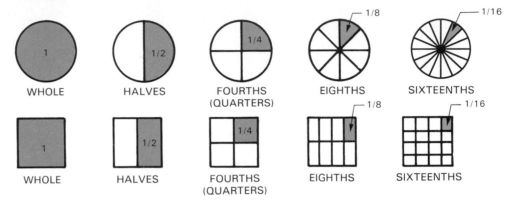

Fig. 2-1. A circle and square divided into fractional parts.

The size of the denominator

The **denominator** is the bottom number of a fraction. It states into how many parts a whole unit is divided. The more divisions there are, the smaller the size of each part. As an example, picture in your mind a fresh, blueberry pie, to be shared by either four or eight people. In which case does everyone get the larger piece? If the piece is cut into four pieces, each piece is a fourth (1/4). If it was cut into eight pieces, each piece would be smaller. Each would be an eighth (1/8). Therefore, a larger denominator has smaller pieces, and a smaller denominator has larger pieces. The following are typical fractions, arranged from larger to smaller:

$$\frac{1}{2} \quad \frac{1}{3} \quad \frac{1}{4} \quad \frac{1}{5}$$

The size of the numerator

The **numerator** is the top number of a fraction. It states how many parts are concerned. As an example, if our blueberry pie was cut into 8 pieces, and 5 pieces were eaten, 5/8 of the pie is gone. Also, 3/8 of the pie, or 3 of the 8 pieces, are saved for midnight snacks. With the denominator constant (the same), as the size of the numerator increases, the value of the fraction increases. The following are typical fractions, arranged from smaller to larger:

$$\frac{1}{8} \quad \frac{2}{8} \quad \frac{5}{8} \quad \frac{11}{8}$$

If the numerator is larger than the denominator, the fraction is larger than 1 whole. Imagine 2 blueberry pies. Each is cut into eight pieces and 11 of those pieces are eaten—all of the first pie and 3 pieces from the second (11/8.)

Practice Problems

Complete the following problems on a separate sheet of paper.

Identify the numerator and denominator in each of the following fractions.

1. $\dfrac{3}{5}$

2. $\dfrac{2}{7}$

3. $\dfrac{5}{9}$

4. $\dfrac{1}{2}$

Arrange each set of fractions in ascending order (using the size of the numerators and denominators).

5. $\dfrac{1}{7}$ $\quad$ $\dfrac{1}{12}$ $\quad$ $\dfrac{1}{10}$ $\quad$ $\dfrac{1}{4}$ $\quad$ $\dfrac{1}{6}$

6. $\dfrac{15}{16}$ $\quad$ $\dfrac{7}{16}$ $\quad$ $\dfrac{2}{16}$ $\quad$ $\dfrac{3}{16}$ $\quad$ $\dfrac{22}{16}$

7. $\dfrac{1}{9}$ $\quad$ $\dfrac{1}{16}$ $\quad$ $\dfrac{1}{12}$ $\quad$ $\dfrac{1}{5}$ $\quad$ $\dfrac{1}{2}$

8. $\dfrac{1}{5}$ $\quad$ $\dfrac{7}{5}$ $\quad$ $\dfrac{4}{5}$ $\quad$ $\dfrac{3}{5}$ $\quad$ $\dfrac{2}{5}$

9. $\dfrac{1}{5}$ $\quad$ $\dfrac{1}{18}$ $\quad$ $\dfrac{1}{14}$ $\quad$ $\dfrac{1}{3}$ $\quad$ $\dfrac{1}{7}$

10. $\dfrac{10}{12}$ $\quad$ $\dfrac{17}{12}$ $\quad$ $\dfrac{6}{12}$ $\quad$ $\dfrac{7}{12}$ $\quad$ $\dfrac{12}{12}$

NUMBERS WITH AND WITHOUT FRACTIONS

Numbers can take different forms, including the forms of whole numbers, proper fractions, improper fractions, mixed numbers, and decimals. Decimals will be discussed in the next chapter.

A **whole number** is a complete unit, containing no fractional parts. The numbers 0, 1, 2, 3, 4, etc., are whole numbers.

A **proper fraction** is a fraction with the numerator smaller than the denominator. A proper fraction has a value of less than 1 (in magnitude). The numbers 1/2, 3/5, 5/7, and 7/8 are examples of proper fractions.

An **improper fraction** is a fraction with the numerator the same or larger than the denominator. An improper fraction has a value equal to or greater than 1 (in magnitude). The numbers 9/4, 5/5, 3/1, and 8/2 are examples of improper fractions.

A **mixed number** contains both a whole number and a fraction. The numbers 3 6/7, 1 4/8, 4 2/3, and 8 10/6 are examples of mixed numbers.

Practice Problems

On a separate sheet of paper, identify each of the following numbers as a whole number, a proper fraction, an improper fraction, or a mixed number.

1. $\dfrac{2}{3}$

2. $4\dfrac{3}{5}$

3. 5

4. $\dfrac{7}{5}$

5. $7\dfrac{3}{2}$

6. 10

7. $\dfrac{7}{16}$

8. $1\dfrac{1}{3}$

9. $\dfrac{12}{12}$

10. $\dfrac{1}{2}$

CHANGING FRACTIONS FROM ONE FORM TO ANOTHER

Fractions can be written in different forms and still represent the same value. Fig. 2-2 uses a ruler to demonstrate the equivalent value of mixed numbers and improper fractions. Dimensions are used to help visualize the relative size of the fractional numbers.

With different arithmetic operations, fractions are written as either mixed numbers or improper fractions, depending on the particular arithmetic operation. Follow the rules of improper fractions or mixed numbers when changing from one form to the other.

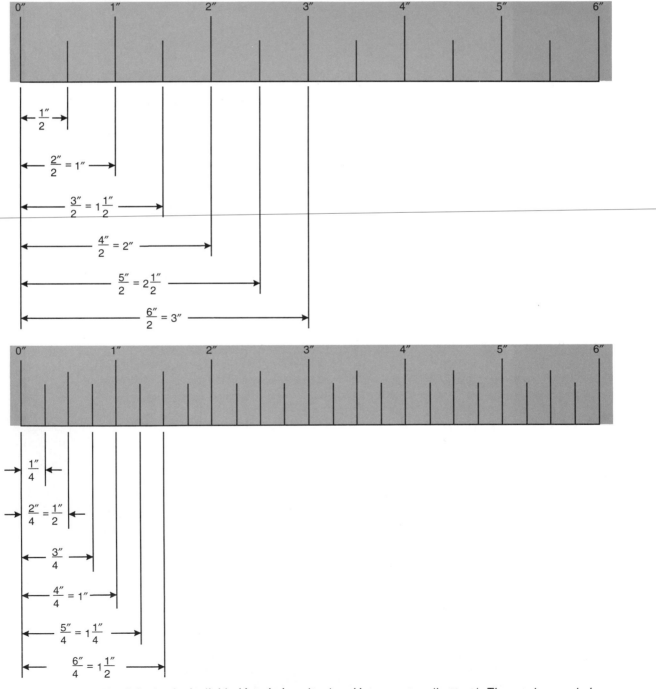

Fig. 2-2. A ruler is divided into halves (top) and into quarters (bottom). These rulers are being used to show that improper fractions have mixed number equivalents.

RULES OF IMPROPER FRACTIONS AND MIXED NUMBERS

- To change a mixed number to an improper fraction—multiply the whole number times the denominator. Add this product to the numerator. Then, write the result over the original denominator.

Examples of changing mixed numbers to improper fractions:

$$2\frac{3}{4} = \frac{11}{4} \; (2 \times 4 = 8), \; (8 + 3 = 11)$$

$$4\frac{1}{6} = \frac{25}{6} \; (4 \times 6 = 24), \; (24 + 1 = 25)$$

$$5 = \frac{5}{1} \; \text{(Whole numbers are written over 1.)}$$

- To change an improper fraction to a mixed number—divide the numerator by the denominator. This division produces a whole number or a whole number and a remainder. The remainder, placed over the original denominator will be the fractional part.

Examples of changing improper fractions to mixed numbers:

$$\frac{11}{4} = 2\frac{3}{4} \; (11 \div 4 = 2 \text{ with remainder of } \frac{3}{4})$$

$$\frac{23}{6} = 3\frac{5}{6} \; (23 \div 6 = 3 \text{ with remainder of } \frac{5}{6})$$

$$\frac{15}{3} = 5 \; (15 \div 3 = 5 \text{ with no remainder})$$

Practice Problems

Complete the following problems on a separate sheet of paper.

Change the following mixed numbers and whole numbers to improper fractions.

1. $5\frac{1}{3}$

2. $4\frac{3}{4}$

3. $3\frac{1}{2}$

4. 6

5. $9\frac{2}{5}$

6. $1\frac{4}{3}$

7. 10

8. $2\frac{4}{7}$

9. $7\frac{2}{1}$

10. $8\frac{5}{9}$

Change the following improper fractions to mixed numbers or whole numbers.

11. $\frac{12}{7}$

12. $\frac{13}{5}$

13. $\frac{10}{4}$

14. $\frac{25}{5}$

15. $\frac{3}{2}$

16. $\frac{7}{6}$

17. $\frac{31}{6}$

18. $\frac{20}{8}$

19. $\frac{15}{12}$

20. $\frac{17}{10}$

EQUIVALENT FRACTIONS

Examine Fig. 2-3. Notice how the different markings can be represented by different fractions. As the whole unit is divided into smaller segments, the denominator of the fraction changes. However, some divisions of the smaller segments overlap divisions of the larger segments. This results in different fractions representing the same value, or **equivalent fractions**. Even though numerators and denominators have different numbers, equivalent fractions have the same value.

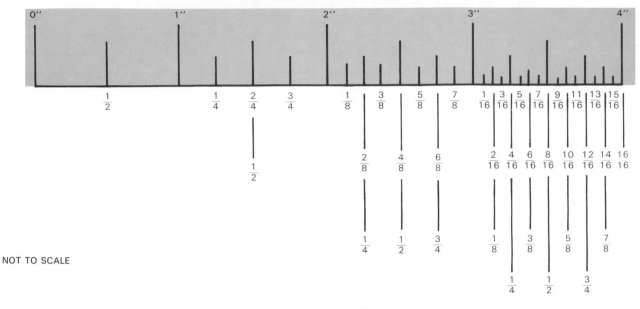

NOT TO SCALE

Fig. 2-3. A ruler is divided into fractional parts. The segments show equivalent fractions.

RULE OF EQUIVALENT FRACTIONS

Equivalent fractions are found by multiplying or dividing both numerator and denominator by the same number. This is the same as multiplying by 1 and does not change the value of the fraction.
Examples:

$$\frac{1}{2} \times \frac{4}{4} = \frac{4}{8} \qquad\qquad \frac{4}{8} \div \frac{4}{4} = \frac{1}{2}$$

$$\frac{3}{8} \times \frac{2}{2} = \frac{6}{16} \qquad\qquad \frac{6}{16} \div \frac{2}{2} = \frac{3}{8}$$

Reducing to lowest terms

Whenever writing a fraction, the final answer must be *reduced* to lowest terms. The purpose is to set a standard for everyone to follow. A fraction reduced to lowest terms is an equivalent fraction using the smallest possible numerator and denominator. Examples of fractions reduced to lowest terms are:

$$\frac{3}{6} \rightarrow \frac{1}{2} \qquad\qquad \frac{9}{12} \rightarrow \frac{3}{4}$$

The most efficient way to reduce a fraction is to factor the numerator and the denominator into their prime factors. **Factors** are any set of numbers that will equal a given number when multiplied together. **Prime factors** are the factored set of which all numbers are prime numbers. A **prime number** is a positive whole number that can be exactly divided (no remainder) only by itself and the number 1. The first 10 prime numbers are: 1, 2, 3, 5, 7, 11, 13, 17, 19, 23.

When factoring a number, review a list of prime numbers. See which ones will exactly divide into the number. Continue factoring until the number has been factored into all prime numbers.

All factors that are common to both numerator and denominator can be "canceled." The remaining factors will result in a fraction reduced to lowest terms. Canceling is possible because any number written over itself as a fraction is equal to 1.

RULES FOR REDUCING FRACTIONS WITH PRIME NUMBERS
1. Find the prime factors.
2. Cancel factors common to numerator and denominator.
3. Rewrite as reduced fraction.

Sample Problem 2-1.

Reduce $\frac{24}{36}$ using prime factors.

Step 1. Find the prime factors.

$$\frac{24}{36} = \frac{2 \times 2 \times 2 \times 3}{2 \times 2 \times 3 \times 3}$$

Step 2. Cancel factors one for one, common to both numerator and denominator.

$$\frac{24}{36} = \frac{2 \times 2 \times 2 \times 3}{2 \times 2 \times 3 \times 3}$$

Step 3. Rewrite as reduced fraction. (Factors remaining here are the reduced fraction.)

$$\frac{24}{36} = \frac{2}{3}$$

Sample Problem 2-2.

Reduce $\frac{60}{75}$ using prime factors.

Step 1. Find the prime factors.

$$\frac{60}{75} = \frac{2 \times 2 \times 3 \times 5}{5 \times 3 \times 5}$$

Step 2. Cancel common factors.

$$\frac{60}{75} = \frac{2 \times 2 \times 3 \times 5}{5 \times 3 \times 5}$$

Step 3. Rewrite as a reduced fraction.

$$\frac{60}{75} = \frac{4}{5}$$

Common denominators

Arithmetic operations of addition and subtraction require changing fractions to equivalents with a common denominator. Up until now, the discussion of equivalent fractions has focused on single fractions. A **common denominator** is a denominator that is *common* to all concerned fractions. When all fractions in a problem have the same denominator, they have a *common denominator*.

Common denominators are also used when it is necessary to compare the value of different fractions. Using a common denominator, a numerator will show the value of a fraction in relation to other fractions.

To change fractions to equivalents with a common denominator, decide by inspection on a common denominator. Then,

(a) divide the common denominator by the original denominators.

(b) multiply the respective numerator *and* denominators by these results.
Examples:

Change to a denominator of 16:

$$\frac{1}{2} \rightarrow \text{a) } 16 \div 2 = 8 \quad \text{b) } \frac{1 \times 8}{2 \times 8} = \frac{8}{16}$$

$$\frac{3}{4} \rightarrow \text{a) } 16 \div 4 = 4 \quad \text{b) } \frac{3 \times 4}{4 \times 4} = \frac{12}{16}$$

$$\frac{5}{8} \rightarrow \text{a) } 16 \div 8 = 2 \quad \text{b) } \frac{5 \times 2}{8 \times 2} = \frac{10}{16}$$

Lowest common denominator. The previous example is one way of changing fractions to an equivalent with a common denominator. Perhaps you were working with a ruler. You knew that 16 is very often a common denominator of fractional inches. By inspection, you could see that this was the case for the fractions 1/2, 3/4, and 5/8. There are times, however, when the new denominator will not be so obvious, as you will soon see.

A common denominator can always be found by multiplying the numbers in the denominator together. For instance, in the fractions of the example, $2 \times 4 \times 8 = 64$. It is best, however, to work with the smallest common denominator because the numbers are easiesr to work with. The common denominator that is smallest in value is called the **lowest common denominator (LCD)**.

At times, the LCD can be determined at a glance. It is easy to see the LCD of 1/2, 3/4, and 5/8 is 8. However, what about the LCD of the fractions 1/3, 2/7, 2/9, and 1/18? The LCD is found as follows:

1. Find the prime factors of all denominators.
   Denominators: 3, 7, 9, 18
   Prime factors:
   $3 = 3$
   $7 = 7$
   $9 = 3 \times 3$
   $18 = 3 \times 3 \times 2$

The least common denominator will consist of all the prime factors of the largest number and any others that are unique.

$$\text{LCD} = 3 \times 3 \times 2 \times 7 = 126$$

Arranging fractions in order of size. To determine the size, or value, of fractions in comparison to others, it is necessary to change all fractions to the same denominator. For example: A wrench set ranges from 1/4 inch to 1 inch. It is spaced every 1/16 inch. The wrenches are marked with the fraction reduced to lowest terms. One way to tell which wrench is larger or smaller would be to change the sizes marked to equivalent fractions with a common denominator. In this way, the wrench having the greatest numerator would be the largest.

Sample Problem 2-3.

Arrange these fractions in ascending order:

$$\frac{1}{2} \quad \frac{3}{8} \quad \frac{3}{4} \quad \frac{7}{16} \quad \frac{5}{8}$$

Step 1. Change all fractions to a common denominator. The common denominator will very often be the largest denominator in the list. All denominators must be able to divide into the common denominator exactly. Sometimes, it is necessary to use one larger than any in the list.
Common denominator is 16.

$$\frac{1}{2} = \frac{1 \times 8}{2 \times 8} = \frac{8}{16}$$

$$\frac{3}{8} = \frac{3 \times 2}{8 \times 2} = \frac{6}{16}$$

$$\frac{3}{4} = \frac{3 \times 4}{4 \times 4} = \frac{12}{16}$$

$$\frac{7}{16} = \frac{\text{(no}}{\text{change)}} = \frac{7}{16}$$

$$\frac{5}{8} = \frac{5 \times 2}{8 \times 2} = \frac{10}{16}$$

Step 2. Arrange the fractions in ascending order of size by comparing their numerators.

$$\frac{6}{16} \qquad \frac{7}{16} \qquad \frac{8}{16} \qquad \frac{10}{16} \qquad \frac{12}{16}$$

Step 3. Complete the problem by rewriting the answer using the original fractions.

$$\frac{3}{8} \qquad \frac{7}{16} \qquad \frac{1}{2} \qquad \frac{5}{8} \qquad \frac{3}{4}$$

Practice Problems

Complete the following problems on a separate sheet of paper.

Change the fraction on the left to an equivalent fraction with the denominator shown on the right.

1. $\frac{5}{12} = \frac{}{24}$ 2. $\frac{1}{4} = \frac{}{16}$

3. $\frac{4}{5} = \frac{}{20}$ 4. $\frac{15}{16} = \frac{}{32}$

5. $\frac{12}{16} = \frac{}{8}$ 6. $\frac{6}{12} = \frac{}{6}$

7. $8 = \frac{}{4}$ 8. $5 = \frac{}{12}$

9. $\frac{2}{3} = \frac{}{15}$ 10. $\frac{5}{7} = \frac{}{28}$

Reduce the following fractions to lowest terms.

11. $\frac{7}{14}$ 12. $\frac{4}{12}$

13. $\frac{6}{8}$ 14. $\frac{9}{12}$

15. $\frac{8}{10}$ 16. $\frac{4}{16}$

17. $\frac{12}{4}$ 18. $\frac{15}{4}$

19. $4\frac{6}{4}$ 20. $3\frac{10}{16}$

Arrange the following sets of fractions in ascending order, from smallest to largest. Show the lowest common denominator.

21. $\frac{1}{3} \qquad \frac{5}{12} \qquad \frac{5}{6} \qquad \frac{11}{24}$

22. $\frac{3}{4} \qquad \frac{15}{16} \qquad \frac{7}{8} \qquad \frac{29}{32}$

23. $\dfrac{1}{2}$ $\dfrac{7}{10}$ $\dfrac{3}{5}$ $\dfrac{13}{20}$

24. $\dfrac{2}{5}$ $\dfrac{3}{15}$ $\dfrac{1}{3}$ $\dfrac{7}{30}$

25. $1\dfrac{1}{16}$ $\dfrac{33}{32}$ $1\dfrac{1}{64}$ $\dfrac{9}{8}$

ADDITION OF FRACTIONS

Addition of fractions requires the use of common denominators. Also, the final answer must be reduced to lowest terms. When selecting a common denominator, it is best to select the LCD to avoid difficult reducing.

RULES FOR ADDITION OF FRACTIONS

1. Convert all fractions to a common denominator.
2. Add the *numerators*. Keep the same denominator. Add whole numbers separately.
3. Reduce answer to lowest terms.
   Example:

$$\frac{1}{3} + \frac{1}{6} = \frac{2}{6} + \frac{1}{6} = \frac{3}{6} = \frac{1}{2}$$

Addition with mixed numbers

When adding mixed numbers, the fractions will be added independently from the whole numbers. Then, the results will be combined and reduced to lowest terms.

Sample Problem 2-4.

Add the following using mixed numbers:

$$2\frac{3}{4} + 3\frac{2}{5}$$

Step 1. Change fractions to the lowest common denominator.
LCD = 20.

$$2\frac{3}{4} = 2\frac{3 \times 5}{4 \times 5} = 2\frac{15}{20}$$

$$3\frac{2}{5} = 3\frac{2 \times 4}{5 \times 4} = 3\frac{8}{20}$$

Step 2. Add the numerators. Keep the same denominator. Add the whole numbers separately.

$$2\frac{15}{20}$$
$$+3\frac{8}{20}$$
$$\overline{5\frac{23}{20}}$$

Step 3. Reduce to lowest terms.

$$5\frac{23}{20} = 5 + 1\frac{3}{20} = 6\frac{3}{20}$$

Sample Problem 2-5.

Add the following using mixed numbers.

$$8 \frac{5}{6} + 3 \frac{1}{2}$$

Step 1. Change to the lowest common denominator.
LCD = 6.

$$8 \frac{5}{6} = \qquad\qquad 8 \frac{5}{6}$$

$$3 \frac{1}{2} = 3 \frac{1 \times 3}{2 \times 3} = 3 \frac{3}{6}$$

Step 2. Add the numerators. Keep the same denominator.
Add whole numbers separately.

$$8 \frac{5}{6}$$

$$+ \; 3 \frac{3}{6}$$

$$\overline{11 \frac{8}{6}}$$

Step 3. Reduce to lowest terms.

$$11 \frac{8}{6} = 11 + 1 \frac{2}{6} = 12 \frac{2}{6} = 12 \frac{1}{3}$$

Addition with improper fractions

Rather than adding with mixed numbers, fractions can be added using improper fractions. Change all mixed numbers and whole numbers to improper fractions. Then, change all fractions in the problem to a common denominator and add. Reduce the answer to lowest terms.

Sample Problem 2-6.

Add the following, using improper fractions.

$$4 \frac{2}{3} + \frac{1}{2} + 6 + \frac{25}{2}$$

Step 1. Change mixed numbers and whole numbers to improper fractions. Change to the lowest common denominator.
LCD = 12.

$$4 \frac{2}{3} = \frac{14}{3} = \frac{14 \times 4}{3 \times 4} = \frac{56}{12}$$

$$\frac{1}{2} = \frac{1}{2} = \frac{1 \times 6}{2 \times 6} = \frac{6}{12}$$

$$6 = \frac{6}{1} = \frac{6 \times 12}{1 \times 12} = \frac{72}{12}$$

$$\frac{25}{6} = \frac{25}{6} = \frac{25 \times 2}{6 \times 2} = \frac{50}{12}$$

Step 2. Add the numerators. Keep the same denominator.

$$\frac{56}{12}$$

$$\frac{6}{12}$$

$$\frac{72}{12}$$

$$+ \ \frac{50}{12}$$

$$\overline{\frac{184}{12}}$$

Step 3. Reduce to lowest terms.

$$\frac{184}{12} = 15\frac{4}{12} = 15\frac{1}{3}$$

Practice Problems

Complete the following problems on a separate sheet of paper.

Add the following fractions.

1. $\frac{1}{3} + \frac{1}{6}$

2. $\frac{1}{2} + \frac{3}{8}$

3. $\frac{2}{3} + \frac{3}{4}$

4. $\frac{5}{12} + \frac{1}{4}$

5. $\frac{2}{7} + \frac{7}{12}$

6. $\frac{1}{12} + \frac{7}{16}$

Add using mixed numbers.

7. $5\frac{5}{6} + 2\frac{3}{5}$

8. $1\frac{5}{9} + 3\frac{7}{10}$

9. $9 + 4\frac{3}{1} + 6\frac{1}{3}$

10. $2\frac{2}{4} + 3\frac{3}{6} + 8\frac{16}{32}$

11. $6\frac{3}{5} + 7\frac{7}{6} + 8\frac{4}{3}$

12. $\frac{5}{8} + 2\frac{9}{12} + 3\frac{4}{15}$

Add using improper fractions.

13. $\frac{12}{5} + 2\frac{2}{3} + 4\frac{3}{4}$

14. $6 + \frac{2}{6} + 3\frac{35}{3} + 4\frac{18}{9}$

15. $1\frac{1}{2} + 5\frac{5}{7} + 3\frac{9}{14} + 7\frac{10}{1}$

16. $2\frac{12}{12} + 4 + 5\frac{6}{2} + 4\frac{3}{3}$

17. $4\frac{7}{16} + 2\frac{1}{8} + 4\frac{3}{4}$

18. $5\frac{2}{3} + 8\frac{3}{14} + 15$

Solve the following word problems. Be sure to include units of measure (inches, feet, minutes, dollars, etc.).

19. The outside wall of a house is constructed as shown in Fig. 2-4. What is the total thickness of the wall?

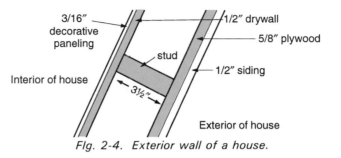

Fig. 2-4. Exterior wall of a house.

20. The pattern for a garment calls for 1/3 yard of material for the sleeves, 1/2 yard for the back, 5/8 yard for the front, and 1/16 yard for trim. If an additional 1/4 yard material is allowed for waste, how much material should the tailor plan on buying?

SUBTRACTION OF FRACTIONS

Subtraction, like addition, can be performed using either mixed numbers or improper fractions. Both methods will be shown. The rules for subtraction are similar to those of addition.

RULES FOR SUBTRACTION OF FRACTIONS

1. Convert all fractions to a common denominator.
2. Subtract the *numerators*. Keep the same denominator. Subtract whole numbers separately.
3. Reduce answer to lowest terms.

Example:

$$\frac{1}{3} - \frac{1}{6} = \frac{2}{6} - \frac{1}{6} = \frac{1}{6}$$

Subtraction with mixed numbers

When subtracting mixed numbers, it will be necessary to "borrow" in a manner similar to subtracting whole numbers. Borrowing uses a digit from the column of the next higher place value.

For example:

$$\begin{array}{r} 41 \\ -\ 7 \\ \hline 34 \end{array}$$

The number 41 is the same as 40 + 1. To borrow, the 40 + 1 is changed to 30 + 11. Then, 7 is subtracted from the 11, resulting in 4. The 30 is brought down, giving a final answer of 34.

When borrowing with mixed numbers, the whole number will be split into additive parts to make the fraction larger.

For example:

$$3\,\frac{1}{4} = 2 + \frac{4}{4} + \frac{1}{4} = 2\,\frac{5}{4}$$

Sample Problem 2-7.

Subtract using mixed numbers.

$$6\frac{1}{3} - 4\frac{1}{2}$$

Step 1.   Change to the lowest common denominator.
LCD = 6.

$$6\frac{1}{3} = 6\frac{1 \times 2}{3 \times 2} = 6\frac{2}{6}$$

$$4\frac{1}{2} = 4\frac{1 \times 3}{2 \times 3} = 4\frac{3}{6}$$

Step 2.   Arrange problem for borrowing. Subtract the numerators. Keep the same denominator. Subtract whole numbers separately.

$$6\frac{2}{6} = 5 + \frac{6}{6} + \frac{2}{6} = 5\frac{8}{6}$$

$$-4\frac{3}{6}$$
$$\overline{1\frac{5}{6}}$$

Step 3.   Reduce to lowest terms.
This number is already in lowest terms.

Answer: $1\frac{5}{6}$

Subtraction with improper fractions

When the numbers in the subtraction problem are changed to improper fractions, the need for borrowing is eliminated.

Sample Problem 2-8.

Subtract using improper fractions.

$$5\frac{3}{8} - 2\frac{3}{4}$$

Step 1.   Change to the lowest common denominator.
LCD is 8.

$$5\frac{3}{8} = = 5\frac{3}{8}$$

$$2\frac{3}{4} = 2\frac{3 \times 2}{4 \times 2} = 2\frac{6}{8}$$

Step 2.   Change to improper fractions. Subtract numerators. Keep the same denominator.

$$5\frac{3}{8} = \frac{43}{8}$$

$$-2\frac{6}{8} = \frac{22}{8}$$
$$\overline{\frac{21}{8}}$$

Step 3.   Reduce to lowest terms.

$$\frac{21}{8} = 2\frac{5}{8}$$

Practice Problems

Complete the following problems on a separate sheet of paper. Be sure your answers are reduced to lowest terms.

Subtract the following fractions.

1. $\dfrac{3}{4} - \dfrac{1}{2}$

2. $\dfrac{5}{6} - \dfrac{2}{3}$

3. $\dfrac{5}{8} - \dfrac{7}{16}$

4. $\dfrac{5}{12} - \dfrac{1}{3}$

5. $\dfrac{4}{5} - \dfrac{4}{7}$

6. $\dfrac{7}{10} - \dfrac{5}{12}$

Subtract using mixed numbers.

7. $4\dfrac{2}{3} - 2\dfrac{1}{2}$

8. $5\dfrac{3}{4} - 2\dfrac{1}{8}$

9. $7\dfrac{1}{6} - 3\dfrac{1}{3}$

10. $9\dfrac{1}{4} - \dfrac{2}{3}$

11. $15 - 3\dfrac{3}{5}$

12. $5\dfrac{1}{2} - 4$

Subtract using improper fractions.

13. $4\dfrac{1}{2} - 2\dfrac{1}{3}$

14. $8\dfrac{2}{9} - 7\dfrac{2}{3}$

15. $6\dfrac{1}{3} - 2$

16. $12\dfrac{3}{16} - \dfrac{12}{8}$

17. $5\dfrac{12}{16} - \dfrac{48}{64}$

18. $12\dfrac{6}{4} - 3\dfrac{2}{3}$

Solve the following word problems. Be sure to include units of measure (inches, feet, minutes, dollars, etc.).

19. A plumber cuts 48 3/16 inches from an 8-foot (96-inch) length of pipe. How much of the pipe remains?

20. Determine the size of the opening allowed for the refrigerator in Fig. 2-5.

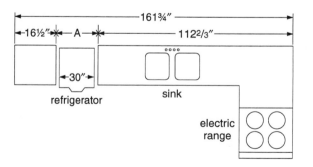

Fig. 2-5. Space requirements in a kitchen.

MULTIPLICATION OF FRACTIONS

Multiplication of fractions requires multiplying the numerators *and* denominators. The answer is then reduced to lowest terms. If the problem contains mixed numbers or whole numbers, they must first be changed to improper fractions.

Prior to multiplying, if possible, the numbers should be reduced by canceling prime factors. As before, cancellation is performed by removing factors that are common to both the numerator and denominator. Factors may be canceled across different fractions within the same multiplication problem.

> ## RULES FOR MULTIPLICATION OF FRACTIONS
> 1. Change mixed numbers and whole numbers to improper fractions.
> 2. Arrange problem for multiplication. Factor and cancel like terms. Common denominators are *not* required for multiplication.
> 3. Multiply numerators; multiply denominators.
> 4. Reduce answer to lowest terms.

Sample Problem 2-9.

Multiply these numbers.

$$\frac{2}{5} \times 2\frac{3}{8} \times 6$$

Step 1. Change mixed numbers and whole numbers to improper fractions.

$$\frac{2}{5} = \frac{2}{5} \text{ (no change)}$$

$$2\frac{3}{8} = \frac{19}{8}$$

$$6 = \frac{6}{1}$$

Step 2. Arrange problem for multiplication. Factor and cancel like terms.

$$\frac{2}{5} \times \frac{19}{8} \times \frac{6}{1}$$

Factoring for cancellation:

$$\frac{2}{5} \times \frac{19}{2 \times 2 \times 2} \times \frac{2 \times 3}{1}$$

Cancellation:

$$\frac{\cancel{2}}{5} \times \frac{19}{\cancel{2} \times 2 \times 2} \times \frac{2 \times 3}{1}$$

Result of cancellation:

$$\frac{1}{5} \times \frac{19}{2} \times \frac{3}{1}$$

Step 3. Multiply numerators; multiply denominators.

$$\frac{1 \times 19 \times 3}{5 \times 2 \times 1} = \frac{57}{10}$$

Step 4. Reduce to lowest terms.

$$\frac{57}{10} = 5\frac{7}{10}$$

Note: Reducing in this step is much easier because the factors were canceled in Step 2.

Practice Problems

Complete the following problems on a separate sheet of paper.

Multiply the following. Use cancellation whenever possible.

1. $\frac{1}{2} \times \frac{4}{7}$

2. $\frac{3}{5} \times \frac{3}{4}$

3. $\frac{7}{10} \times \frac{7}{12}$

4. $\frac{2}{3} \times \frac{4}{5}$

5. $2\frac{5}{16} \times \frac{4}{25}$

6. $3\frac{2}{7} \times 4\frac{14}{28}$

7. $1\frac{12}{6} \times 4\frac{2}{14}$

8. $0 \times 3\frac{4}{7}$

9. $\frac{32}{20} \times 5\frac{15}{8}$

10. $6\frac{35}{5} \times 1\frac{10}{14}$

Solve the following word problems. Be sure to include units of measure (inches, feet, minutes, dollars, etc.).

11. A recipe for cookies calls for 2 2/3 cups of sugar. If three batches of cookies are to be made, how much sugar is needed?
12. A sheet of plywood weighs 70 5/8 pounds. If 15 sheets of plywood are loaded into the back of a pickup truck, what would the load weigh? Does this load exceed the truck's load capacity rating of 1000 pounds?
13. A milling machine is adjusted to remove 3/32 inch of material with each pass over the surface of a steel plate. How much material is removed from the surface of the plate with 5 passes? Is this more or less than 1/2 inch?
14. Fig. 2-6 shows the dimensions of a wooden box ordered by a customer. Fifteen of these boxes are to be built. What is the total length of wood needed for the long sides? What is the total length needed for the short sides?

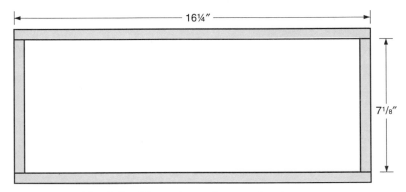

Fig. 2-6. Dimensions of a wooden box.

DIVISION OF FRACTIONS

Division of fractions is very similar to multiplication, with one added step. The **divisor**, or the number by which you divide, must be inverted. Then, the two fractions are multiplied following the rules for multiplication. Cancellation must be performed *after* the divisor has been inverted.

RULES FOR DIVISION OF FRACTIONS

1. Change mixed numbers and whole numbers to improper fractions.
2. Arrange problem for division—*invert* the divisor; change sign to multiplication.
3. Factor and cancel like terms. Multiply numerators; multiply denominators.
4. Reduce to lowest terms.

Sample Problem 2-10.

Divide the following fractions.

$$\frac{2}{5} \div 3\frac{5}{8}$$

Step 1. Change the mixed numbers to improper fractions.

$$\frac{2}{5} = \frac{2}{5} \text{ (no change)}$$

$$3\frac{5}{8} = \frac{29}{8}$$

Step 2. Arrange problem for division—invert the divisor; change sign to multiplication.

$$\frac{2}{5} \times \frac{8}{29}$$

Step 3. Factor and cancel like terms. Multiply numerators; multiply denominators.

$$\frac{2}{5} \times \frac{2 \times 2 \times 2}{28} = \frac{16}{145}$$

Note: These numbers cannot be canceled.

Step 4. Reduce to lowest terms.

Note: This number cannot be reduced.

$$\frac{16}{145}$$

Practice Problems

Complete the following problems on a separate sheet of paper. Reduce your answers to lowest terms and include units of measure (inches, feet, minutes, dollars, etc.) where applicable.

Divide the following. Use cancellation whenever possible.

1. $\frac{1}{2} \div \frac{1}{6}$ 2. $\frac{1}{4} \div \frac{1}{2}$

3. $\frac{3}{4} \div \frac{1}{3}$ 4. $\frac{5}{8} \div \frac{3}{4}$

5. $7 \div 2$ 6. $5 \div 8$

7. $2\frac{5}{12} \div 3\frac{5}{6}$ 8. $4\frac{7}{16} \div 1\frac{32}{8}$

9. $8\frac{7}{10} \div 9\frac{5}{6}$ 10. $5\frac{5}{9} \div 2\frac{5}{3}$

11. How many 15 1/4-inch-long pieces can a 83 3/5-inch-long pipe be divided into?
    Note: If the answer is not a whole number, the remaining fraction is scrap.

12. Determine dimension A for the three equally spaced shelves shown in Fig. 2-7.

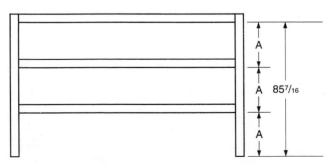

Fig. 2-7. Bookcase with three equally spaced shelves.

SUMMARY OF RULES

Fig. 2-8 is provided as a reference. It is a brief summary of the rules of fractions.

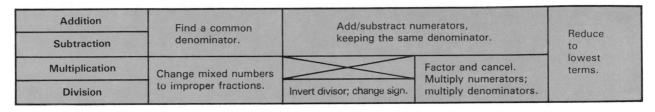

Addition	Find a common denominator.	Add/substract numerators, keeping the same denominator.		Reduce to lowest terms.
Subtraction				
Multiplication	Change mixed numbers to improper fractions.		Factor and cancel. Multiply numerators; multiply denominators.	
Division		Invert divisor; change sign.		

Fig. 2-8. Summary of rules of fractions.

TEST YOUR SKILLS

Do *not* write in this book. Use a separate sheet of paper to complete the following problems. Show your work and your final answer.

CHANGING FRACTIONS FROM ONE FORM TO ANOTHER

Change the mixed numbers and whole numbers to improper fractions. Change improper fractions to mixed numbers.

1. $\dfrac{29}{6}$

2. $5\dfrac{2}{3}$

3. 9

4. $\dfrac{56}{7}$

5. $2\dfrac{5}{9}$

6. $\dfrac{37}{5}$

7. $\dfrac{49}{6}$

8. $3\dfrac{3}{8}$

9. $4\dfrac{3}{4}$

10. $\dfrac{28}{3}$

EQUIVALENT FRACTIONS

Change the fraction on the left to an equivalent fraction with the denominator shown on the right.

11. $\dfrac{3}{4} = \dfrac{}{12}$

12. $\dfrac{3}{5} = \dfrac{}{10}$

13. $6 = \dfrac{}{12}$

14. $\dfrac{8}{5} = \dfrac{}{30}$

15. $3\dfrac{3}{7} = 3\dfrac{}{21}$

16. $7\dfrac{2}{9} = 7\dfrac{}{81}$

17. $10 = \dfrac{}{10}$

18. $\dfrac{32}{16} = \dfrac{}{8}$

19. $\dfrac{3}{16} = \dfrac{}{48}$

20. $\dfrac{25}{50} = \dfrac{}{100}$

Reduce the following to lowest terms.

21. $\dfrac{6}{12}$

22. $\dfrac{6}{18}$

23. $\dfrac{8}{32}$

24. $\dfrac{10}{18}$

25. $\dfrac{48}{32}$

26. $\dfrac{50}{225}$

27. $\dfrac{12}{3}$

28. $3\dfrac{12}{8}$

29. $\dfrac{100}{1000}$

30. $\dfrac{7}{13}$

Arrange the following sets of fractions in descending order. Show the lowest common denominator.

31. $\dfrac{2}{3}$ $\dfrac{5}{6}$ $\dfrac{8}{24}$ $\dfrac{9}{12}$

32. $\dfrac{1}{4}$ $\dfrac{3}{8}$ $\dfrac{3}{16}$ $\dfrac{7}{32}$

33. $\dfrac{3}{6}$ $\dfrac{7}{18}$ $\dfrac{11}{36}$ $\dfrac{35}{72}$

34. $\dfrac{7}{9}$ $\dfrac{13}{18}$ $\dfrac{43}{54}$ $\dfrac{2}{3}$

35. $\dfrac{8}{11}$ $\dfrac{15}{22}$ $\dfrac{17}{22}$ $\dfrac{31}{44}$

ADDITION OF FRACTIONS

Add the following fractions. Reduce the answer to lowest terms. Show your work.

36. $\dfrac{1}{2} + \dfrac{1}{2}$

37. $\dfrac{5}{6} + \dfrac{3}{6}$

38. $\dfrac{3}{4} + \dfrac{5}{8}$

39. $3\dfrac{3}{5} + 2\dfrac{7}{10}$

40. $6 + 5\dfrac{2}{9}$

41. $\dfrac{32}{8} + \dfrac{4}{32}$

42. $2\dfrac{2}{3} + 3\dfrac{3}{4} + 4\dfrac{4}{5}$

43. $1\dfrac{2}{6} + 2\dfrac{13}{7}$

44. $\dfrac{11}{13} + \dfrac{9}{11}$

45. $5 + 0 + \dfrac{15}{32}$

SUBTRACTION OF FRACTIONS

Subtract the following. Reduce the answer to lowest terms. Show your work.

46. $\dfrac{1}{2} - \dfrac{1}{4}$

47. $\dfrac{3}{4} - \dfrac{2}{3}$

48. $\dfrac{3}{5} - \dfrac{1}{10}$

49. $1\dfrac{2}{3} - \dfrac{1}{6}$

50. $5\dfrac{1}{4} - 2\dfrac{1}{2}$

51. $7\dfrac{1}{3} - 3\dfrac{3}{5}$

52. $8 - 2\dfrac{2}{3}$

53. $5\dfrac{1}{6} - 3$

54. $15\dfrac{1}{8} - 14\dfrac{4}{4}$

55. $6\dfrac{3}{5} - \dfrac{5}{6}$

MULTIPLICATION OF FRACTIONS

Multiply the following. Reduce the answer to lowest terms. Show your work.

56. $\frac{1}{3} \times \frac{2}{3}$

57. $\frac{3}{4} \times \frac{2}{5}$

58. $\frac{4}{6} \times \frac{3}{8}$

59. $\frac{5}{10} \times \frac{25}{50}$

60. $3\frac{3}{5} \times \frac{3}{4}$

61. $\frac{2}{3} \times 4\frac{1}{2}$

62. $\frac{7}{8} \times 2\frac{5}{6}$

63. $5\frac{3}{5} \times 2\frac{5}{6}$

64. $15 \times 3\frac{1}{5}$

65. $2\frac{1}{9} \times 3$

DIVISION OF FRACTIONS

Divide the following. Reduce the answer to lowest terms. Show your work.

66. $\frac{1}{2} \div \frac{1}{4}$

67. $\frac{1}{4} \div \frac{2}{3}$

68. $\frac{3}{5} \div \frac{5}{6}$

69. $\frac{5}{12} \div \frac{6}{10}$

70. $4 \div 3$

71. $2\frac{7}{8} \div 3\frac{3}{4}$

72. $8\frac{1}{2} \div 2$

73. $7\frac{1}{9} \div 3\frac{6}{9}$

74. $0 \div \frac{2}{3}$

75. $4\frac{5}{6} \div 6\frac{5}{4}$

WORD PROBLEMS WITH FRACTIONS

Solve the following word problems. Show all of the steps involved. Be sure answer is reduced to lowest terms and include units with the final answer.

76. In a stack of 147 boards of dimensional lumber, 3/7 of the boards are 2 x 4s. How many of the boards are other sizes?
77. A HVAC technician worked 6 1/3 hours each of the first three days of last week. On the fourth day, the technician worked 12 1/4 hours. How many hours were worked the fifth day to complete a 40-hour week?
78. A length of copper pipe 14 3/4 feet long is cut into 5 equal pieces. How long is each piece?
79. Barrett is 8 1/4 years old. Mark is 6 1/2 years old. Eric is 2 2/3 years old. Bettina is 9 3/4 years old. Diane is 7 1/3 years old. What is the average of their ages?
80. If a pound of nails cost 24 cents per pound, what is the cost of 3 1/2 pounds?
81. A piping contractor installs a 12-inch-nominal pipe. Given the dimensions shown in Fig. 2-9, calculate the inside diameter (ID).

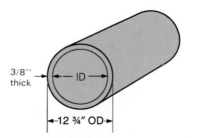

3/8″ thick

← ID →

←12 ¾″ OD→

Fig. 2-9. Schedule 40, galvanized steel pipe.

82. A finished steel casting weighs 45 2/3 pounds. Prior to finish, it weighed 54 1/4 pounds. What will be the total steel lost if 150 castings are produced?

83. Three pieces of copper wire, used for wiring an electrical fixture, will run through 13 3/4 feet of conduit. If 18 inches of extra wire are needed at each end to terminate each wire, how many feet of wire are required?

84. Three additional, equally spaced holes are to be drilled between A and B in Fig. 2-10. What will be the distance between the centers of the holes?

85. In machining operations that use a rotating workpiece, such as lathe turning, the reduction in diameter of the workpiece is twice the depth of cut. If a round aluminum shaft is reduced in diameter by 7/16 inch, what is the total depth of cut?

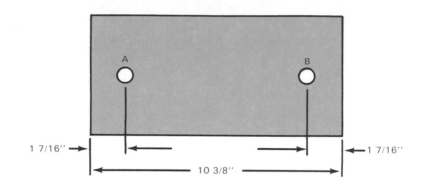

Fig. 2-10. Add three 1/2 inch holes to this part.

PROBLEM-SOLVING ACTIVITIES

Activity 2-1.
Arrange a wrench set by size
   Objective: To use equivalent fractions and a common denominator to arrange a wrench set in ascending order.
   Instructions:
   1. Use a set of wrenches marked in fractions.
   2. Write the fractions for each wrench and convert these fractions to an LCD.
   3. Using the fractions, arrange the wrenches in order of size.

Activity 2-2.
Adding fractions on a ruler
   Objective: To use a ruler to verify the addition of fractions.
   Instructions:
   1. Add the following using the rules of fractions:

$$6 \frac{1}{2} + 12 \frac{1}{4} + 18 \frac{3}{4} + 4 \frac{5}{8} + 9 \frac{3}{8}$$

   2. Measure 6 1/2 inches from a wall and place a piece of tape on the floor.
   3. Measure 12 1/4 inches from the tape and place another piece of tape on the floor. Continue this process for each of the numbers in the addition problem.
   4. Measure from the wall to the last piece of tape. The total distance should equal the sum found in step 1. Note: Slight variations may result due to inaccurate placement of the tape.

Activity 2-3.
Subtracting fractions on a ruler
   Objective: To use a ruler to verify the subtraction of fractions.
   Instructions:
   1. Subtract 21 1/8 inches from 44 3/8 inches and record your answer.
   2. Place a piece of masking tape on the floor 44 3/8 inches from a wall.

3. Place a second piece of tape 21 1/8 inches from the wall.
4. Measure the distance between the pieces of tape and compare the measurement to your calculations.

Activity 2-4.
Material to make a picture frame

> **Objective:** To determine the exact length of the material needed to make a picture frame.

> **Instructions:**
1. Measure the outside dimensions of a picture frame. The outside dimension of each side is the length of the material needed for the side.
2. Add the measurements of the sides to find the total length of material needed to construct the frame.

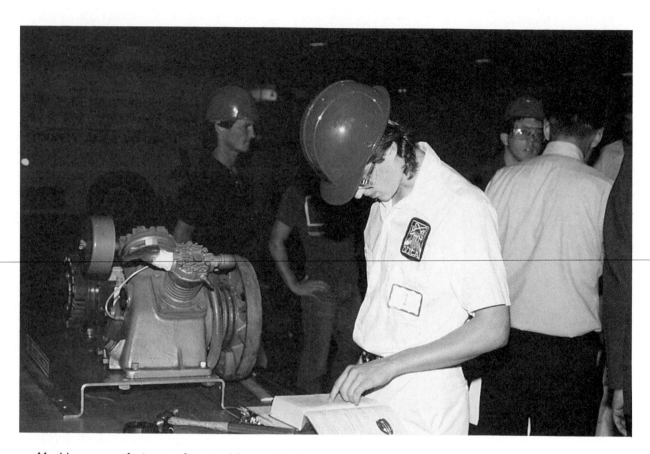

Machinery manufacturers often provide specifications for part sizes, wear tolerances, clearances, gaps, etc. These specifications are commonly expressed as decimal fractions. (Vocational Industrial Clubs of America)

Chapter 3

Decimal Fractions

OBJECTIVES

After studying this chapter, you will be able to:
- Explain basic terminology and functions of decimals.
- Convert fractions to decimals.
- Convert decimals to fractions.
- Identify significant figures of numbers.
- Round off numbers to indicated degrees of accuracy or precision.
- Perform the operations of arithmetic on decimals.

A **decimal fraction**, or **decimal number**, or just **decimal**, is used to represent a fractional part of a whole. Decimals are similar to fractions, but both have their own applications. This chapter will discuss the rules for using decimals.

VALUE OF DECIMAL NUMBERS

A decimal number is made up of digits and a *decimal point* (.). Each digit is assigned a value that is a multiple of 10—the definition of the **decimal system**—depending on its place, or location, in the complete number. The quantity to the right of the decimal point is, in reality, a fraction written in a different form. The quantity to the left is a whole number.

Reading decimal numbers

In explaining decimal numbers, it is necessary to include written numbers in the text. Before any attempt to read and know what decimals are, it is logical that you know *how* to read (and say) them. Methods of reading decimal numbers are: reading them like fractions; or saying "point" (meaning decimal point) with the numbers. Examples:

0.056 = "fifty-six thousandths"
      = "point zero fifty-six"
      = "point zero five six"
0.75 = "seventy-five hundredths"
     = "point seventy-five"
     = "point seven five"
0.3 = "three-tenths"
   = "point three"

If the decimal number contains a whole number, read the whole number first. Then, add the decimal using the word "and" or, simply, by saying "point." Example:

2.89 = "two and eighty-nine hundredths"
     = "two point eighty-nine"
     = "two point eight nine"

Digits

Digits are any of the numerals 0 through 9. Digits of *any given number* are the *individual numerals* in that number. For example, digits of the number 359 are the numerals 3, 5, and 9. A number may be quantified in terms of the number of digits it has. For example, the number 359 is a 3-digit number.

The decimal point

A decimal point separates a number into a whole number and a fraction. The whole number appears to the left. The fraction, which is written in decimal form, appears to the right. The value is less than 1 (in magnitude). Digits to the right of the decimal point, including zeros, are referred to in terms of **decimal places**. For example, 5 is in the third *decimal place* of the decimal number 1.375.

Unwritten decimal point

Numbers written without fractions or decimal points, as you have learned, are whole numbers (or integers). Such numbers have no fractional parts. Whole numbers can be converted to decimal numbers by placing a decimal point after the number. For example, 250 and 250. have the same value. Neither number has fractional parts. However, 250 is an integer and 250. is a decimal number.

Place values

The value of the location of a digit in a number, called the **place value**, determines the value of the digit. For example, in the number 359, the location of the digit 3 has a *place value* of 100. The 3 indicates there are 3 – 100s. The value of the digit is 300. Likewise, 5 in the 10s place (place value = 10) has a value of 50. The 9 in the units (ones) place has a value of 9.

Fig. 3-1 shows a place value chart. Notice that the values are multiples of 10. Each place moving left of the ones place is represented by the next higher multiple

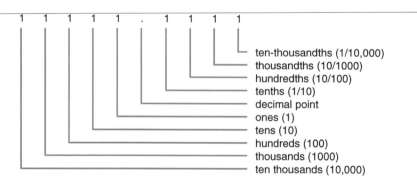

Fig. 3-1. Place value chart.

of 10. Each place moving right of the decimal point is represented by the next smaller multiple of 1/10. A complete number is formed by adding the values of each digit multiplied by its place value. The following are examples of place values:

56,304 =

$$
\begin{array}{rcrcr}
5 & \times & 10,000 & = & 50,000 \\
6 & \times & 1,000 & = & 6,000 \\
3 & \times & 100 & = & 300 \\
0 & \times & 10 & = & 00 \\
4 & \times & 1 & = & +\quad 4 \\
\hline
 & & & & 56,304
\end{array}
$$

4,081 =

$$
\begin{array}{rcrcr}
4 & \times & 1,000 & = & 4,000 \\
0 & \times & 100 & = & 000 \\
8 & \times & 10 & = & 80 \\
1 & \times & 1 & = & +\quad 1 \\
\hline
 & & & & 4,081
\end{array}
$$

$$0.302 =$$

$$3 \times \frac{1}{10} = \frac{3}{10} = \frac{300}{1000}$$

$$0 \times \frac{1}{100} = \frac{0}{100} = \frac{0}{1000}$$

$$2 \times \frac{1}{1000} = \frac{2}{1000} = + \frac{2}{1000}$$

$$\frac{302}{1000} = .302$$

When a fraction has a denominator that is already a multiple of 10, the fraction can be written as a decimal number by placing a decimal point in front of the numerator and removing the denominator. Caution: It is necessary to maintain the place values by using zeros to hold places that are not used. For example, the fraction 1/1000 is written as .001 (not as .1) in decimal form. The zeros placed before the 1 in the decimal form hold the places that are not used, i.e., the tenths and hundredths places.

The following are examples of changing fractions to decimal fractions:

$$\frac{506}{1000} = .506$$

$$\frac{92}{1000} = .092$$

$$\frac{76}{100} = .76$$

$$\frac{8}{100} = .08$$

$$\frac{5}{10} = .5$$

Zeros in a decimal number

Although a zero has no value, it is extremely important to the value of the complete number. Follow the rules of zeros in decimal numbers.

RULES OF ZEROS IN DECIMAL NUMBERS

- A zero is used in a number as a placeholder, to keep the other digits in their proper places. A zero placed between numbers or between a number and decimal point changes the value of the number.
  Example: 56 ≠ 506; .506 ≠ .0506
- A zero placed on the left side of a whole number does not change its value.
  Example: 35 = 035.
- A zero placed to the right of a decimal number does not change its value.
  Example: .68 = .680.
- A zero placed to the left of the decimal point does not change the value. It is common to write the zero this way.
  Example: .59 = 0.59.

Practice Problems

Complete the following problems on a separate sheet of paper.

Multiply each digit in the following whole numbers by its place value.

1. 209	2. 3,567
3. 10,451	4. 150
5. 872	6. 10
7. 2,083	8. 235,710
9. 1,045,102	10. 26,821

Multiply each digit in the following decimal numbers by its place value.

11. 3.52	12. 8.068
13. 9.125	14. 15.8715
15. 21.38512	16. 4.27
17. 4.0067	18. 2.05
19. 0.85	20. 0.20871

Change each of the following decimal numbers into a fraction. Do not reduce.

21. 0.25	22. 0.17
23. 0.3	24. 0.5
25. 0.125	26. 0.371
27. 0.007	28. 0.0035
29. 0.00002	30. 0.00000105

SIGNIFICANT FIGURES AND ROUNDING

Significant figures are the figures of a number that begin with the first nonzero figure to the left and end with the last figure to the right that is not zero or is a zero that indicates a number to be exact and not approximate. **Rounding** is giving a close approximation of a number. A number is rounded to a specified **accuracy** (number of significant digits) or to a certain **precision** (place value of last digit). Sometimes rounding is arbitrarily done depending on the degree of importance of a digit to the complete value of a number. For example, if a person has $999.01, the $0.01, in many cases, could be thought insignificant in comparison to the complete number. The number could be rounded to $999.

Most and least significant figures

The significance of figures is always with *most* significant on the left and *least* significant on the right, regardless of the location of the decimal point. Refer to the following examples:

- 3056 — 3 is most, 6 is least.
- 5.4409 — 5 is most, 9 is least.
- 0.921 — 9 is most, 1 is least.

RULES OF SIGNIFICANT FIGURES

- Significant figures follow in consecutive place value (next to each other), from left to right.
- All nonzero numbers (digits 1-9) are considered significant.
- Zeros are significant when used as placeholders between two nonzero numbers.
- Zeros at the end of a number and to the right of a decimal point indicate exactness of a number and are, therefore, significant.
- Zeros appearing in front of all nonzero numbers are not significant.
- Zeros appearing at the end of nonzero numbers and ending to the left of a decimal point may be assumed not significant.

The following examples of significant figures should aid your understanding.

- 27 has 2 significant figures.
- 0.56 has 2 significant figures.
- 100 has 1 significant figure.
- 0.09 has 1 significant figure.
- 80.9 has 3 significant figures.
- 6750.00 has 6 significant figures.
- 0.3040 has 4 significant figures.
- 0.00356 has 3 significant figures.
- 0.0002109 has 4 significant figures.

Rounding numbers

Sometimes a number is written that contains so many digits that the number is difficult to deal with. The number can be *rounded* (rounded off) to a number that is easier to deal with, and is roughly the same value. Numbers can be rounded to a specified amount of significant figures or to a certain place value.

RULES OF ROUNDING NUMBERS

- Decide on the desired number of significant figures or to which decimal place the number is to be rounded. The digit to the right of the rounded digit determines the outcome of rounding.
- All of the digits to the right of the rounded digit will be dropped (and replaced with zeros as placeholders where necessary).
- If the digit to the right is *5 or more,* change the rounded digit to *1 digit higher.* Note: A 9 changes to a 10.
- If the digit to the right is *less than 5,* the rounded digit does not change.

The following are examples of number rounding:

Rounding to three significant figures:
- 87,321 = 87,300
- 570,802 = 571,000
- 608.29 = 608
- 0.006237 = 0.00624
- 5.5555 − 5.56
- 110.99 = 111

Rounding to the nearest thousandth (third decimal place):
- 0.03653 = 0.037
- 0.90085 = 0.901
- 2.00110 = 2.001
- 56.0007 = 56.001

Practice Problems

Complete the following problems on a separate sheet of paper.

Identify the most and least significant figures in each of the following numbers.

1. 102	2. 512
3. 30.1	4. 8.15
5. 0.125	6. 0.521
7. 0.00308	8. 0.0000905
9. 308.1	10. 951.7

Round the following decimal numbers to 3 significant figures.

11. 875.012	12. 281.872
13. 35.2893	14. 10.90881
15. 9.99855	16. 3.333333
17. 0.66666	18. 0.555555
19. 18.1818	20. 999.9999

Round the following decimal numbers to the nearest hundredth (2 decimal places).

21. 0.1950812	22. 3.0036712
23. 1.06211	24. 2.15512
25. 2.00001258	26. 9.0008971
27. 0.89091	28. 1.00802010
29. 5.061501	30. 890.0981199

ADDITION AND SUBTRACTION OF DECIMAL NUMBERS

Addition and subtraction of decimal numbers are performed using the same rules and methods as used for whole numbers. However, there is one additional rule: decimal points must be aligned in a column so that you add or subtract numbers with the same place value.

With addition, zeros may be added optionally to the right of the decimal to fill missing places in the columns.

With subtraction, zeros *must* be added to fill the missing places in the columns to the right of the decimal. Example will follow.

RULES FOR ADDITION AND SUBTRACTION OF DECIMALS

1. Align decimal points in a column.
2. Add/subtract numbers in the same place value column. Add zeros to the number (optional for addition). Decimal point in the answer remains in the same location.

Sample Problem 3-1.

Add the following:

$$2.3 + 13.05 + 12 + 0.704 + 0.0039$$

Step 1. Align numbers, placing decimal points in a column.

```
 2.3
13.05
12.
 0.704
 0.0039
```

Step 2. Use zeros to fill out place value columns (optional). Add the numbers in each column. Decimal point remains in the same location.

```
  2.3000
 13.0500
 12.0000
  0.7040
+ 0.0039
 28.0579
```

Sample Problem 3-2.

Subtract the following:

$$5.1 - 2.96$$

Step 1. Align decimal points.

```
5.1
2.96
```

Step 2. Use zeros to fill in blank spaces. (A zero is required to allow subtraction.) Subtract and keep decimal point in the same location.

```
  5.10
- 2.96
  2.14
```

Practice Problems

Complete the following problems on a separate sheet of paper.

Perform the indicated operations.

1. $2.5 + 3.05 + 10$
2. $0.11 + 1.2 + 55.89$
3. $0.44 + 5.06 + 6.619$
4. $12 + 1.5 + 0.015 + 100.59$
5. $21.9 - 9.8$
6. $0.0145 - 0.0029$
7. $0.875 - 0.0382$
8. $0.022 - 0.01255$
9. $6.1 + 4.55 - 2.93$
10. $12.06 - 5.678 + 2.5658$

Solve the following word problems. Be sure to include units of measure (inches, feet, minutes, dollars, etc.).

11. What is the total thickness if four steel plates with the following thicknesses are welded together: 0.0125 inches, 0.0250 inches, 0.0125 inches, and 0.0085 inches?
12. A surveyor made the following length measurements on a section of a road: 100.8 feet, 89.05 feet, 65.175 feet, 231.1 feet. What is the total length of this section of road?
13. An employee's gross weekly pay is $450.50 (before deductions). How much is brought home after the following deductions have been made: $125.83 for taxes, $25.21 for FICA, $12.15 for health insurance, and $10.32 miscellaneous items?
14. Find the total length of the object shown in Fig. 3-2.

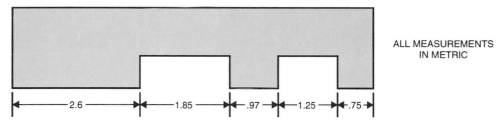

ALL MEASUREMENTS IN METRIC

Fig. 3-2. Find the total length.

15. Find dimension "A" in Fig. 3-3.

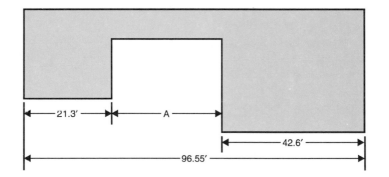

Fig. 3-3. Determine the missing dimension A.

MULTIPLICATION AND DIVISION OF DECIMAL NUMBERS

Multiplication and division with decimal numbers is performed much the same as with whole numbers. The rules demonstrated in this section explain how to deal with the decimal point.

Multiplication of decimals

It is important that you know how to multiply decimals. Many everyday problems require this knowledge. Follow the rules for multiplication for the proper procedure.

RULES FOR MULTIPLICATION OF DECIMALS

1. Set up the problem. The decimal points do not need to be aligned in a column.
2. Perform the multiplication. Ignore the decimal points.
3. Count the total number of places to the right of each decimal point in the numbers multiplied. Place the decimal point in the answer by counting from right to left, the total decimal places. If you count more places than you have numbers in your answer, fill places with zero. Example: $0.1 \times 0.1 = 0.01$.

Sample Problem 3-3.

Multiply the following:

$$3.245 \times 2.3$$

Step 1. Set up the problem as if multiplying whole numbers. Note that decimal points are not aligned.

$$\begin{array}{r} 3.245 \\ \times \quad 2.3 \\ \hline \end{array}$$

Step 2. Perform the multiplication. Ignore the decimal points.

$$\begin{array}{r} 3.245 \\ \times \quad 2.3 \\ \hline 9735 \\ 6490 \quad \\ \hline 74635 \end{array}$$ (decimal point not placed yet)

Step 3. Place the decimal point by counting the total number of places in the numbers multiplied. Then, place the decimal point the total number of places, counting from right to left.
- 3.245 has 3 decimal places.
- 2.3 has 1 decimal place.
- Total is 4 decimal places.

- Answer: 7.4635

Division of decimals

Using decimals allows for dividing numbers that would not divide exactly. The remainder that is formed when dividing such numbers is a fractional part. It can be written as a fraction, or the division can be continued as many decimal places as desired.

Keep in mind there are three formats to use when writing a division problem. Each of these (following) read "A divided by B."

$$A \div B \quad \text{or} \quad \frac{A}{B} \ \text{(also seen as A/B)} \quad \text{or} \quad B \overline{\smash{)}A} \ \text{("bracket" format)}$$

Referring to the three formats, remember that A is the *dividend* (number being divided), and B is the *divisor* (number by which the dividend is divided). The answer is the *quotient*. If the number in the divisor contains a decimal point, you will need to adjust the location of decimal points in the dividend and the divisor. The rules for division of decimals provide methods for dealing with the decimal point.

RULES OF DIVISION OF DECIMALS

- If the dividend and divisor do not contain decimal points, and the division does not come out exactly:
  a. Place a decimal point to the right of the dividend. Place a decimal point in the answer, aligned with the dividend decimal point.
  b. Add zeros to the dividend. Continue dividing to as many decimal places as necessary.
  Example:

$$\begin{array}{r} 4+ \\ 8\,\overline{\smash{)}35} \end{array} = \begin{array}{r} 4.375 \\ 8\,\overline{\smash{)}35.000} \end{array}$$

- If the dividend contains a decimal point and the divisor does not:
  a. Divide as usual.

b. Place a decimal point in the answer aligned with the dividend decimal point.
Example:

$$\frac{.5}{5\overline{)2.5}}$$

- If the *divisor* contains a decimal point:
  a. Move the decimal point to the far right of the divisor. Count the places moved.
  b. Shift the decimal point of the dividend right the same number of places. (If the dividend does not have a decimal point, first add one at the far right.)
  c. Use zeros as placeholders, if necessary.
  d. Divide as usual.
  e. Place a decimal point in the answer, aligned with shifted decimal point in the dividend.
  Example:

$$2.4\,\overline{)0.51}$$

$$2.4.\,\overline{)0.5.1}\quad \text{(move decimal point 1 place)}$$

$$\frac{.2125}{24.\,\overline{)51000}}\quad \text{(perform division)}$$

Sample Problem 3-4.

Divide the following:

$$29 \div 4$$

Step 1. Arrange the problem using the bracket format. Perform division and see if it divides exactly.

$$\frac{7+}{4\overline{)29}}$$
$$\frac{28}{1}$$

Step 2. Place a decimal point. Add zeros as needed. Bring the zeros down in the long division steps for use in subtracting. Decimal point in the answer is above the decimal point inside the bracket (dividend).

$$\frac{7.25}{4\overline{)29.00}}$$
$$\frac{28}{10}$$
$$\frac{8}{20}$$
$$\frac{20}{}$$

Sample Problem 3-5.

Divide the following:

$$1.92 \div 0.6$$

Step 1. Arrange the problem using the bracket format. Do *not* divide until the outside decimal point (divisor decimal point) is moved.

$$0.6\,\overline{)1.92}$$

Step 2. Move the decimal point in the divisor to the far right. The decimal point in the dividend is moved the same number of places.

$$0.\underset{\curvearrowright}{6.} \overline{\smash{\big)}1.\underset{\curvearrowright}{9.2}}$$

Step 3. Perform the division. Decimal point in the answer (quotient) is placed directly over the shifted decimal point of the dividend.

$$
\begin{array}{r}
3.2 \\
6 \overline{\smash{\big)}19\!2} \\
\underline{18} \\
12 \\
\underline{12}
\end{array}
$$

Practice Problems

Complete the following problems on a separate sheet of paper.

Perform the indicated operations. Round division problems to 3 significant figures.

1. 4.5×2.3
2. 66.125×20
3. 0.358×0.12
4. 0.6635×0.0021
5. 50×0.657
6. 9.5×4.505
7. $16 \div 4.2$
8. $21 \div 3.6$
9. $4.5 \div 32$
10. $7.4 \div 36$
11. $0.025 \div 0.125$
12. $0.375 \div 0.48$
13. $5.60 \div 0.85$
14. $6.80 \div 0.4$
15. $71 \div 3.2$

Solve the following word problems. Be sure to include units of measure (inches, feet, minutes, dollars, etc.).

16. Find the total length of material needed for the steel posts shown in Fig. 3-4.

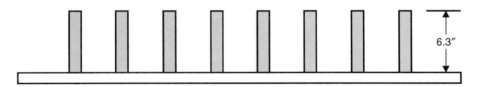

6.3″

Fig. 3-4. Find the total length of material needed for the steel posts.

17. If three 3.5-inch-thick boards are nailed together to form a base, what is the total thickness of the base?
18. If 285 guests are expected at a banquet, how many tables are needed if each table seats 8 people? Round your answer to the next whole number.
19. A 250-foot roll of electrical wire is to be cut into pieces measuring 10.4 feet each. How many pieces can be cut from the roll?
20. If one sheet of paper is 0.0015 inches thick, how thick is a 250-page book made from the same paper (without the cover)?

CONVERTING DECIMALS AND FRACTIONS

The system for units of measurement used in the United States uses both fractions and decimals. For example, a ruler measures inches, with fractions between the whole numbers; a micrometer measures in decimals to thousandths of an inch. In everyday use, it is often necessary to convert from one to the other.

ADDITION AND SUBTRACTION OF DECIMAL NUMBERS

Perform the indicated operations. Show your work. Do *not* round answers.
Example:

2.35 + 46.1 + 0.009

$$
\begin{array}{r}
2.35 \\
46.1 \\
+\ 0.009 \\
\hline
48.459
\end{array}
$$

26. 53.46 + 3.5 + 2.901
27. 0.0012 + 5.409 + 9.2
28. 100.0 + 10.001 + 3.059
29. 56.05 − 23.9
30. 0.0035 − 0.0009
31. 0.01 − 0.001
32. 1000 + 60.00 + 0.001
33. 1.001 − 0.0005
34. 5020.09 + 604.107 − 32.87
35. 10 + 100 − .01 + .001 − 01.01

MULTIPLICATION AND DIVISION OF DECIMAL NUMBERS

Multiply. Do *not* round answers.
Example: 3.4 × 6.1

$$
\begin{array}{r}
3.4 \\
\times\ 6.1 \\
\hline
34 \\
204\ \\
\hline
20.74
\end{array}
$$

36. 12.2 × 8   37. 10.9 × 5.6
38. 0.01 × 0.05   39. 25.05 × 20.5
40. 10.6 × 0.0005

Divide. Round to 3 decimal places, if necessary.
Example: 1 ÷ 3

$$
\begin{array}{r}
.3333 \\
3\,\overline{\smash{)}\,1.000} \\
\end{array} = \rangle\ .333
$$

$$
\begin{array}{r}
9 \\
\hline
10 \\
9 \\
\hline
10 \\
9 \\
\hline
1
\end{array}
$$

41. 49 ÷ 0.7   42. 5.6 ÷ 0.8
43. 7.2 ÷ 12   44. 0.125 ÷ 0.5
45. 0.66 ÷ 3   46. 0.857 ÷ 0.8
47. 9.36 ÷ 0.005   48. 875 ÷ 0.6
49. 4.5 ÷ 0.008   50. 67 ÷ 0.525

RULES FOR CHANGING DECIMALS TO FRACTIONS

1. Place numbers that are left of the decimal point to the left of the fraction.
2. Place numbers to the right of the decimal point in the numerator of the fraction.
3. Make the denominator of the fraction the same as the place value of the least significant figure (on the far right). There will be as many zeros as there are decimal places.
4. Reduce to lowest terms.
   Examples:

$$0.2 = \frac{2}{10} = \frac{1}{5}$$

$$0.25 = \frac{25}{100} = \frac{1}{4}$$

$$0.125 = \frac{125}{1000} = \frac{1}{8}$$

$$3.1875 = 3\,\frac{1875}{10,000} = 3\,\frac{3}{16}$$

RULES FOR CHANGING FRACTIONS TO DECIMALS

1. Place whole numbers to the left of the decimal point.
2. *Divide* the numerator by the denominator.
3. If it does not divide exactly, round off to 3 decimal places, unless otherwise specified.
   Examples:

$$\frac{1}{2} = 0.5$$

$$\frac{2}{3} = 0.667$$

$$5\,\frac{3}{8} = 5.375$$

Practice Problems

Complete the following problems on a separate sheet of paper.

Convert the following decimals to fractions. Reduce the fractions to lowest terms.

1. 0.6   2. 0.8
3. 0.25   4. 0.75
5. 0.035   6. 0.040
7. 4.004   8. 3.125
9. 5.375   10. 2.002

Convert the following fractions to decimals. Round the decimals to 3 significant figures.

11. $\frac{1}{2}$ 12. $\frac{3}{4}$

13. $\frac{2}{5}$ 14. $\frac{5}{8}$

15. $3\,\frac{3}{9}$ 16. $6\,\frac{5}{16}$

17. $5 \frac{9}{16}$ 18. $1 \frac{7}{16}$

19. $1 \frac{8}{12}$ 20. $6 \frac{7}{10}$

TEST YOUR SKILLS

Do *not* write in this book. Use a separate sheet of paper to complete the following problems. Show your work and your final answer.

PLACE VALUES—Whole Numbers

Write each digit times its place value.
Example:

$$506 = 5 \times 100$$
$$0 \times 10$$
$$6 \times 1$$

1. 1,324   2. 90,027
3. 819   4. 2,560,409

PLACE VALUES—Decimal Numbers

Write as a fraction, with the denominator being the place value of the least significant figure. Do *not* reduce.
Example:

$$0.00045 = \frac{45}{100,000}$$

5. 0.03   6. 0.005
7. 0.000079   8. 0.106
9. 0.5   10. 0.00204
11. 0.049   12. 0.0006

ROUNDING NUMBERS—3 Significant Figures

Round to 3 significant figures.
Example: 28.7956 = 28.8

13. 29.05   14. 1.561
15. 0.07392   16. 87,345
17. 560.6   18. 3.45531
19. 0.7269

ROUNDING NUMBERS—3 Decimal Places

Round to 3 decimal places.
Example: 56.035862 = 56.036

20. 2.500256   21. 0.005324
22. 29.59999   23. 0.0092306
24. 0.03551   25. 0.021903

Loan officers must use percentages to calculate loan payments and interest charges. (Jack Klasey)

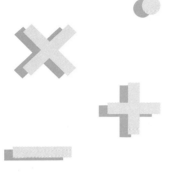

Chapter 4

Percentage

OBJECTIVES

After studying this chapter, you will be able to:
■ Explain the basic terminology and functions of percentage.
■ Express numerical values in terms of percentage.
■ Change percent values to decimals and fractions.
■ Compute bases, rates, or portions of numbers.
■ Determine maximum and minimum allowable measurements given percent tolerances.

Percentage is an application of fractional parts. With percentage, 100 percent (%) is used to describe 1 whole unit. Some examples of where percentage is used in everyday life are store discounts and markups, taxes, and paycheck deductions. Other examples arc tolerance of accuracy, and interest charges and earnings.

RELATIONSHIP OF PERCENTAGE TO FRACTIONS AND DECIMALS

With fractions, the denominator represents how many times a whole is divided. The numerator describes how many of the fractional parts are being used. Using fractions, a whole can be divided into any number of parts.

With decimals, a denominator is not needed. This is because the number of decimal places used indicates how many times a whole is divided. With decimals, the divisions of the whole get smaller in size with each place to the right a digit moves of the decimal point. The size of the division decreases by a factor of 1/10 with each decimal place.

Percentage is yet another way of expressing a fractional amount. The word "percent" means *per 100*. A **percentage** is a quantity expressed on a basis of 100. For example, if 50 of 200 parts produced at a factory one day were bad, then 25 *per 100*, or 25 percent, were bad. (50/200 = 25/100 = 0.25 = 25%) If 2 out of 8 people were off sick that day, then 25 percent of the workforce was absent. (2/8 = 0.25 = 25%)

From the examples, you should be able to see the relationship of percent to fractions and decimal fractions. A number expressed as a percentage is simply a decimal fraction multiplied by 100.

Converting fractions, decimals, and percent

To convert fractions to decimals, divide the numerator by the denominator. If a mixed fraction, place the whole number left of the resulting decimal point. To convert decimals to fractions, leave the number left of the decimal point alone. Write the number right of the decimal point over its place value. Reduce to lowest terms. (Refer to Chapter 3.)

To convert fractions to percent or percent to fractions, first change to a decimal. Then, follow applicable rule for converting.

RULES OF CONVERTING DECIMALS TO PERCENT

Move the decimal point two places to the right. Attach the "%" sign.
Note: The percent sign symbolizes the number has been multiplied by 100.
Examples: 0.25 = 25%
0.333 = 33.3%
0.5   = 50%
1.0   = 100%

RULES OF CONVERTING PERCENT TO DECIMALS

Remove the "%" sign. Move the decimal point two places to the left.
Examples: 45% = 0.45
75.8% = 0.758
2% = 0.02

Values greater than 100%

Since 100 percent represents a whole of something, a percentage value greater than 100 represents more than the whole. For example, 200 percent is 2 times 100 percent, or 2 times a whole. If a whole is represented by 10,000 gallons, 200 percent of it is 2 times 10,000 gallons, or 20,000 gallons. If a whole is represented by 0.25 pounds, 200 percent of it is 0.5 pounds. Converting the percentage to a decimal will tell you how many *times* larger than a whole a number is.

For example:

200% = 2.00 times larger than whole
150% = 1.50 times larger than whole
358% = 3.58 times larger than whole
675% = 6.75 times larger than whole

It is very important to identify exactly what the "whole" is. Beware if a problem is asking for percentage of increase, raise, gain, or similar term. This can be confusing. For example, $100 is *200% of* $50, or *2 times larger* than $50. However, $100 is a *100% increase* over $50. In this case, the increase is $50 ($100 − 50), and $50 is 100% of $50. A 200% increase would be 2 times the $50 increase, or a $100 increase. This is only the amount of increase, however. To get the new figure, add the increase to the original amount: $50 + 100 = $150.

In another example, a class of 100 students now has 150 students. The new class size is 150% of the original, or 1.5 times as large. However, this class *gained* in size by 50%, not 150%. The gain was 50 students (150 students − 100), and 50 is 50% of the original 100. Had the class gained in size by 150%, it would now have 250 students.

Practice Problems

On a separate sheet of paper, complete the following chart by expressing the given value in two other forms. Reduce fractions to lowest terms. Round decimals to 3 significant figures.

	Percent	Decimal	Fractions
1.	25%	?	?
2.	?	0.15	?
3.	?	?	1/8
4.	65.5%	?	?
5.	?	0.078	?
6.	?	?	3/5
7.	133.3%	?	?
8.	?	2.875	?
9.	?	?	5/16
10.	25.5%	?	?

SOLVING BASIC PERCENT PROBLEMS

To solve a problem containing percentage, you must first identify quantities given in the problem. The following are the quantities you want to identify:
- **Base (B)** — the original or total amount.
- **Portion (P)** — a fractional part (larger or smaller) of the base.
- **Rate (R)** — the figure in the problem that carries the % sign.

After identifying the quantities, use appropriate formula and find the unknown quantity. To solve a problem, you must know two of three quantities.

RULES FOR SOLVING BASIC PERCENT PROBLEMS

1. Identify the quantities. Select the proper formula below.
2. Substitute the quantities. Perform the arithmetic.

KNOWN	TO FIND	USE FORMULA	CONVERT R TO
Base and Rate	Portion	$P = B \times R$	Decimal
Portion and Base	Rate	$R = \dfrac{P}{B}$	Percent
Portion and Rate	Base	$B = \dfrac{P}{R}$	Decimal

Note: In a formula, a fraction is used to represent division.

Sample Problem 4-1.

Find 65% of 120.

Step 1. Identify the quantities. Select the proper formula.

> Base = 120
> Rate = 65%
> Portion = unknown
> Formula: $P = B \times R$

Step 2. Change the rate to a decimal and substitute the quantities into the formula. Perform the arithmetic.

$$P = 120 \times 0.65 = 78$$

Sample Problem 4-2.

What percent of 9 is 16?

Step 1. Identify the quantities. Select the proper formula.

> Portion = 16
> Base = 9
> Rate = unknown
> $R = \dfrac{P}{B}$

Step 2. Substitute the quantities into the formula. Perform the arithmetic. Change the rate to percent.

$$R = \frac{16}{9} = 1.7778 = 177.78\%$$

Sample Problem 4-3.

Nine (9) is 15% of what number?

Step 1. Identify quantities. Select proper formula.

$$
\begin{aligned}
\text{Portion} &= 9 \\
\text{Rate} &= 15\% \\
\text{Base} &= \text{unknown}
\end{aligned}
$$

Formula: $B = \dfrac{P}{R}$

Step 2. Change the rate to a decimal. Substitute quantities into the formula. Perform the arithmetic.

$$15\% = 0.15$$

$$B = \frac{9}{0.15} = 60$$

Sample Problem 4-4.

In a new house, 18 windows have been installed; 15 await installation. What percentage of windows have yet to be installed?

Step 1a. Find the base, which is the total amount, by adding the two portions together.

$$
\begin{array}{r}
15 \\
+\ 18 \\
\hline
33 \ \text{windows}
\end{array}
$$

Step 1b. Identify the quantities. Select proper formula.

$$
\begin{aligned}
\text{Portion} &= 15 \\
\text{Base} &= 33 \\
\text{Rate} &= \text{unknown}
\end{aligned}
$$

Formula: $R = \dfrac{P}{B}$

Step 2. Substitute quantities into the formula. Perform the arithmetic. Change the rate to percent.

$$R = \frac{15}{33} = 0.455 = 45.5\%$$

Practice Problems

Complete the following problems on a separate sheet of paper. For each problem:
a. identify the base, rate, and portion.
b. show the formula used.
c. substitute the values into the formula.
d. show the final answer.

1. What is 60% of 210?
2. What percent of 96 is 16?
3. Three (3) is 20% of what number?
4. What is 30% of 84?
5. What percent of 150 is 25?
6. Sixty (60) is 250% of what number?
7. What is 150% of 40?
8. What percent of 90 is 30?
9. Twenty-five (25) is 100% of what number?
10. If each question on a 50-question math test has the same point value, what is a student's score if 15 questions were answered incorrectly?
11. In an applied math class, there are 15 males and 8 females. What is the percentage of males in the class?
12. In a high school parking lot, 75% of the cars belong to students. The remaining cars belong to school employees. If there are 248 cars in the lot, how many belong to school employees?

13. A grade of 70% is received on a 20 question test. If all questions have the same point value, how many questions were answered correctly?
14. If 1 out of every 8 students in the senior class of a community college does not graduate, what percentage of the senior class graduates?
15. Before oil is added to a certain road surface material, the other ingredients are mixed. The mixture contains 45 percent 3/4-inch stone and 32 percent 1/2-inch stone. The balance of the mixture is made up of sand. If the mixture weighs 15,000 pounds before the oil is added, how many pounds of each ingredient are in the mixture?

PERCENT TOLERANCE

A tolerance is allowed on measurements because it is impossible to be consistently perfect. The **nominal** value is the exact or ideal measurement given in specifications. The **tolerance** is the acceptable range of accuracy. The tolerance can be expressed as either a specific amount (± example 0.001") or as a percentage. This section deals with percent tolerance.

Tolerance is stated as plus or minus (±), where + is the maximum and − is the minimum measurement allowed. The tolerance range allows for inaccuracies. If the actual measurement is within this range, it is acceptable. As examples:

- 21 inches ± 5% = 21 + 5% to 21 − 5%.
- 120 ohms ± 10% = 120 + 10% to 120 − 10%.
- $35,000 ± 2% = 35,000 + 2% to 35,000 − 2%.

RULES FOR SOLVING PERCENT TOLERANCE

1. Find the range of tolerance by multiplying the nominal by the percentage.
2. Find the maximum by *adding* the range to the nominal.
3. Find the minimum by *subtracting* the range from the nominal.

Sample Problem 4-5.

Determine the maximum and minimum allowable measurements of 3 inches ± 5%.

Step 1. Find the range by multiplying the nominal by the percentage. Percent must first be changed to decimal form.

$$5\% = 0.05$$
$$3 \times 0.05 = 0.15 \text{ in.}$$

Step 2. Find the maximum by adding the range to the nominal.

$$\begin{array}{r} 3. \\ + 0.15 \\ \hline 3.15 \text{ in. maximum} \end{array}$$

Step 3. Find the minimum by subtracting the range from the nominal.

$$\begin{array}{r} 3.00 \\ - 0.15 \\ \hline 2.85 \text{ in. minimum} \end{array}$$

Practice Problems

Complete the following problems on a separate sheet of paper. Include units of measure (inches, feet, minutes, dollars, etc.) where applicable.

Determine the maximum and minimum values based on the percent tolerance. Do not round answers.

1. 350 ± 5%   2. 120 ± 10%
3. 48 ± 20%   4. 24 ± 3%

5. 0.36 ± 1%
6. 0.9 ± 2%
7. 0.082 ± 0.5%
8. 0.016 ± 1.2%
9. 2.9 ± 2.5%
10. 5.6 ± 0.2%

11. A resistor used in an electronic circuit has a rated value of 180 ohms ± 10%. The actual measured value of this resistor is 190 ohms. What are the maximum and minimum values? Is the measured value within the acceptable range?
12. The drawings for a steel pin call for a diameter of 1.50 inches ±1%. What are the maximum and minimum sizes? The actual measurement, which was taken with a micrometer, is 1.436 inches. Is this steel pin within the acceptable range?
13. What is the range if the accuracy of a voltmeter is ±10% when measuring 15 volts?
14. A telephone survey had an error range of 3%. If 260 people were contacted for the survey, and 38% voted yes, 48% said no, and the remainder had no opinion, what is the range with no opinion?
15. A car loan requires a 15% down payment. How much money is needed for a down payment when purchasing a $14,000 automobile?

TEST YOUR SKILLS

Do *not* write in this book. Use a separate sheet of paper to complete the following problems. Show your work and your final answer.

RELATIONSHIP OF PERCENTAGE TO FRACTIONS AND DECIMALS

Express the given values in terms of the other two forms. Reduce fractions to lowest terms.

	Percent	Decimal	Fraction
1.	?	?	$\frac{4}{5}$
2.	$6\frac{1}{2}\%$	?	?
3.	45%	?	?
4.	?	0.03	?
5.	?	?	$\frac{5}{8}$

BASIC PERCENT PROBLEMS

For each problem:
    a. Identify the base, rate and portion.
    b. Show the formula used.
    c. Substitute the values into the formula.
    d. Show the final answer.
6. What is 55% of 280?
7. What percent of 120 is 36?
8. Eighteen (18) is 70% of what number?
9. Find 40% of 80.
10. What percent of 50 is 125?
11. What percent of 90 is 30?
12. Seventy-five (75) is 75% of what number?

WORD PROBLEMS WITH PERCENT

13. To buy a $15,000 car, a 25% down payment is required. How much money is required?
14. A machine produced 60 parts. If the length of 3 of them is out of tolerance, what is the percentage of defective pieces?
15. A box of 300 screws has been used to put up drywall. If the job is 60% complete, how many more screws will be required to finish the job?
16. A quenching pit had 30,000 gallons of water in it. After more water was added, it had 33,000 gallons. What percent of increase does this represent?
17. Motor 1 has 370 watts of power available at its output when power is applied at the rate of 500 watts. Motor 2 has 418 watts of power available when 550 watts are applied to it. Which motor is most efficient? Justify the answer with a percentage comparison.
18. A painting contractor adds 15 percent to the estimate of paint for waste. If 29 gallons are estimated, excluding waste, how many 5-gallon pails must be purchased for the job, including waste?
19. At what percent of capacity is a truck carrying 11,525 pounds if full capacity is 14,000 pounds?
20. A recycling truck has 1200 cubic feet of space for recyclable material—newspaper, aluminum, and plastic. The truck is filled to capacity. With plastic being 50 percent by volume, and paper 35 percent, how many cubic feet of space does the aluminum occupy?

PERCENT TOLERANCE

Determine the maximum and minimum amount based on the percent tolerance. Do *not* round answers.

21. 160 feet ± 10%.
22. 470 ohms ± 5%.
23. 0.25 ounces ± 2%.
24. 0.005 inches ± 1.3%.
25. 0.125 millimeters ± 0.5%.

PROBLEM-SOLVING ACTIVITIES

Activity 4-1.
Percentage of male and female students in class

> **Objective:** To determine the percentage of males and females in your math class.
> **Instructions:**
> 1. Count the total number of students in your classroom.
> 2. Count the number of males and the number of females.
> 3. Determine the percentage of each.

Activity 4-2.
Percentage of two-door cars

> **Objective:** To determine the percentage of two-door cars in a section of your school parking lot.
> **Instructions:**
> 1. Select a section of the parking lot, such as the front row, and count the total number of cars in the section.
> 2. Count the two-door models in the section.
> 3. Determine the percentage of two-door cars in the chosen section of the lot.

Part II

Applied Basic Math

Chapter 5

Personal Finances

OBJECTIVES

After studying this chapter, you will be able to:
- Explain basic terminology used with personal finance.
- Perform arithmetic with dollars and cents.
- Estimate net income.
- Determine simple or compound interest on a sum of money.
- Estimate monthly loan payments.
- Apply percentages to figure merchandise pricing.

Everyone needs to know how to apply math to everyday problems of money. Not only is money used to purchase merchandise, it is also used for payment of labor and services. Banks specialize in money matters, with loans and savings and checking accounts being the most familiar.

DOLLARS AND CENTS

In the United States and Canada, the units of measure for money are dollars and cents. Although these units are often used consecutively when an amount of money is stated, they represent different units of measure. The difference as to whether units of a given amount (preceded by a dollar sign) are stated as dollars or cents depends on which side of the decimal point numbers appear.

The dollar sign

The dollar sign ($) is the symbol of the dollar. It is written in front of a number. With no decimal point, numbers are whole dollars. For example, $64 is 64 whole dollars—no fractional parts.

When a monetary amount is a decimal number, numbers to the right of the decimal point, as usual, represent fractional parts. The dollar system is based on 100 fractional parts. Each part represents 1 cent. With no numbers left of the decimal point, a fraction of a dollar noted with a dollar sign is generally read in cents. For example, $0.78 or 78 hundredths dollars, is read "78 cents." This fractional part of a dollar is represented by 78 whole cents.

When numbers are present both left and right of the decimal, a monetary amount is represented by whole dollars and fractional parts of a dollar. The fractional parts are, again, read as cents. For example, $1.25 is "1 dollar and 25 cents."

A number to the right of a decimal can be more than two decimal places. When this is the case, the number can be said to represent dollars, cents, and fractional parts of a cent. For example, $2.875 is "2 dollars and 87.5 cents."

The cent sign

The cent sign (¢) is the symbol of the cent. A cent is one hundredth of a dollar. To convert dollars to cents, move the decimal point 2 places right, drop the $ sign, and add a ¢ sign. (This is similar to converting decimals to percent.) As examples:

$0.50 = 50¢
$0.75 = 75¢
$0.02 = 2¢
$0.125 = 12.5¢
$3.80 = 380¢

Caution: A common mistake is to write both the dollar sign and cents sign at the same time.
For example:

- $.25¢ — This is wrong!
- $1.36¢ — This is wrong!

Arithmetic with dollars and cents

When performing arithmetic with dollars and cents, it is important to change all of the numbers in the problem to the same unit of measure. You cannot mix units of measure in the same problem. Either symbol can be used; however, it is usually easier to change cents to dollars.

Sample Problem 5-1.

What is the sum of $3.45 and 25¢?
Step 1. Change to the same units.

25¢ = $0.25

Step 2. Perform the arithmetic.

$$\begin{array}{r} \$3.45 \\ + \quad 0.25 \\ \hline \$3.70 \end{array}$$

Sample Problem 5-2.

It took 49.3 gallons of fuel to make a certain trip. If the average cost was $1.25 per gallon, how much was spent on fuel?
Step 1. Identify the operation.
Multiplication is used.
Step 2. Perform the arithmetic. Include the units. Round to the nearest cent.

$$\begin{array}{r} 49.3 \\ \times \quad 1.25 \\ \hline \$61.625 = \$61.63 \end{array}$$

Step 3. Check the answer.
The units are correct. The answer looks reasonable.

Practice Problems

Complete the following problems on a separate sheet of paper.

Write each of the following expressions as a number with a dollar sign and a decimal point.

1. One dollar and twenty-five cents.
2. Fifty dollars.
3. Ten and 6/100 dollars.
4. Two hundred and sixty cents.
5. Seven dollars and fifty and one-half cents.
6. Nine hundred and sixty-three dollars and one cent.
7. Eleven cents.

8. Twelve dollars and thirty-five and three-fourths cents.
9. Eight hundred and seventy-one cents.
10. Thirty-two cents.

Write each of the following numbers as a written expression.
11. $4.32
12. $53.90
13. $.85
14. $56
15. $3.06
16. $0.55
17. $409.85
18. $235.02
19. $21.00
20. $.251

Write each of the following expressions as a number with a cents sign.
21. Sixty-three cents.
22. Five cents.
23. One hundred and thirty cents.
24. Eighty-one and one-half cents.
25. Nine cents.
26. Eighty-five cents.
27. Two hundred and one cents.
28. Two dollars and fifteen cents.
29. Twelve dollars and thirty-nine cents.
30. Twenty-eight and one-quarter cents.

Write each of the following numbers as a written expression. Include only the word "cents," not dollars.
31. 33¢
32. 18¢
33. $0.15
34. 235¢
35. 873¢
36. $.48
37. 12.5¢
38. $3.25
39. $9.12
40. 25.75¢

Perform the indicated arithmetic. Include units in the answers.
41. What is the sum of $45.30, $0.21, $1.82, $0.75, 45¢, 632¢, five-dollars, and thirty-five cents?
42. What is the difference between $0.85 and 35¢?
43. What is the sum of 56¢ and four-dollars and ten cents?
44. If one book costs $22.95, how much will two books cost?
45. If one gallon of gasoline costs $1.25, how much will 10.8 gallons cost?
46. A meal costs $10.50, a beverage costs 65¢, dessert costs $3.70, tax is seventy-five cents, and the tip is one dollar. What was the total spent?
47. Six members of a band earn $255 for an evening. If they split the money evenly, how much does each get paid?
48. If you are paid $15 for cutting the neighbor's lawn and your lawnmower burns $0.35 worth of gasoline during the job, what is your profit?
49. A customer at a computer repair center is charged $45.50 per hour for 5.25 hours of labor. She is also charged $125.65 for parts. What is the total cost of the repair?
50. Which is a better buy (lowest cost per ounce): a 20-ounce box of cereal for $3.15 or a 15-ounce box for $2.90? What is the difference per ounce?

GROSS INCOME

The amount of money a person earns while working is not the same amount of money in the paycheck received at the end of the week. There is federal income tax, state income tax, social security tax, health insurance, union dues, etc., that must first be paid. These items, called *payroll deductions,* are subtracted from the total amount earned.

Gross income

The money that a person earns prior to payroll deductions is **gross income**. If you earn $5 per hour, with a 40-hour workweek your gross income for the week will be $200. Gross income, or gross pay, can be given as hourly, weekly, monthly, or yearly.

When working for an hourly wage, an employee is usually paid straight time for the normal 40-hour week. *Straight time* is pay for a standard work period. It is paid at a rate equal to an employee's base pay. *Base pay* is the rate or amount of pay prior to payroll deductions and exclusive of extra payments or allowances. For time worked in excess of 40 hours, *overtime* is generally paid. Overtime is commonly paid at time and a half, or 1 1/2 times the base pay. An employee may have extra pay for such things as hazardous pay, high time, shift differential, and holiday pay.

Sample Problem 5-3.

A help-wanted ad shows an opening for a carpenter. Pay is $8.50 per hour plus time and a half for overtime. The standard workweek is 40 hours plus 8 hours overtime. Determine the weekly gross pay.

Step 1. Identify the operation.

Multiply to find the straight time, overtime pay rate, and pay. Add these amounts together for the weekly gross pay.

Step 2. Perform the arithmetic. Include the units. Calculate the pay for the first 40 hours.

$$\begin{array}{r} \$8.50 \\ \times \quad 40 \\ \hline \$340.00 \end{array}\text{ straight-time pay}$$

Calculate the pay for 8 hours overtime, at 1 1/2 times the normal hourly rate.

$$\begin{array}{r} \$8.50 \\ \times \quad 1.5 \\ \hline \$12.75 \end{array}\text{ per hour overtime} \qquad \begin{array}{r} \$12.75 \\ \times \quad 8 \\ \hline \$102.00 \end{array}\text{ overtime pay}$$

Add the straight time to the overtime for a total gross pay.

$$\begin{array}{r} \$340.00 \\ + \quad 102.00 \\ \hline \$442.00 \end{array}\text{ gross income}$$

Step 3. Check the answer.

The units are correct. The answer looks reasonable.

Practice Problems

Complete the following chart on a separate sheet of paper. Annual income is for 52 weeks. Overtime pay (hours over 40 per week) is calculated as 1 1/2 times the base hourly rate. Round your answers to the nearest cent.

	Weekly Hours	Base Hourly Rate	Regular Time Pay	Overtime Pay	Combined Weekly Income	Annual Gross Income
1.	40	$6.25	?	?	?	?
2.	35	?	?	?	?	$18,000
3.	40	?	?	?	$340.00	?
4.	50	$10.25	?	?	?	?
5.	48	$16.00	?	?	?	?
6.	40	?	?	?	$650.00	?

	Weekly Hours	Base Hourly Rate	Regular Time Pay	Overtime Pay	Combined Weekly Income	Annual Gross Income
7.	35	?	?	?	?	$28,028
8.	60	$5.10	?	?	?	?
9.	45	$14.30	?	?	?	?
10.	50	$8.60	?	?	?	?

NET INCOME AND PAYROLL DEDUCTIONS

Net income is the amount of money received after payroll deductions. The amount in a paycheck is net income. Net income is also called *take-home pay.*

An *employer* pays someone to work. An *employee* is the worker. Many of the deductions an employer uses to calculate net pay for an employee are from books of tables, supplied by the IRS (Internal Revenue Service). Payroll deductions for taxes are generally a percentage of gross income. These vary from person to person depending on personal situations, such as number of dependents. The tables supplied in the books make it easy for the employer to determine the amount to *withhold* (deduct).

As an employee, you will not likely have IRS tax tables at hand. However, for any amount of hours you work, you can figure, roughly, your take-home pay. This can be done by finding the percentage that payroll deductions are of your gross income. To do this, you will need one of your old pay stubs, which was at one point attached to your paycheck. The *pay stub* shows payroll deductions. Once the percentage is figured, you can estimate your take-home pay prior to receiving your paycheck.

In Chapter 4, percentages were demonstrated using the concepts of base, rate, and portion. These same concepts are used again, where base is the *gross pay,* rate is the *applied deduction percentage,* and portion is the dollar amount of the *deduction.*

The following are usual deductions to gross income for taxes and are referred to as *tax withholdings.* (This list is not all-inclusive. In some locales, for example, city and/or county taxes are also withheld.)

- *Federal income tax* is determined by applying a percentage to the gross income. The percentage varies with every employee.
- *FICA — Social security tax* is determined by applying a percentage to the gross income. The percentage is the same for all employees. (The employer also pays an amount equal to the amount each employee pays.) FICA percentages increase slightly from year to year.
- *State income tax* is determined by individual state regulations and varies from state to state. The examples in this book apply a percentage to the gross income.

Very often, there are other deductions made to your gross income. They are for such things as health insurance, employee savings plans, union dues, etc. You should account for these as well when estimating your take-home pay.

Sample Problem 5-4.

Using the listed pay stub amounts, calculate the percentage withheld for federal income tax, FICA, and state income tax. Use the amounts given to determine the net income. Also, determine the net income expected on $1000.00 gross income.

PAY STUB: Gross = $600.00. Federal tax = $72.85. FICA = $44.71. State tax = $16.73.

Federal tax percentage:
Base = $600. Portion = $72.85. Find rate.

$$R = \frac{P}{B} = \frac{72.85}{600} = 0.1214 = 12.14\% \text{ of gross}$$

FICA percentage:
Base = $600. Portion = $44.71. Find rate.

$$R = \frac{P}{B} = \frac{44.71}{600} = 0.07451 = 7.45\% \text{ of gross}$$

State tax percentage:
Base = $600. Portion = $16.73. Find rate.

$$R = \frac{P}{B} = \frac{16.73}{600} = 0.02788 = 2.79\% \text{ of gross}$$

Net income:

$$
\begin{array}{r}
\$72.85 \text{ federal} \\
44.71 \text{ FICA} \\
+ \quad 16.73 \text{ state} \\
\hline
\$134.29 \text{ total deductions}
\end{array}
$$

$$
\begin{array}{r}
\$600.00 \text{ gross} \\
- \quad 134.29 \text{ deductions} \\
\hline
\$465.71 \text{ pay received}
\end{array}
$$

Total tax percentage:
Base = $600. Portion = $134.29. Find rate.

$$R = \frac{P}{B} = \frac{134.29}{600} = 0.2238 = 22.38\% \text{ of gross}$$

Total tax expected:

$$
\begin{array}{r}
\$1000 \\
\times \quad .224 \\
\hline
\$ \ 224
\end{array}
$$

Net income expected:

$$
\begin{array}{r}
\$1000 \\
- \quad 224 \\
\hline
\$ \ 776
\end{array}
$$

Practice Problems

Complete the following chart on a separate sheet of paper. Calculate the *net income* based on the information given. Percentages are calculated on gross income. Show your work and your final answers. Round your answers to the nearest cent.

	Gross Income	Federal Tax	State Tax	FICA	Insurance	Net Income
1.	$680 weekly	$170	$34	$54.40	$26	?
2.	$31,200 annually	35%	5%	8%	$1300	?
3.	$460 weekly	25%	4%	8%	$18	?
4.	$28,600 annually	$8580	$900	$2288	$1200	?
5.	$2320 monthly	$800	$140	$185	$128	?

SALES AND PROPERTY TAXES

In addition to the taxes paid on income, there are other types. Sales tax and property tax are two of these. Each is based on a percentage applied to the value of an item. The percentage varies widely from state to state and within different areas of the same state.

Sales tax

Sales tax is a tax on goods and services. It is paid on items bought in retail stores, on restaurant meals, lodging, and airfare, to name a few. It is based on a percentage applied to the purchase price of such items, is added to the purchase price, and paid when the purchase is made.

Sample Problem 5-5.

With sales tax of 5.5%, what is the total amount for a stereo that costs $357?

Step 1. Determine the sales tax portion.

Basc = $357. Rate = 5.5%. Find portion.

$$P = B \times R = \$357 \times 0.055 = \$19.635$$
$$\text{Tax} = \$19.64 \text{ (round to nearest cent)}$$

Step 2. Add tax to purchase price.

$$\begin{array}{r} \$357.00 \\ + \quad 19.64 \\ \hline \$376.64 \text{ total cost} \end{array}$$

Property tax

Property tax is tax on property—real estate or personal. It may be collected by the local government of the region in which the property is owned. The amount is usually a percentage of the estimated assessed value of the property. The tax rate is called the *mil rate* or millage, which is dollars of tax per thousand dollars of property. The units for mil rate are mils. (1 mil = $1/$1000 = 0.001)

Example: Tax on a $30,000 property at mil rate of 5 mils.

mil rate = 5 mils = 5 × 1 mil = 5 × 0.001 = 0.005
tax = property value × mil rate = $30,000 × 0.005 = $150

Notice that mil rate is changed into a decimal number before being multiplied with property value. As you can see from the example, mils have been changed to decimal form, in effect, by moving the decimal 3 places left. Mil rate may be changed to percentage, then, if desired. As examples:

- 8 mils = 0.008 = 0.8%
- 15 mils = 0.015 = 1.5%
- 36 mils = 0.036 = 3.6%
- 41.5 mils = 0.0415 = 4.15%

You may want to think of a property tax problem as a percentage problem. Mils are changed to percent and used as the rate.

Sample Problem 5-6.

Find the taxes on an $80,000 house in a town with a mil rate of 23.6 mils.

Base = $80,000. Rate = 2.36% (23.6 mils = 2.36%). Find portion.

$$P = B \times R = \$80,000 \times 0.0236 = \$1888$$
$$\text{Tax} = \$1888$$

Practice Problems

Complete the following charts on a separate sheet of paper. Round answers to the nearest cent.

	Price of Item	Sales Tax	Total Purchase Price
1.	$15.00	5%	?
2.	$238.50	6%	?
3.	$46.80	4%	?

	Price of Item	Sales Tax	Total Purchase Price
4.	$0.35	3%	?
5.	$28.95	5.5%	?
6.	$358.00	6.5%	?
7.	$6,500	3%	?
8.	$12,600	4%	?
9.	?	5%	$126.00
10.	?	6%	$42.40

	Value	Mil Rate	Property Tax
11.	$5,600	30	?
12.	$10,800	35	?
13.	$18,400	40.1	?
14.	$125,000	25	?
15.	$235,000	20.3	?
16.	$8,400	32	?
17.	$12,600	38	?
18.	$42,200	41.2	?
19.	$135,000	23	?
20.	$410,000	26.4	?

INTEREST

Interest is a percentage of a sum of money that is saved or loaned. With savings, it is money received by a savings customer. With loans, it is money paid by a borrowing customer. Banks and other financial institutions pay interest on savings accounts as incentive for a customer to keep money in the bank. They charge interest on loans while a customer uses the money. This interest is a source of income for the bank. Interest is calculated based on three items—principal, interest rate, and time. **Principal** is the original amount of a loan or deposit on which interest is paid. **Interest rate** is the percentage applied to the principal. **Time** has to do with duration and the period for interest compounding. (Explanations will follow.)

Simple interest

Simple interest is interest applied only to the principal of a savings account (or loan). At the end of a period of time, simple interest will total the same amount whether accumulated in installments (partial payments) or in a single payment. Simple interest is not used on most savings accounts. Normally, interest is paid on interest, too, not just the principal. This results in more money being paid. However, simple interest demonstrates the concepts of interest payments. It can also provide a quick, approximate calculation of interest compounding when short periods of time are involved.

Interest rates must be given on a one year basis. This basis is called the **annual percentage rate (APR)**. APR is average annual interest divided by outstanding principal.

Annual interest is found by dividing total interest by the total period of time in years. If the period of time involved is not exactly one year, fractions of a year are used. The following are examples of months expressed as such:

$$3 \text{ months: } \frac{3 \text{ months}}{12 \text{ months}} = \frac{1}{4} = 0.25 \text{ year}$$

$$4 \text{ months: } \frac{4 \text{ months}}{12 \text{ months}} = \frac{1}{3} = 0.333 \text{ year}$$

$$8 \text{ months: } \frac{8 \text{ months}}{12 \text{ months}} = \frac{2}{3} = 0.667 \text{ year}$$

$$9 \text{ months: } \frac{9 \text{ months}}{12 \text{ months}} = \frac{3}{4} = 0.75 \text{ year}$$

$$14 \text{ months: } 1\frac{2 \text{ months}}{12 \text{ months}} = 1\frac{1}{6} = 1.167 \text{ year}$$

For a single-payment plan, outstanding principal is the same as the original. For installment plans, outstanding principal is the original principal minus total payments made (excluding interest).

The following problems are examples of finding simple interest and APR.

Sample Problem 5-7.

Determine the simple interest on a $1500 savings account held for 6 months at an APR of 5 1/2%.

Step 1. Change the months to a partial year.

$$6 \text{ months} = \frac{6}{12} = \frac{1}{2} = 0.5 \text{ yr.}$$

Step 2. Determine simple interest for 1 year.
Base = $1500. Rate = 5.5%. Find portion.
$P = B \times R = \$1500 \times 0.055 = \$82.500$
Yearly interest = $82.50

Step 3. Multiply by the partial year for the total interest paid.
$82.50 × 0.5 = $41.25
Interest paid for 6 months = $41.25

Sample Problem 5-8.

Determine the simple interest paid on a $750 single-payment loan after 2 years at an APR of 8%.

Step 1. Determine simple interest for 1 year.
Base = $750. Rate = 8%. Find portion.
$P = B \times R = \$750 \times 0.08 = \$60$
Yearly interest = $60

Step 2. Multiply by the number of years for the total interest paid.
$60 × 2 = $120
Interest paid for 2 years = $120

Compound interest

Compound interest is paid periodically. Each time the interest is paid, the amount is added to the account balance. Interest is paid on the *balance* (principal plus *accrued* [accumulated] interest). This results in a higher amount of interest being paid each period. Compound interest is used on most savings accounts.

Sample Problem 5-9.

Determine the balance in a savings account at the end of one year given the following information:
- Starting balance is $800.
- Annual interest rate is 6.25%.

• Interest is compounded quarterly.

Step 1. Calculate interest the first quarter.

Determine simple interest for 1 year.

Base = $800. Rate = 6.25%. Find portion.

$$P = B \times R = \$800 \times 0.0625 = \$50$$

Yearly interest = $50

Multiply by 0.25 for 1 quarter.

$$\$50 \times 0.25 = \$12.50$$

Interest paid for 1 quarter = $12.50

Add the interest earned to the principal. This is the new balance at the end of the first quarter.

$$
\begin{array}{r}
\$800.00 \\
+\quad 12.50 \\
\hline
\$812.50 \text{ balance}
\end{array}
$$

Step 2. Calculate interest the second quarter.

Determine simple interest for 1 year.

Base = $812.50. Rate = 6.25%. Find portion.

$$P = B \times R = \$812.50 \times 0.0625 = \$50.78$$

Yearly interest = $50.78

Multiply by 0.25 for 1 quarter.

$$\$50.78 \times 0.25 = \$12.70$$

Interest paid for 1 quarter = $12.70

Add the interest earned to the principal. This is the new balance at the end of the second quarter.

$$
\begin{array}{r}
\$812.50 \\
+\quad 12.70 \\
\hline
\$825.20 \text{ balance}
\end{array}
$$

Step 3. Calculate interest the third quarter.

Determine simple interest for 1 year.

Base = $825.20. Rate = 6.25%. Find portion.

$$P = B \times R = \$825.20 \times 0.0625 = \$51.58$$

Yearly interest = $51.58

Multiply by 0.25 for 1 quarter.

$$\$51.58 \times 0.25 = \$12.90$$

Interest paid for 1 quarter = $12.90

Add the interest earned to the principal. This is the new balance at the end of the third quarter.

$$
\begin{array}{r}
\$825.20 \\
+\quad 12.90 \\
\hline
\$838.10 \text{ balance}
\end{array}
$$

Step 4. Calculate interest the fourth quarter.

Determine the interest for 1 year.

Base = $838.10. Rate = 6.25%. Find portion.

$$P = B \times R = \$838.10 \times 0.0625 = \$52.38$$

Yearly interest = $52.38

Multiply by 0.25 for 1 quarter.

$$\$52.38 \times 0.25 = \$13.10$$

Interest paid for 1 quarter = $13.10

Add the interest earned to the principal. This is the new balance at the end of the fourth quarter.

$$
\begin{array}{r}
\$838.10 \\
+\quad 13.10 \\
\hline
\$851.20 \text{ balance}
\end{array}
$$

Estimating monthly loan payments

When thinking about taking out a loan, whether for a car, boat, land, house, or any other item, monthly payments must be considered. For any such large purchase, a down payment is usually required. The loan would be for the amount remaining after the down payment.

To determine the exact monthly payment, a book of monthly payment tables is needed. Banks issue these books to their employees. The tables simplify determination of payments. Actual calculations to figure them are quite complicated. Interest payments are compounded, then reduced with each payment. Most people do not have these books for their own use.

It is possible to estimate monthly payments without the book of tables. Amount of the loan, APR, and term of the loan must be known. (The term of the loan is the time required to pay it off.) The calculations shown here result in a figure that will be close enough to assist in making a decision about taking out the loan.

Sample Problem 5-10.

Estimate the monthly payments on a $5000 loan, with an APR of 12% for 5 years.

Step 1. Estimate total interest paid.

Base = $5,000. Rate = 12%. Find portion.

$$P = B \times R = \$5000 \times 0.12 = \$600$$

Yearly interest = $600

Multiply by the number of years for the total interest paid.

$600 \times 5 = \$3000$ total interest

Interest paid for 5 years = $3000

Step 2. Add the total estimated interest to the loan for the total cost.

$$\begin{array}{r} \$5000 \\ +\ \ \underline{3000} \\ \$8000 \text{ total cost} \end{array}$$

Step 3. Divide by the number of months for the estimated monthly payment.

$8000 \div 60 = \$133$ estimated monthly payment

Practice Problems

Complete the following charts on a separate sheet of paper.

Complete the following chart by calculating the simple interest paid on a single payment loan based on the values given. Round your answers to the nearest dollar.

	Principal	APR %	Time in Months	Interest Paid
1.	$1200	7.0	12	?
2.	$5600	4.0	18	?
3.	$10,500	3.5	24	?
4.	$56,800	6.0	36	?
5.	$480	5.5	6	?
6.	$3300	2.5	15	?
7.	$9800	5.0	20	?
8.	12,000	3.0	48	?
9.	$38,000	4.5	66	?
10.	$250	6.5	9	?

Complete the following chart by calculating the balance in a savings account at the end of the time shown if the interest is compounded as specified. Round your answers to the nearest dollar.

	Starting Balance	APR %	Compounding Period	Time	Balance
11.	$200	5	quarterly	1 year	?
12.	$600	6	monthly	3 months	?
13.	$3000	4	quarterly	9 months	?
14.	$8000	3	monthly	3 months	?
15.	$12,000	5	quarterly	6 months	?

Complete the following chart by calculating the estimated monthly payment for the loans shown in the following chart. Round your answers to the nearest dollar.

	Loan	APR %	Time	Monthly Payment
16.	$600	5	1 year	?
17.	$800	6	6 months	?
18.	$2600	7	1 1/2 years	?
19.	$3200	12	4 years	?
20.	$15,000	10	5 years	?
21.	$400	4	4 months	?
22.	$700	8	1 year	?
23.	$1800	14	2 1/2 years	?
24.	$6400	18	3 years	?
25.	$24,000	12	10 years	?

MERCHANDISE PRICING USING PERCENTAGE

The *retail price* is the amount charged to consumers in retail stores. Many different methods are used by stores to determine the selling price of merchandise. This section will demonstrate using percentages to calculate the retail price.

Wholesale cost is the price a store pays to buy an item. In addition, the store has operating expenses. Rent, utilities, and payroll are major operating expenses. **Markup** is an amount added to wholesale cost. Wholesale cost plus markup is retail price. The markup allows the store to pay operating expenses. It enables the store to make some profit, to reinvest, and to prepare for future cost increases.

Markup using percentage

Markup percentage rate varies widely from one store to the next. Often, different items in the same store have different markups. A faster-selling item, for example, may have a higher markup than an item that sits on the shelf. Stores will take less profit, if necessary, in order to speed the turnover of slower items. Competition is also a major factor in setting the markup percentage of an item. To determine values where markup is a factor, follow rules of percent markup.

RULES OF PERCENT MARKUP

- Retail price is always the result of *adding* markup to wholesale cost.

    retail $ = wholesale $ + markup $
    retail % = wholesale % + markup %

Note: Wholesale % is 100%; retail is higher than 100%.

- Generally, solving markup problems is a two-step process: one step using the base/rate/portion formulas, the other step using addition/subtraction.

Relationships in Base/Rate/Portion Formulas

Substitute for:	Retail	Markup
Base	Wholesale $	Wholesale $
Rate	Retail %	Markup %
Portion	Retail $	Markup $

Sample Problem 5-11.

The wholesale cost of an item is $25. The markup percentage rate is 85%. Find the retail cost.

Step 1. Use 100% for wholesale %. Add markup % to find retail %.

$$100\%$$
$$+ \ \ 85\%$$
$$185\% \text{ retail percentage}$$

Step 2. Since you know wholesale $ and retail %, use basic percent formula and solve for retail $.

Base (wholesale $) = $25.
Rate (retail %) = 185%.
Find portion (retail $).
P = B × R = $25 × 1.85 = $46.25
Retail price = $46.25

This problem may be worked another way (as is generally the case). Note Sample Problem 5-12.

Sample Problem 5-12.

Determine the retail cost of the item in Sample Problem 5-12 using an alternate method.

Step 1. Find the dollar value of the markup.

Base (wholesale $) = $25.
Rate (markup %) = 85%.
Find portion (markup $).
P = B × R = $25 × 0.85 = $21.25
Markup = $21.25

Step 2. Determine the retail $ by adding the markup $ to the wholesale $.

$$\$25.00$$
$$+ \ \ 21.25$$
$$\$46.25 \text{ retail price}$$

Sample Problem 5-13.

Find the wholesale cost of an item with a retail price of $79.50 and a 95% markup.

Step 1. Add the markup % to wholesale % (100%) to find the retail %.

$$\begin{array}{r} 100\% \\ +\ \ 95\% \\ \hline 195\% \end{array} \text{ retail percentage}$$

Step 2. Use the retail $ in the percent formulas.

Portion (retail $) = $79.50.
Rate (retail %) = 195%.
Find base (wholesale $).

$$B = \frac{P}{R} = \frac{\$79.50}{1.95} = \$40.769$$

Wholesale cost = $40.77

Sample Problem 5-14.

Determine the percent markup of an item with a retail price of $159.25 if $91.00 is the wholesale cost.

Step 1. Determine the retail % using the percent formulas.

Portion (retail $) = $159.25.
Base (wholesale $) = $91.00
Find rate (retail %).

$$R = \frac{P}{B} = \frac{\$159.25}{\$91.00} = 1.75$$

Retail percentage = 175%.

Step 2. Find the markup % by subtracting wholesale % (100%) from the retail %.

$$\begin{array}{r} 175\% \\ -\ 100\% \\ \hline 75\% \end{array} \text{ markup percentage}$$

Discount using percentage

Discount is an amount subtracted from the retail price, resulting in a lower, *sale price*. To determine values where discount is a factor, follow rules of percent discount.

RULES OF PERCENT DISCOUNT

- Sale price is always the result of *sutracting* the discount from the retail price.

      sale $ = retail $ − discount $
      sale % = retail % − discount %

  Note: Retail % is 100%; sale is lower than 100%.

- Generally, solving discount problems is a two-step process: one step using the base/rate/portion formulas, the other step using addition/subtraction.

Relationships in Base/Rate/Portion Formulas

Substitute for:	Sale	Discount
Base	Retail $	Retail $
Rate	Sale %	Discount %
Portion	Sale $	Discount $

Sample Problem 5-15.

An item with a retail price of $12.50 is on sale for 30% off. Find the sale price.

Step 1. Determine the sale % by subtracting the discount % from 100%.

$$100\%$$
$$-\ 30\%$$
$$\overline{70\% \text{ sale percentage}}$$

Step 2. Since you know retail $ and sale %, use basic percent formula and solve for sale $.

Base (retail $) = $12.50.
Rate (sale %) = 70%.
Find portion (sale $).
$$P = B \times R = \$12.50 \times 0.70 = \$8.75$$
Sale price = $8.75

This problem may be worked another way. Note Sample Problem 5-16.

Sample Problem 5-16.

Determine the sale price of the item in Sample Problem 5-16 using an alternate method.

Step 1. Determine the dollar value of the discount.

Base (retail $) = $12.50.
Rate (discount %) = 30%.
Find portion (discount $).
$$P = B \times R = \$12.50 \times 0.30 = \$3.75$$
Discount = $3.75

Step 2. Find the sale $ by subtracting the discount $ from the retail $.

$$\$12.50$$
$$-\ \ \ 3.75$$
$$\overline{\$\ 8.75 \text{ sale price}}$$

Sample Problem 5-17.

The sale price marked on an item is $7.80. It is advertised at 35% off. What was the retail price before the sale?

Step 1. Determine the sale % by subtracting the discount % from the retail % (100%).

$$100\%$$
$$-\ 35\%$$
$$\overline{65\% \text{ sale percentage}}$$

Step 2. Find the retail $ by using the sale % as the rate in the percent formulas.

Portion (sale $) = $7.80.
Rate (sale %) = 65%.
Find base (retail $).

$$B = \frac{P}{R} = \frac{\$7.80}{0.65} = \$12$$

Retail price = $12.00

Sample Problem 5-18.

Determine the percent discount if the sale price marked on an item is $25.50 and $30.00 was the retail price.

Step 1. Find the sale % using the percent formulas.

Base (retail $) = $30.00.
Portion (sale $) = $25.50.
Find rate (sale %).

$$R = \frac{P}{B} = \frac{\$25.50}{\$30.00} = 0.85$$

Sale percentage = 85%

Step 2. Percent discount is found by subtracting the sale % from retail % (100%).

$$\begin{array}{r} 100\% \\ - \quad 85\% \\ \hline 15\% \text{ discount} \end{array}$$

Practice Problems

Complete the following charts on a separate sheet of paper. Include units of measure. When necessary, round your answers to nearest cent.

	Wholesale Cost	Retail Price	Percent Markup
1.	$32.50	$42.25	?
2.	$19.00	?	50%
3.	?	$50.00	120%
4.	$125.00	$175.00	?
5.	$250.50	?	75%
6.	?	$30.50	80%
7.	?	$680.00	150%
8.	$25.00	$75.00	?
9.	$12.00	?	50%
10.	?	$0.50	200%

	Retail Price	Sale Price	Percent Discount
11.	$50.00	$35.00	?
12.	$6.50	?	25%
13.	?	$62.00	20%
14.	$180.00	$90.00	?
15.	$360.00	?	1/3
16.	?	$18.00	30%
17.	?	6.00	0%
18.	$36.00	$21.60	?
19.	$75.00	?	25%
20.	?	$120.00	40%

TEST YOUR SKILLS

Do *not* write in this book. Use a separate sheet of paper to complete the following problems. Show your work and your final answer. Round answers to the nearest cent (two decimal places).

DOLLARS AND CENTS

1. What is the cost of 18 gallons of gasoline with a price of $1.279 per gallon?
2. If Bettina bought 2 books for $8.95 each, how much change would be received from a $20 bill?

3. At a certain time of day, the telephone rates from Boston to Seattle are 87¢ for the first 3 minutes. Thereafter, they are 1/6 that amount for each minute. What would be the cost for an 18-minute call?

4. If 3 1/2 pounds of roofing nails cost $2.87, what is the cost per pound?

5. Barrett pays 49¢ per pound for 8d (eightpenny) nails and 87¢ per pound for 16d nails. How much change will he receive from a $20 bill if he buys 6 pounds of 8d and 12 pounds of 16d?

GROSS INCOME, NET INCOME, AND THE PAYCHECK

6. What is the weekly gross pay of an office clerk who is paid an annual (52 weeks) salary (gross income) of $12,560?

7. An electronic technician is paid $9.60 per hour. Based on 40 hours per week, what is this technician's annual salary?

8. What is the weekly gross pay of a machinist earning $10.75 per hour working 48 hours per week? (Overtime is time and a half for hours over 40.)

9. A maintenance electrician is paid $15.70 per hour, plus a 25% bonus for each hour worked during the second shift. What is the gross pay for a 40-hour week working the second shift?

10. Listed is Robert's weekly pay stub. Use the information given to determine the total percentage withheld. All items are a percentage of gross pay.
PAY STUB: Gross = $560.00. Federal tax = $78.40. FICA = $40.04. State tax = $12.54.

11. Robert, from Problem 10, is expecting a 15% raise. Using the same percentages from the old pay stub, estimate his net pay after the raise.

12. Joyce earns $13.75 per hour. She receives a paycheck (net pay) of $430 for a 40-hour week. What is the total percentage deducted?

13. An employer hires a mechanic at $7.50 per hour for a 40-hour week. The employer has additional employee expenses as follows: matching FICA payment at a rate of 7.13%; health insurance premium at $12.60 per week; worker's compensation insurance at $6.80 per week; unemployment insurance at $9.70 per week. What is the hourly cost to the employer for a 40-hour week?

14. A salesperson earns $5.60 per hour, plus 3% commission on all sales. If the total sales for a 35-hour week is $15,000, what is the gross pay?

15. Using the pay stub listed, determine the total percent deductions. All items are a percentage of gross pay.
PAY STUB: Gross = $430.00. Federal tax = $64.35. FICA = $35.04. Retirement = $8.58.

SALES AND PROPERTY TAXES

16. Sales tax on a car is figured on the difference between trade-in value and purchase price. What is the sales tax (6.5% rate) on a new pickup truck costing $12,600, with a trade-in of a vehicle valued at $2,800?

17. The price of a candy bar in a vending machine is 60¢. Sales tax is 4.5%. Determine the tax collected for each candy bar sold. Hint: Base is the price of candy without tax. Vending price % is 100% + tax %.

18. Plywood costs $15.50 per sheet. When 4 sheets are purchased, the total cost is $66.00, including tax. What is the percentage rate of sales tax?

19. A motor home is valued at $12,000 in a town with a property tax rate of 23 mils. If property tax is paid on a motor home, what is the tax on the vehicle?

20. With a property tax rate of 43 mils, what is the yearly tax on a house valued at $86,000?

INTEREST

21. Calculate the simple interest earned in 4 months on a $600 savings account at a 7% interest rate.

22. How long must a $1500 investment be held to earn $225 given a simple interest rate of 10%?
23. Determine the balance at the end of 3 months if 6% interest is compounded monthly on a savings account that started with $300.
24. What is the total interest earned on a savings account in 18 months with 8% interest, compounded semiannually, and a starting balance of $1000?
25. Someone deposits $50 each week into a savings account offering 7% interest compounded quarterly. What is the balance at the end of 1 year? (Note: 1 quarter = 13 weeks)

ESTIMATING MONTHLY LOAN PAYMENTS

Find the estimated monthly payment for these loans. Round to the nearest dollar.
26. Loan = $12,000, APR = 9%, time = 4 years.
27. Loan = $4000, APR = 15%, time = 36 months.
28. Loan = $600, APR = 18%, time = 1/2 year.
29. Loan = $2500, APR = 10%, time = 2 years, 6 months.
30. Loan = $8000, APR = 7 1/2%, time = 5 years.

MERCHANDISE PRICING USING PERCENTAGE

Complete the following tables:

	Wholesale Cost	Retail Price	Percent Markup
31.	$10.00	$15.00	?
32.	$18.00	?	75%
33.	?	$125.00	80%
34.	$24.60	?	130%
35.	$3.50	$12.50	?

	Retail Price	Sale Price	Percent Discount
36.	$64.80	?	40%
37.	?	$44.50	50%
38.	?	$72.00	33.3%
39.	$1.90	$1.00	?
40.	$50.00	$30.00	?

PROBLEM-SOLVING ACTIVITIES

Activity 5-1.
Combined pocket change

> **Objective:** To find the total amount of change in the pockets of a class of students.
> **Instructions:**
> 1. Ask each student in your classroom to count the amount of the change in his/her pockets.
> 2. Write the individual totals on the board and calculate the total amount of change carried by all the students in the classroom.

Activity 5-2.
Comparison shopping

> **Objective:** To determine the better buy based on the cost per unit (ounce, pound, quart, etc.).
>
> **Instructions:**

1. Calculate the cost per unit measure for two identical items that are in different size packages.
2. Determine which is the better buy based on price per unit.

Activity 5-3.
Gross income from help-wanted ads

> **Objective:** To determine the annual gross pay from help-wanted ads.
>
> **Instructions:**

1. Locate a help-wanted ad from a local newspaper. Be sure the ad includes the hourly, weekly, or monthly rate of pay.
2. Based on the information given in the ad, calculate the annual rate of pay. If hours are not given, assume a 40-hour work week.

Activity 5-4.
Purchase price, including sales tax

> **Objective:** To determine the total cost to purchase an item, including sales tax.
>
> **Instructions:**

1. Look through a sales advertisement and select at least three items you'd like to purchase.
2. Determine the sales tax for your area. If there is no sales tax, use 5% for this exercise.
3. Calculate the total price to buy the three items, including the sales tax.

Activity 5-5.
Property tax based on mil rate

> **Objective:** To determine the property tax for a piece of property.
>
> **Instructions:**

1. Look through a real estate advertisement and select a piece of property that you would like to purchase.
2. Determine the property tax mil rate for your area. If there is none, use 20 mils for this exercise.
3. Calculate the property tax based on the value of the property you have selected.

Activity 5-6.
Percent discount

> **Objective:** To calculate the percent discount of selected items of merchandise.
>
> **Instructions:**

1. Using a merchandise sale paper, select different items that are advertised with both the sale price and the original price. Some papers advertise the amount taken off the original price.
2. Using the original and sale prices, determine the percent discount.

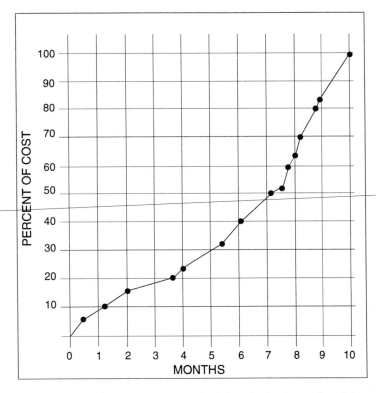

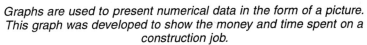

Graphs are used to present numerical data in the form of a picture. This graph was developed to show the money and time spent on a construction job.

Chapter 6

Graphs

OBJECTIVES

After studying this chapter, you will be able to:
- Explain basic terminology of graphs.
- Read line, bar, and circle graphs.
- Prepare line, bar, and circle graphs from tabular data.

Numerical data is the backbone of technology. It comes from a variety of sources. *Theoretical data* comes from theoretical calculations. *Empirical data* comes from actual measurement or experience. Tied to this is *statistical data*. It comes from statistics. **Statistics** are collections of numbers. They are derived through records or sampling (i.e., statistics on weather or on faulty parts). Statistics are used for the purpose of supplying information. All of the above are types of numerical data.

Data is generally recorded in tabular form. Given large amounts, it can be hard to extract information from data presented this way. Graphs are used for this purpose. They present data in the form of a "mathematical picture." Graphs organize the information and help the reader to "see." This chapter will present three different types of graphs—line graphs, bar graphs, and circle graphs.

LINE GRAPHS/BAR GRAPHS

Line and bar graphs present data quite differently—line graphs are depicted by lines; bar graphs by bars. They are very similar, however, in their labeling and, often, in the type of data they present. (Circle graphs present a different type of information and will be presented later in this chapter.) The information of this section applies to both line and bar graphs.

Graph paper

Graph paper is used for drawing graphs. A variety is available. The paper you will be using here has equally-spaced, horizontal and vertical lines drawn on it. Except in certain instances, the lines on the paper are used to represent numerical values. These lines have no specific value until assigned. The graph to be drawn dictates what their value will be. Pay attention to rules of graph paper lines when plotting data. (**Plotting data** is locating numerical data points on graph paper.) Quite often, graphs do not appear on graph paper. A grid merely enables points to be plotted accurately.

RULES OF GRAPH PAPER LINES

- It is not necessary to label every line on the paper.
- Each line will represent a specific value.

- If the graph paper has some lines drawn heavier than others, these become major divisions. They should be labeled. The lighter lines between are minor divisions. They are not labeled.
- On graph paper with major/minor divisions, *incremental value* of minor divisions (proportional amount by which divisions increase) is found by dividing incremental value of major divisions by number of blocks (not lines) between them. Example:

  Given—Major divisions are labeled in increments of 5 (i.e., 0, 5, 10, etc.).
  Number of blocks between major divisions is 5.

  Found—Incremental value of minor divisions is 5 ÷ 5 = 1.

Fig. 6-1 shows two different styles of graph paper. Fig. 6-1A shows graph paper with major and minor divisions. The major horizontal lines shown here are labeled on the left, every 10 degrees. Minor lines are in 2 degree increments since there are 5 blocks between major divisions. The major vertical lines are labeled across the bottom, every 1 hour. Minor lines are in 0.2 hour (12 minute) increments.

With no major or minor divisions, the value of lines of Fig. 6-1B will change by the same amount. The horizontal lines shown here are labeled in dollars per thousand board feet. Lines are in 50 dollar increments. The vertical lines shown here are not labeled with numerical values. The lines themselves are not labeled at all. This type of graph, when complete, will be a bar graph. (You will learn about these shortly.) The vertical lines are used here to set the bounds of the bar. The area within the bounds will be shaded. Each bar will represent a wood type and will be labeled as shown.

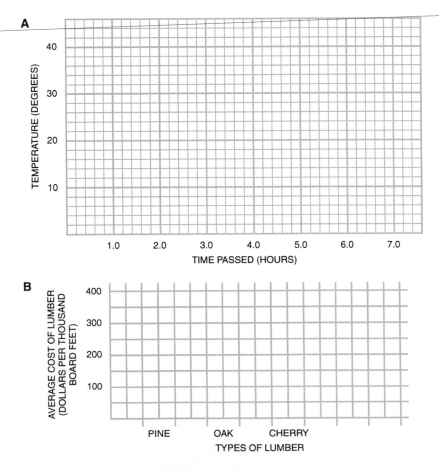

Fig. 6-1. Styles of graph paper. A—Graph paper with major and minor divisions. B—Graph paper with all divisions equal.

Basic terminology

The labeled, horizontal line on the bottom of a graph is an *axis* of the graph. Specifically, it is the **x-axis**. The x-axis is sometimes called the *axis of abscissas*. The vertical line to the left of a graph is the **y-axis**. It is sometimes called the *axis of ordinates*. Together, they are *coordinate axes*. The point at which coordinate axes meet is the **origin**.

Independent and dependent variables

The quantities that are represented by a graph are **variable quantities,** or variables. This means they are subject to change. (Symbols used to denote variable quantities are also called variables. These are introduced in Chapter 11.) A graph is a plot of two variable quantities. One of the variables is independent. It is the controlling factor. The other variable is dependent. Its value depends on the value of the other. The **independent variable** is generally represented by the x-axis; the **dependent variable** by the y-axis. For example, outside temperature might be recorded every hour for a 24-hour period. At the end of the period, the data will show how the temperature varied. The independent variable in this case is time. The dependent variable is temperature. In other words, temperature depended on the time of day. The time of day did not depend on temperature. Time is independent of temperature. If this information were placed on a graph, the x-axis would be labeled in units of time. The y-axis would be labeled in units of temperature.

Referring to graphs

When graphs are presented, they are commonly referred to in one of two ways. These are as graphs of:
- a dependent variable *versus* an independent variable.
- a dependent variable *as a function of* an independent variable.

Of these, the former applies to any line or bar graph. Take the foregoing example. This graph would be referred to as the graph of outside temperature versus time. Take another example. Suppose you were to plot highest temperature recorded each week in a certain location. Here, temperature is the dependent variable. Week number would be the independent variable. The graph that resulted would be referred to as the graph of maximum temperature versus week number. Other examples include graphs of gas mileage versus average speed; gas mileage versus tire pressure; tire pressure versus temperature.

If the relationship between dependent and independent variables can be defined by a formula, it is a *functional relationship*. You are somewhat familiar with formulas. In Chapter 4, you used basic percent formulas. A **formula** is a mathematical statement of a real-life situation. This statement expresses one quantity in terms of one or more other quantities. A formula will have dependent and independent variables. The dependent variable is said to be *a function of* the independent variable. Thus, if a functional relationship exists between variables of a graph, it may be referred to accordingly.

For example, a functional relationship exists between air pressure and altitude. The two can be defined by a formula with air pressure the dependent variable; altitude the independent variable. A graph of the two could be referred to as the graph of air pressure versus altitude or the graph of air pressure as a function of altitude. You would not refer to the graph of maximum temperature versus week number in the latter terms. This is because a functional relationship does not exist between the two. They cannot be defined by a formula.

Interpreting and plotting

To interpret a graph, the exact data value can be taken from the graph at the point where the bar ends, or where the line passes through. Extend horizontal and vertical lines from the point to the axes. The value read where the lines cross the axes gives the value of the variables. When stating the values, give the independent variable first.

Trends are the general direction the graph is going. For example, take a student's class grades. There will be a variation from grade to grade, but the *trend* could be up, down, or the same. The trend can be seen on a graph by placing a straightedge through the average of data points in a desired segment. The general slope is then examined. In Fig. 6-2, for example, the slope of the straightedge is positive (uphill slope). Although the first six quiz scores get worse, the general trend is positive. In other words, the scores are progressively improving.

Regarding the plotting of graphs, convention governs how graphs are plotted. Most graphs follow the rules for plotting a graph.

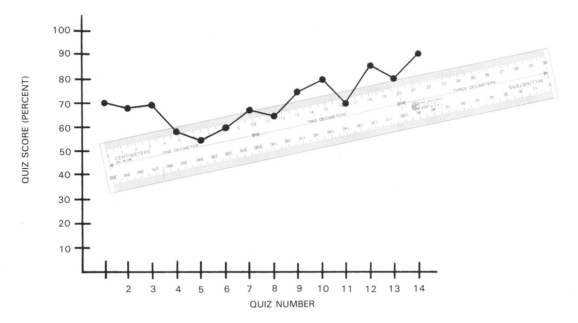

Fig. 6-2. Quiz Score versus Quiz Number.

RULES FOR PLOTTING A GRAPH

1. Determine the independent variable (controlling factor). Determine the dependent variable.
2. Select suitable graph paper.
3. Label axes—independent variable usually along the bottom; dependent variable along the left side.
4. Assign values to the horizontal and vertical lines. Label only enough lines to establish pattern. Major lines should always be in increments that are easy to work with (i.e., increments of 1, 2, 5, 10, etc.). You would not want to increment lines by, for example, 4.5 or 7.2, or 16, etc.

Line graphs

A **line graph** is a graph of data depicted by a line. The line may connect each data point, resulting in a kind of zigzag effect. Otherwise, it may be drawn as the average of the data points, resulting in a smooth line or curve. The former type is used when the data cannot be tied to a formula. As previously discussed, this is the case when no functional relationship exists (graph of maximum weekly price per share of a corporate stock, for example). A smooth line or curve is used when a functional relationship exists (graph of gas mileage as a function of average speed, for example). A line graph works well in showing change. For example, ups and downs of sales at a car dealership is clearly depicted by a line graph.

In plotting data for a line graph, a dot is placed at the point where the value of the dependent and independent variables coincide. This point is found by extending horizontal and vertical lines from the corresponding value along the axes. The point is located where the lines cross.

For an example of a line graph, refer to Fig. 6-3. This graph shows the changing effects of speed on gas mileage. The independent variable is speed in miles per hour (mph). The dependent variable is gas mileage in miles per gallon (mpg). The dots are data points of gas mileage at the various speeds. Notice that the "mathematical picture" produced allows you to see which speed gives the best gas mileage at a glance. You do not have to labor through all of the numbers that are given.

The location of a data point is given by its **coordinates**. The *x-coordinate* is given first. It is given by the value along the x-axis. The *y-coordinate* is given second. It is given by the value along the y-axis. In Fig. 6-3, for example, the point of peak gas mileage is given by the coordinates (30, 28).

miles per gallon	0	18	24	28	26	22	18	15
miles per hour	0	10	20	30	40	50	60	70

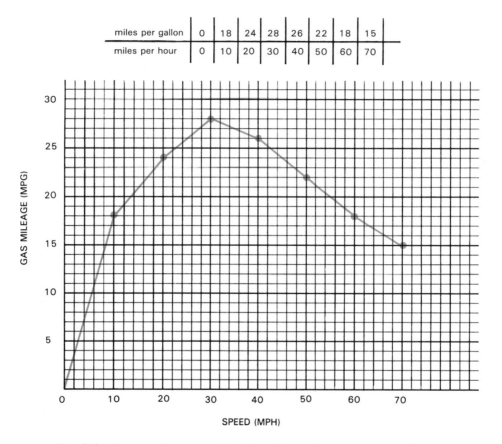

Fig. 6-3. Line graph representing the effect of speed on gas mileage.

Practice Problems

Complete the following problems on a separate sheet of paper.

(Refer to Fig. 6-3 when answering questions 1-4.)
1. At what speed is gas mileage the highest?
2. What is the trend for MPG as the speed increases past 40 MPH?
3. What is the gas mileage at 55 MPH?
4. What is the percentage decrease in gas mileage at 55 MPH compared to the maximum gas mileage?
5. Plot a line graph based on the following student's grades. Refer to Fig. 6-2 as a guide when making the axes.
   Student's quiz grades:

#1 = 85%	#4 = 75%	#7 = 90%
#2 = 95%	#5 = 70%	#8 = 85%
#3 = 90%	#6 = 80%	#9 = 80%

(Use the graph created in question 5 to answer questions 6-10.)
6. What is the student's overall average?
7. Draw a straight line representing the student's average on the graph created in question 5.
8. What is the general trend?
9. If the class average is 78%, how many of this student's grades are above average? How many student's grades are below the average?
10. How many percentage points is this student's average from the class average?

Bar graphs

A **bar graph** is a graph of data depicted by bars, or heavy lines. Bar graphs, or bar charts as they are sometimes called, are used chiefly in the presentation of statistical data. They display relative size of a group of statistics. Axes of bar graphs, like line graphs, are labeled with specific information. The bar graph gives the value of the statistic on one axis. On the other, it gives the statistical group. At times, bar graphs do not show axes. Even so, they are still bar graphs.

One type of bar graph that is used very often for statistical data is the histogram. A **histogram** is a graph of frequency (number of occurrences) by groups. See Fig. 6-4. You can see from the histogram that the greatest number of tanks, by group, are 5 to 10 (9.99) years old. Examining this graph should give you a good idea about the age of all tanks in the population (i.e., all 1000 tanks).

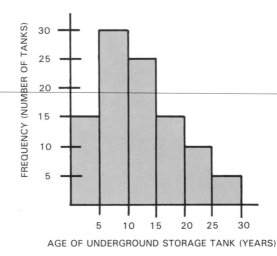

Fig. 6-4. A sample histogram. (Random sample of 100 tanks out of a population of 1000.)

When plotting a bar graph, first place your statistical groups. (These are your independent variables.) Decide how thick a bar you want and how much space you want between each. Then, place your labels. For each bar, locate the dependent variable value. If desired, shade in the area chosen for the bar. This area starts at the axis and ends at the located value. The result will be a bar representing the relative size of the statistic. The bars on a graph can be either horizontal or vertical. The most common form is vertical.

For another example of a bar graph, refer to Fig. 6-5. This graph shows mean (average) temperature by month. The table gives the numerical data. The bottom line, the independent variable, is months of the year. The left side, the dependent variable, is the mean value of the temperature.

Practice Problems

Complete the following problems on a separate sheet of paper. Include units of measure.

(Refer to Fig. 6-5 when answering questions 1-5.)
1. Which month has the highest mean temperature?

Months	J	F	M	A	M	J	J	A	S	O	N	D
Degrees	15	20	25	40	60	70	80	90	75	50	35	25

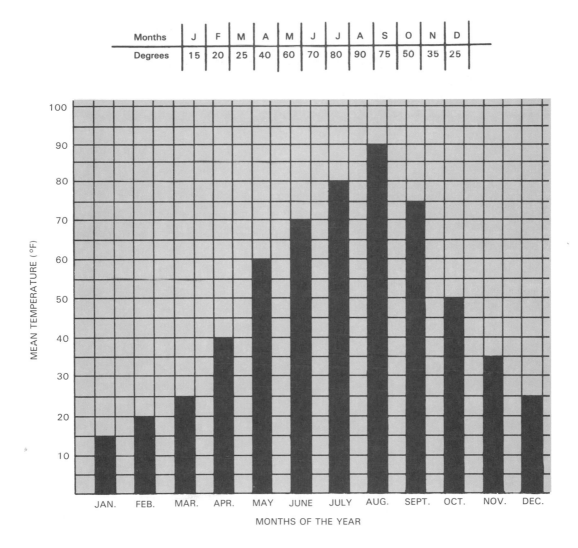

Fig. 6-5. Bar graph of mean temperatures.

2. Which month has the lowest mean temperature?
3. What is the yearly average temperature?
4. What is the mean temperature for the month of April?
5. What is the difference in mean temperature between the months of April and October?
6. Make a bar graph based on the following wattage estimates of selected household appliances.

Appliance	Wattage
Hair dryer	380 watts
Coffeemaker	900 watts
Room air conditioner	860 watts
Vacuum cleaner	650 watts
Microwave oven	1450 watts
Toaster	1200 watts

(Use the graph created in question 6 to answer questions 7-10.)
7. Which appliance listed uses the most electrical wattage? Which appliance uses the least wattage?
8. What is the average wattage of the appliances listed?

9. Which appliances are above the average?
10. Which appliances are below the average?

CIRCLE GRAPHS

A **circle graph** is a graph of data depicted by a circle. Circle graphs are also called pie graphs, or pie charts. Values are represented by "slicing" a circle into pie-shaped pieces called **sectors.** The sectors show portions in relation to the total. Usually, values are given as percentages. The size of each sector is determined by the percentage. Fig. 6-6 is a circle graph. It shows the distribution of expenses for a certain factory. Total expenses, including profit, equal 100 percent of the wholesale cost. The percentages of the sectors add up to 100 percent.

Plotting a circle graph requires at least some knowledge of circles. The first thing to do is draw a circle. Lightly mark the center. (Use a compass or a circle template. These will allow you to identify the center.)

You will now need to figure how big each sector should be. Each sector is some percentage of the whole circle. Its shape is defined by two radii (plural of **radius** — a line between the center of a circle and the circle itself) and an arc of the circle. An **arc** is a portion of the circle. It has the form of a curved line. You will know how big to make the sector by finding the angle between the radii. (Otherwise, use percentage grid paper or a percentage protractor.)

To find the angle between the radii, use the basic percent formula, $P = B \times R$. The base is the total number of degrees in a circle (360°). The rate is the percentage given by the sector. The portion is the angle of the sector.

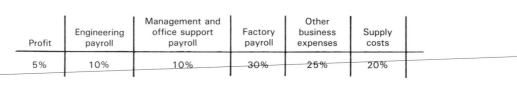

Profit	Engineering payroll	Management and office support payroll	Factory payroll	Other business expenses	Supply costs
5%	10%	10%	30%	25%	20%

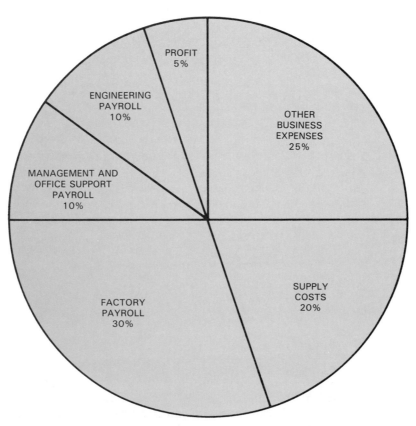

Fig. 6-6. Circle graph used to show the distribution of expenses for a certain factory.

Sample Problem 6-1.

Find the angle of the sector showing "profit" in Fig. 6-6.
Base = 360°. Rate = 5%. Find portion.

$$P = B \times R = 360° = \times 0.05 = 18°$$

Angle of sector = 18°

The angle between the radii of the sector in Sample Problem 6-1 is 18°. Assume that you wanted to show this sector on your circle graph. Using a protractor, you would draw in two radii, 18° apart. The area of this sector would be 5% of the whole circle. The arc of this sector would be 5% of the **circumference**, or distance around the circle. (Circumference is also the boundary of a circle.)

Practice Problems

Complete the following problems on a separate sheet of paper. Include units of measure.

(Refer to Fig. 6-6 when answering questions 1-5.)

1. Calculate the number of degrees for the angle of each sector: engineering payroll, management payroll, factory payroll, supply costs, other, and profit.
2. If the company's total yearly expenses were $875,000, how much money was spent on supplies?
3. With the budget shown in question 2, determine the average yearly salary for the 3 members of the engineering department.
4. With the budget shown in question 2, determine the average yearly salary for the 40 members of the factory department.
5. What two sectors combined equal 50% of the total budget?
6. The budget of a selected rural town is shown below. Make a circle graph using this information.

> Annual town budget = $6.4 million
>   Education = 35%
>   Road maintenance = 25%
>   Fire/police = $640,000
>   Administration = 15%
>   Miscellaneous = Remainder

(Use the graph made in question 6 to answer questions 7-10.)

7. What percentage of the budget is allowed for miscellaneous expenses? What percentage is allowed for fire/police protection?
8. What is the dollar amount allotted for education?
9. If half the police/fire budget is used for salaries, what is the average salary with 20 people in the department?
10. If the road maintenance department exceeds its budget by 5% due to an unusually snowy winter, how many dollars must be taken from the miscellaneous account?

TEST YOUR SKILLS

Do *not* write in this book. Use a separate sheet of paper to complete the following problems. Show your work and your final answer.

LINE GRAPHS

(For questions 1 through 6, refer to Fig. 6-7.)

1. Determine the total sales for the 6-month reporting period.
2. What is the average monthly sales? (Round to nearest million.)
3. Sales during the month of October are what percentage of the average? (Round to the nearest whole percent.)
4. What percentage did sales during October drop from average?

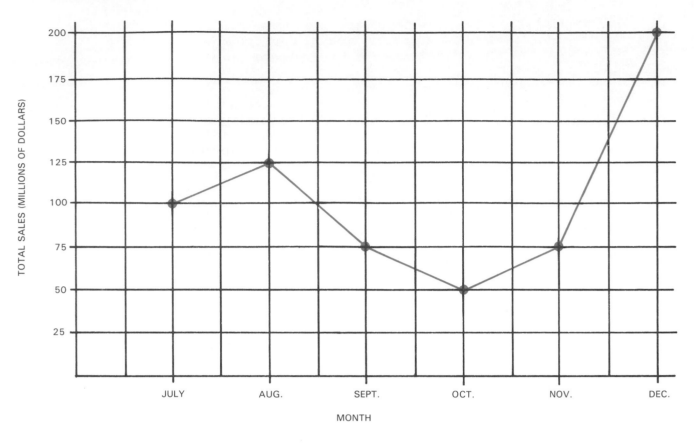

Fig. 6-7. Santa's Toy Company. Total monthly sales for 6-month reporting period. Sales data shown represents wholesale prices.

5. If the retail market has an 85% markup, what is the total retail value for December?
6. If the cost of labor is equal to 30% of the wholesale price, what was the cost of labor during August?
7. Use the following information to plot a line graph.
   Number of defect-free parts manufactured by day of the week:

   Monday = 15
   Tuesday = 38
   Wednesday = 65
   Thursday = 50
   Friday = 30

(For questions 8 through 12, refer to line graph of question 7.)

8. What was the total number of defect-free parts produced that week?
9. Draw a straight line across the graph equal to the average.
10. How many days are above average?
11. Defect-free parts made on Monday are what percentage of average? (Round to nearest tenth of a percent.)
12. Each day represents 8 hours working time. This worker is paid $6.50 per hour plus a bonus of 25¢ per piece above average. What is the worker's gross pay at the end of the week?

BAR GRAPHS

(For questions 13 through 17, refer to Fig. 6-8.)

13. How many times did monthly income exceed monthly expenses?
14. Find the total income (within 1 M$) for the 6-month period. Use a straightedge to read correct amount.
15. Find the total expenses for the 6-month period. Again, use a straightedge.

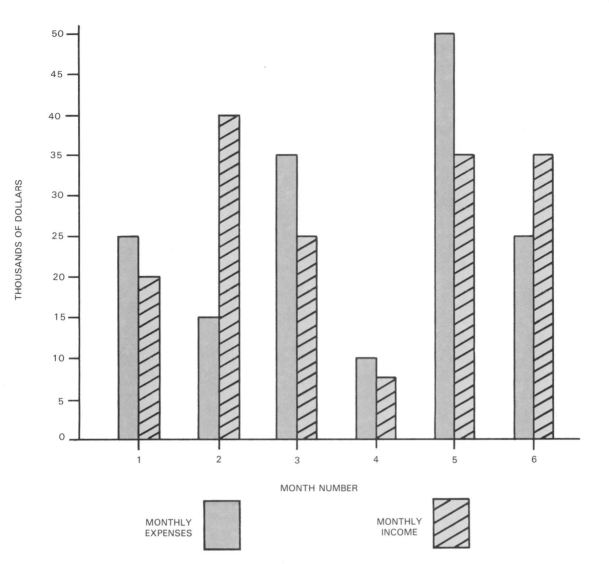

Fig. 6-8. A 6-month financial summary comparing income and expenses.

16. Determine the profit netted for the 6-month period.
17. With regard to the second month only, what percentage of the income was profit?
18. Use the following information to plot a bar graph.
    Gas mileage of various cars:
    Car #1 = 23 miles per gallon
    Car #2 = 45 miles per gallon
    Car #3 = 30 miles per gallon
    Car #4 = 15 miles per gallon
    Car #5 = 38 miles per gallon

(For questions 19 through 23, refer to bar graph of Question 18.)

19. Identify the independent variable.
20. Identify the dependent variable.
21. Which car costs least to drive on a long trip?
22. If gasoline costs $1.25 per gallon, what is the cost to drive car #3 a distance of 600 miles?
23. If smaller cars get better gas mileage, which car is largest?

CIRCLE GRAPHS

(For questions 24 through 28, refer to Fig. 6-9.)
    Total monthly budget = $1200
    Rent = 25% — block A

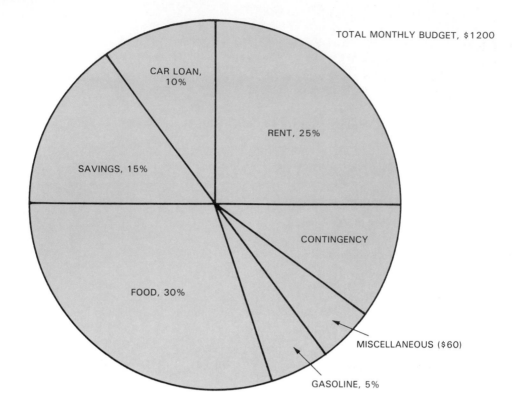

TOTAL MONTHLY BUDGET, $1200

CAR LOAN, 10%

RENT, 25%

SAVINGS, 15%

CONTINGENCY

FOOD, 30%

MISCELLANEOUS ($60)

GASOLINE, 5%

Fig. 6-9. A pie graph of a monthly budget. (The item of "contingency" is often presented with budgets. It is for unexpected costs such as car repairs, medical emergency, etc.)

Car loan = 10% — block B
Savings = 15% — block C
Food = 30% — block D
Gasoline = 5% — block E
Misc. = $60 — block F
Unexpected — block G

24. How much, in dollars, is the monthly rent?
25. How much, in dollars, is the car loan payment?
26. What is the percentage for miscellaneous items?
27. What is the percentage for contingency?
28. How much, in dollars, is for contingency?
29. Use the information supplied to draw a circle graph.
    Types of jobs available:
    Total = 800
    Technical = 15%
    Clerical = 20%
    Fast-food = 10%
    Services = 30%
    Other = ?

(For questions 30 through 34, refer to circle graph of question 29.)

30. How many jobs are available for technical people?
31. If 80 jobs are filled in the clerical area, how many clerical positions will remain unfilled?
32. Suppose fast-food jobs pay an average of $7 per hour, 40 hours per week. How much money must be available every week for payroll?
33. What percentage of jobs are classified as *other*?
34. How many jobs are available in the *other* category?

PROBLEM-SOLVING ACTIVITIES

Activity 6-1.
Students absent by day of week

Objective: To draw a line graph showing the number of students absent from your school by the day of the week.

Instructions:

1. Collect the absentee list for your school for the past few weeks.
2. Plot a line graph showing the number of students absent on each day. The bottom axis of the graph should be the day of the week (Monday to Friday), and the side axis should be the number of students absent each day.
3. The more weeks shown on this graph, the easier it will be to detect a trend.

Activity 6-2.
Individual student's quiz scores

Objective: To plot a line graph showing your grades for each quiz or test in a given class.

Instructions:

1. Make a list of your quiz and test scores for a specific class.
2. Plot a line graph showing your grades for each quiz or test. The bottom axis should be the number (name) of the quiz or test, and the side axis should be the grade received for the quiz or test.
3. Use a straightedge to determine the trend of the grades.

Activity 6-3.
Number of students per grade level

Objective: To use a bar graph to compare the number of students in your school in each grade level.

Instructions:

1. Obtain the number of students registered for each grade level at your school from the school office.
2. Draw a bar graph showing the number of students in each grade level.
3. Use the graph to compare the number of students in each grade.

Activity 6-4.
Vehicles by categories

Objective: To use a circle graph to show the relative number of vehicles by selected categories, such as 2-door, 4-door, hatch-back, mini-van, etc.

Instructions:

1. Count the cars in a parking lot (or a selected section of the lot).
2. Tabulate the number of cars in each of the selected categories.
3. Determine the percentage in each category.
4. Draw the circle graph showing the percentage of vehicles in each category.

Activity 6-5.
School budget

Objective: To use a circle graph to display the various categories in your school's budget.

Instructions:

1. Ask the school administration for a copy of the school's budget.
2. Make a list of categories each budget item falls into, such as salaries for administrators, teachers, and support staff; utilities; building maintenance; textbooks; supplies; equipment; miscellaneous; etc.
3. Determine the percentage of the budget in each category.
4. Draw a circle graph showing the percentage in each category.

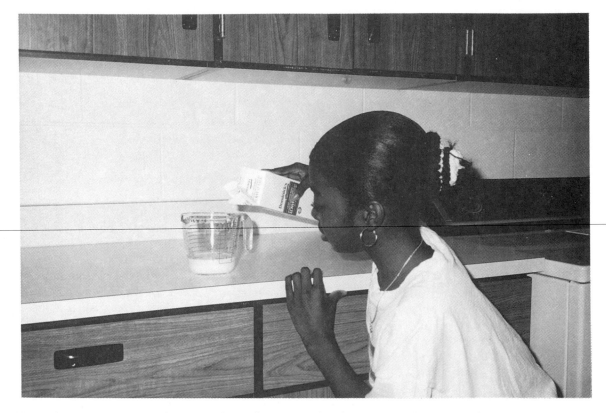

Units of measure give meaning to numbers. Common units of measure include units of length, area, volume, capacity, and weight. Units of capacity, such as liters, ounces, and cups, are often specified in food recipes.

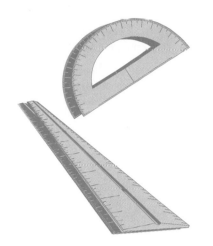

Chapter 7

Units of Measure

OBJECTIVES

After studying this chapter, you will be able to:
- Explain base units and derived units.
- Convert units using a conversion bar.
- Convert units using a factor.
- Solve word problems using a conversion bar.
- Perform operations of arithmetic on denominate numbers.
- Compute travel time between time zones.

Units of measure are used to give values in relation to known standards. They give meaning to numbers. For example, if you said that it was 25 to the store, it would be unclear as to what you meant. Did you mean 25 feet, 25 miles, 25 seconds? What?

Units and "subunits" exist for a wide variety of measurements. Gallons, years, tons, and kilograms are all different units of measure. Subunits of these are quarts, days, pounds, and grams, respectively. Dollars, a monetary unit of measure, have subunits of cents.

There are a few accepted systems of measurement. The United States operates on the *U.S. Conventional System*. This system is based on the English system of measure. It has such units as feet and pounds. Ironically, the English operate only partially on this system. For the most part, they follow what is known as the *SI Metric System*. SI Metric is the international system of measure. The metric system uses such units as meters and kilograms. Both systems use seconds for units of time.

Regardless of the system, most units are derived from the few base quantities that exist. Among these few are quantities of length, time, temperature, and electric charge. In U.S. Conventional, force or weight, in pounds, is included. In SI Metric, mass, in kilograms, is also included. (Weight and mass are thought by most of the general public to be the same thing. They are not. Weight varies with the force of gravity, whereas mass does not. More than this is beyond the scope of this book. From here on, however, the two will be treated as by the general public. The difference, for the most part, will be ignored.)

Units of the base quantities are base units. They include, for example, feet, kilograms, and seconds. Quantities derived from base quantities are derived quantities; their units, derived units. Area and volume, for example, are derived quantities. They are derived from length. Units of velocity are derived from quantities of length and time. Miles per hour is an example of a derived unit of velocity. Some units are given a specified name. Volts, hertz, ohms, and watts are examples of these. Still, these units are all derived from base units.

CATEGORIES OF MEASUREMENT

Five basic categories of measurement are given below. These categories are from a table of "weights and measures." Each features a different quantity, base or derived.

These are some of the most common, and principles relating to them can be applied to any quantity. Units in both U.S. Conventional and SI Metric are given as examples of each quantity.

- **Length**—the base quantity of distance.
    U.S. Conventional—inch, foot, yard, mile.
    Metric—centimeter, meter, kilometer.
- **Area**—the 2-dimensional measurement of a region.
    U.S. Conventional—square inch, square foot, acre.
    Metric—square centimeter, square meter, square kilometer.
- **Volume**—the measurement of space occupied by a 3-dimensional object, such as a closed, cardboard box.
    U.S. Conventional—cubic inch, cubic foot.
    Metric—cubic centimeter, cubic meter.
- **Capacity**—the measure of what is contained within an object, such as a liquid in a bottle. Also, the measure of how much an object can hold. Capacity is a measure of volume, in reality. Some references do not split the two categories. In general, volume refers to dry measure and capacity to fluid measure.
    U.S. Conventional—fluid ounce, cup, pint, quart, gallon.
    Metric—milliliter, liter, kiloliter.
- **Weight**—the measure of how heavy something is.
    U.S. Conventional—ounce, pound, ton.
    Metric—milligram, gram, kilogram.

Practice Problems

Answer the following questions on a separate sheet of paper.

Place each of the following units of measure into the proper category: length, area, volume, capacity, or weight. Also, state if the unit is U.S. Conventional or SI Metric.

1. acre
2. centimeter
3. cubic centimeter
4. cubic inch
5. cubic foot
6. cubic meter
7. cup
8. fluid ounce
9. foot
10. inch
11. gallon
12. gram
13. kilogram
14. kilometer
15. kiloliter
16. liter
17. meter
18. mile
19. milligram
20. milliliter
21. ounce
22. pint
23. pound
24. quart
25. square centimeter
26. square foot
27. square inch
28. square kilometer
29. square meter
30. ton

CONVERTING UNITS

Units have various sizes that are given by subunits. Using appropriate subunits can make numbers easier to work with. For example, it would be impractical to give the distance from Chicago to Texas in inches. To give your height as a fraction of a mile would be equally impractical. It is often necessary to change from one unit size to another. You may need to change pints into gallons, or feet into miles. Generally, except for categories of volume and capacity, units must always be in the same category in order to convert them. You could not convert seconds to pounds, for example.

Conversion factors

The **conversion factor** is a relationship used to change from one unit to another (i.e., 1 foot = 12 inches). A conversion factor is sometimes written in multiplication form. Expressed this way, number of inches is number of feet times 12. Tables of conversion factors can be found in Technical Section B, at the back of this book.

Units can be converted in a manner similar to converting equivalent fractions. When converting fractions, a number is selected by which to multiply or divide both the numerator and denominator. This number could be thought of as a conversion factor. Notice that it is written as a fraction and is equal to 1. Consider the following example:

$$\frac{2}{3} \times \frac{4}{4} = \frac{8}{12}$$

Here, the fraction 2/3 is changed to an equivalent fraction of 8/12. The conversion factor, you might say, is 4/4.

Conversion bar

The *conversion bar* makes the use of conversion factors simple by writing the factor as a fraction equal to 1. The conversion bar can be expanded when a problem requires more than one step in converting. Use of the conversion bar will make it clear whether the arithmetic involved is multiplication or division. The following are examples of the conversion bar giving fractions equal to 1:

$$\frac{3 \text{ feet}}{1 \text{ yard}} \qquad \frac{1 \text{ yard}}{3 \text{ feet}} \qquad \frac{5280 \text{ feet}}{1 \text{ mile}}$$

Notice that, regarding value, it does not matter which number is in the numerator and which is in the denominator. These factors have a value of 1 when written as a fraction regardless of which way they are written. All units, except those in the final answer, will be canceled as demonstrated in the sample problems to follow.

RULES FOR USING THE CONVERSION BAR

1. Identify appropriate conversion factors.
2. Write down number being converted.
3. Write conversion factors as fractions. Arrange so units cancel (except units of the final answer).
4. Cancel the units. Perform the arithmetic. Units in the final answer *must* make sense. If not, not only are units wrong, but number value is probably off, too.

Converting simple units — Part I

The following sample problems demonstrate converting of simple (basic) units.

Sample Problem 7-1.

How many feet are in 180 yards?

Step 1. Conversion factor:

1 yard = 3 feet

Step 2. Write the number you are trying to convert (180 yards), first. This will let you know which way you should write the conversion factor. Write the conversion factor as a fraction. The denominator has the same units as the first number (i.e., yards). Now the units will cancel. Parentheses are drawn to separate the different fractions in the conversion problem.

$$\left(\frac{180 \text{ yards}}{*}\right) \left(\frac{3 \text{ feet}}{1 \text{ yard}}\right)$$

*A "1" is implied but need not be shown (typical throughout this chapter).

Step 3. Cancel units that appear in *both* numerator and denominator. Perform arithmetic following the rules of fractions.

$$\left(\frac{180 \text{ yd.}}{}\right) \left(\frac{3 \text{ ft.}}{1 \text{ yd.}}\right) = \frac{180 \times 3 \text{ ft.}}{} = \frac{540 \text{ ft.}}{}$$

Answer: 540 feet

Sample Problem 7-2.

How many gallons do 16 pints equal?

Step 1. Conversion factors:

1 gallon = 4 quarts
1 quart = 2 pints

Step 2. Write the number you are trying to convert (16 pints), first. Write the conversion factor that will cancel units of the first number (i.e., pints). Units left will be quarts. Now, write the conversion factor that will cancel units of quarts. When quarts are canceled, units of gallons will remain.

$$\left(\frac{16 \text{ pints}}{}\right) \left(\frac{1 \text{ quart}}{2 \text{ pints}}\right) \left(\frac{1 \text{ gallon}}{4 \text{ quarts}}\right)$$

Step 3. Cancel units. Perform arithmetic following the rules of fractions.

$$\left(\frac{16 \text{ pt.}}{}\right) \left(\frac{1 \text{ qt.}}{2 \text{ pt.}}\right) \left(\frac{1 \text{ gal.}}{4 \text{ qt.}}\right) = \frac{16 \times 1 \times 1 \text{ gal.}}{2 \times 4} = \frac{16 \text{ gal.}}{8} = \frac{2 \text{ gal.}}{}$$

Answer: 2 gallons

Note: A conversion bar can also be used to convert between metric units. However, conversion between SI Metric units can be done quite easily without a conversion bar simply by shifting the decimal point. This is only valid, however, for one-dimensional units (i.e., grams or meters but not meters^2 or meters^3.) Fig. 7-1 will assist you.

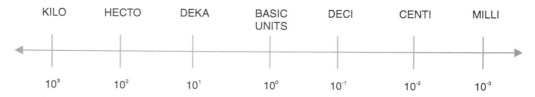

Fig. 7-1. Use this conversion line to convert between one-dimensional metric units.

To use Fig. 7-1, count the number of increments between desired units beginning with the unit to be converted. This gives the number of places to move the decimal point. Move decimal left if counting to the left and vice versa.

Examples:

• Convert 150 milliliters to liters—Count 3 increments left—move decimal point 3 places left: 150 mL = 0.15 L.

• Convert 2 kilometers to centimeters—Count 5 increments right—move decimal point 5 places right: 2 km = 200 000 cm.

Converting simple units — Part II

The following sample problems demonstrate converting of simple units. These problems are slightly harder, but the same rules apply.

Sample Problem 7-3.

How many square meters (m²) in 50 000 square centimeters (cm²)?

Step 1. Conversion factor:

1 square meter = 10 000 square centimeters

Step 2. Set up conversion bar.

$$\left(\frac{50\ 000\ \text{square centimeters}}{}\right) \left(\frac{1\ \text{square meter}}{10\ 000\ \text{square centimeters}}\right)$$

Step 3. Cancel units. Perform arithmetic.

$$\left(\frac{50\ 000\ \cancel{\text{cm}^2}}{}\right) \left(\frac{1\ \text{m}^2}{10\ 000\ \cancel{\text{cm}^2}}\right) = \frac{50\ 000 \times 1\ \text{m}^2}{10\ 000}$$

$$= \frac{50\ 000\ \text{m}^2}{10\ 000} = \frac{5\ \text{m}^2}{}$$

Answer: 5 square meters

Sample Problem 7-4.

Show another way to convert units of Sample Problem 7-3.

Step 1. Conversion factor:

1 meter = 100 centimeters

Step 2. Set up conversion bar. The conversion factor is written as a fraction, as always. In order to get the centimeters of the conversion factor the same as units of the first number (50 000 *square centimeters),* the conversion factor is squared. Squaring the factor does not change its value. It still equals 1. (Squaring a number is multiplying the number by itself; units carry a "raised" 2 [i.e., *cm²*]. This indicates that they are squared.)

$$\left(\frac{50\ 000\ \text{cm}^2}{}\right) \left(\frac{1\ \text{m}^2}{100\ \text{cm}}\right) = \left(\frac{50\ 000\ \text{cm}^2}{}\right) \left(\frac{1\ \text{m}}{100\ \text{cm}}\right) \left(\frac{1\ \text{m}}{100\ \text{cm}}\right)$$

$$= \left(\frac{50\ 000\ \text{cm}^2}{}\right) \left(\frac{1\ \text{m}^2}{10\ 000\ \text{cm}^2}\right)$$

Step 3. Cancel units. Perform arithmetic.

$$\left(\frac{50\ 000\ \cancel{\text{cm}^2}}{}\right) \left(\frac{1\ \text{m}^2}{10\ 000\ \cancel{\text{cm}^2}}\right) = \frac{50\ 000 \times 1\ \text{m}^2}{10\ 000} = \frac{50\ 000\ \text{m}^2}{10\ 000} = \frac{5\ \text{m}^2}{}$$

Answer: 5 square meters

Sample Problem 7-5.

What volume, in cubic feet, must a 50,000 gallon maximum capacity swimming pool have?

Step 1. Conversion factor:

1 gallon = 0.1337 cubic feet

Step 2. Set up conversion bar.

$$\left(\frac{50,000\ \text{gallon}}{}\right) \left(\frac{0.1337\ \text{cubic feet}}{1\ \text{gallon}}\right)$$

Step 3. Cancel units. Perform arithmetic.

$$\left(\frac{50,000\ \cancel{\text{gal.}}}{}\right) \left(\frac{0.1337\ \text{ft.}^3}{1\ \cancel{\text{gal.}}}\right) = \frac{50,000 \times 0.1337\ \text{ft.}^3}{} = \frac{6685\ \text{ft.}^3}{}$$

Answer: 6685 cubic feet

Practice Problems

On a separate sheet of paper, convert the measurement given to the new unit shown. If necessary, use decimal places in the answer.

U.S. Conventional
1. 2 miles = _____ feet
2. 48 fluid ounces = _____ pint
3. 18 inches = _____ feet
4. 12 inches² = _____ feet²
5. 6 yards = _____ inches
6. 36 feet = _____ inches
7. 12 cups = _____ quarts
8. 128 cups = _____ gallons
9. 8 cu. yd. = _____ cu. ft.
10. 5 pounds = _____ ounces

SI Metric
11. 65 meters = _____ centimeters
12. 15 grams = _____ decigrams
13. 350 millimeters = _____ meters
14. 3 kilometers = _____ meters
15. 25 square meters = _____ square decimeters
16. 850 centimeters = _____ meters
17. 5 liters = _____ milliliters
18. 530 square decimeters = _____ square meters
19. 10 kilograms = _____ grams
20. 5000 meters = _____ hectometers

Time
21. 3 weeks = _____ hours
22. 90 hours = _____ minutes
23. 6 days = _____ minutes
24. January = _____ hours
25. 2 hours = _____ seconds
26. 84 days = _____ weeks
27. 2300 minutes = _____ hours
28. 850 hours = _____ days
29. 7200 seconds = _____ minutes
30. 2 years = _____ hours

Converting combined units

As indirectly stated before, different units can be combined in such a way that one unit is compared to another. Miles per hour, for example, measures distance in a unit of time. The term *per* signifies a ratio. It is written mathematically as a fraction. As examples:

- $\dfrac{\text{miles}}{\text{hour}}$ – miles *per* hour (mph).

- $\dfrac{\text{gallons}}{\text{minute}}$ – gallons *per* minute (gpm).

- $\dfrac{\text{pounds}}{\text{square inch}}$ – pounds *per* square inch (psi).

When converting, each unit remains in its category. However, the individual units can be converted to subunits within the same category.

Sample Problem 7-6.

If a car is traveling at 60 miles per hour, how many feet per second is it moving?

Step 1. Conversion factors:

1 mile = 5280 feet
1 hour = 60 minutes
1 minute = 60 seconds

Step 2. Set up conversion bar. Keep in mind which unit will be in the numerator and which will be in the denominator of the final answer.

$$\left(\frac{60 \text{ miles}}{\text{hour}}\right) \left(\frac{5280 \text{ feet}}{1 \text{ mile}}\right) \left(\frac{1 \text{ hour}}{60 \text{ minutes}}\right) \left(\frac{1 \text{ minute}}{60 \text{ seconds}}\right)$$

Step 3. Cancel units. Perform arithmetic.

$$\left(\frac{60 \text{ mi.}}{\text{hr.}}\right) \left(\frac{5280 \text{ ft.}}{1 \text{ mi.}}\right) \left(\frac{1 \text{ hr.}}{60 \text{ min.}}\right) \left(\frac{1 \text{ min.}}{60 \text{ sec.}}\right) =$$

$$\frac{60 \times 5280 \text{ ft.}}{60 \times 60 \text{ sec.}} = \frac{88 \text{ ft.}}{\text{sec.}}$$

Answer: 88 feet per second

Other applications

In Chapter 1, word problems were presented. With some word problems, a "conversion bar" can be used. Its use, at times, can make the solving of problems easier. You are not actually converting units, but the principles are the same.

Sample Problem 7-7.

Weight of a certain pipe is given as 4.3 pounds per linear foot. How much does an 8-foot piece weigh?

Step 1. Set up "conversion bar."

$$\left(\frac{8 \text{ linear feet}}{}\right) \left(\frac{4.3 \text{ pounds}}{\text{linear foot}}\right)$$

Step 2. Cancel units. Perform arithmetic.

$$\left(\frac{8 \text{ lin. ft.}}{}\right) \left(\frac{4.3 \text{ lb.}}{\text{lin. ft.}}\right) = \frac{8 \times 4.3 \text{ lb.}}{} = \frac{34.4 \text{ lb.}}{}$$

Answer: 34.4 pounds

Sample Problem 7-8.

Certain pipe is installed at the average rate of 87 feet per hour. The price of the pipe is $4 per foot. After 10 hours, what is the total value of installed pipe?

Step 1. Set up "conversion bar."

$$\left(\frac{87 \text{ feet}}{\text{hour}}\right) \left(\frac{\$4}{\text{foot}}\right) \left(\frac{10 \text{ hours}}{}\right)$$

Step 2. Cancel units. Perform arithmetic.

$$\left(\frac{87 \text{ ft.}}{\text{hr.}}\right) \left(\frac{\$4}{\text{ft.}}\right) \left(\frac{10 \text{ hr.}}{}\right) = \frac{87 \times \$4 \times 10}{} = \frac{\$3480}{}$$

Answer: $3480

Practice Problems

On a separate sheet of paper, convert the measurement given to the new unit shown. If necessary, use decimal places in the answer.

1. 80 miles per hour = _____ feet/second
2. 90 kilometers per hour = _____ meters/hour
3. 25 miles per gallon = _____ gallons/mile

4. 135 words per minute = _____ words/hour
5. 80 liters per kilometer = _____ liters/meter
6. 35 gallons per hour = _____ quarts/minute
7. 10 pounds per square inch = _____ pounds/square foot
8. 50 feet per second = _____ miles/hour
9. 7 pounds per gallon = _____ pounds/quart
10. 888 grams per minute = _____ kilograms/hour

Complete the following problems on a separate sheet of paper. Show your work and your final answer. Include units of measure.

11. If an airplane burns 7 gallons per hour, how many gallons of fuel will be used in 5 hours?
12. If a typist can type at the rate of 90 words per minute, how long will it take to type 5 documents that average 250 words each?
13. If a chocolate machine can make 40 kilograms of chocolate per minute, how much chocolate can be produced in 2 hours?
14. If a bus travels at an average of 56 miles per hour, how many miles will be traveled in 8 hours?
15. If an economy car can travel 45 miles on one gallon of gasoline, how many gallons will be needed for a 360 mile trip?

CONVERTING BETWEEN U.S. CONVENTIONAL AND SI METRIC

Conversion factors in Technical Section B for converting from one system to another are given in multiplication form. To use this type of conversion, multiply the given units by the factor to obtain the new unit. If the chart does not have a direct conversion between the two desired units, find a multiplication factor with units that can be converted within the same system.

Sample Problem 7-9.

How many meters are in 2 feet?
Step 1. Conversion factor:
(U.S. Conventional to SI Metric)
feet × 0.305 = meters
Step 2. Perform the arithmetic. Final answer will have the new units.
2 × 0.305 = 0.610 m
Answer: 0.610 meters

Sample Problem 7-10.

How many gallons are in 100 liters?
Step 1. Conversion factor:
(SI Metric to U.S. Conventional)
liters × 0.264 = gallons
Step 2. Perform the arithmetic.
100 × 0.264 = 26.4 gal.
Answer: 26.4 gallons

Practice Problems

Convert the measurement given on the left to the new unit shown on the right. If necessary, use decimal places in the answer, rather than compound units.

1. 48 miles = _____ kilometers
2. 80 meters = _____ yards
3. 16 centimeters = _____ inches
4. 24 feet = _____ meters
5. 32 square feet = _____ square meters
6. 18 square meters = _____ square feet

7. 280 ounces = _____ grams
8. 125 pounds = _____ kilograms
9. 48 kilograms = _____ pounds
10. 36 liters = _____ gallons

Complete the following problems on a separate sheet of paper. Show your work and your final answer. Include units of measure.

11. A tourist from the U.S. travels into Canada, where gasoline is sold by the liter. If fill-ups during the trip had been averaging 18 gallons, how many liters would be needed for the same amount of gasoline?
12. A high-school student steps on the scale and finds his weight to be 68 kilograms. What is his weight in pounds?
13. The distance between Chicago, Illinois, and Indianapolis, Indiana, is 289 kilometers. What is the distance between the two cities in miles?
14. The area of a bedroom is 144 square feet. How many square meters of carpet would be needed for the room?
15. The volume of a swimming pool is 20,000 cubic feet. What is the volume of the pool in cubic meters?

ARITHMETIC WITH DENOMINATE NUMBERS

A **denominate number** is a number with units attached — specifically, units of measure. Examples are 50 feet, 50 pounds, and 10 gallons. An **abstract number** is a number with no units. Arithmetic with denominate numbers is somewhat different from abstract numbers because of the units. By now, you are familiar with denominate numbers. In working with the conversion bar, you have had practice in multiplying and dividing denominate numbers. Always follow the applicable rules when working with these numbers.

RULES OF ADDITION/SUBTRACTION WITH DENOMINATE NUMBERS

- Units must be the same.
- If units are not the same, convert so they are.

Sample Problem 7-11.

Add 6 meters and 3 meters.
Since these two already have the same units, align in a column and add.

$$
\begin{array}{r}
6 \text{ m} \\
+\ 3 \text{ m} \\
\hline
9 \text{ m}
\end{array}
$$

Sample Problem 7-12.

Subtract 6 feet from 4 yards.
Step 1. Since the units are not the same, one unit must be converted to the other.

4 yards = 12 feet

Step 2. Subtract.

$$
\begin{array}{r}
12 \text{ ft.} \\
-\ 6 \text{ ft.} \\
\hline
6 \text{ ft.}
\end{array}
$$

RULES OF MULTIPLICATION/DIVISION WITH DENOMINATE NUMBERS

- Units do not have to be the same, but should be if working with area or volume, in particular.
- Multiply/divide numbers together.
- Multiply/divide units together.
  1. If multiplying/dividing denominate number by abstract number, units will be the same as units of denominate number.
     Examples:
     > 3 linear feet $\times$ 3 = 9 linear feet
     > 120 pounds $\div$ 3 = 40 pounds
  2. If multiplying/dividing denominate numbers together, all units carry over.
     Examples:
     > 120 pounds $\times$ 3 feet = 360 foot-pounds
     > 50 pounds $\div$ 2 inches^2 = 25 pounds/inches^2 (25 pounds per square inch)
     > 5 feet $\times$ 5 feet = 25 feet^2
     > 100 feet^3 $\div$ 10 feet = 10 feet^2

Note: When multiplying denominate numbers of the same units together, the units of the answer are raised to a power. The power is determined by adding together the powers. If unwritten, a power is equal to 1. For example:

$$3 \text{ feet} \times 2 \text{ feet} \times 2 \text{ feet} = 3 \text{ feet}^1 \times 2 \text{ feet}^1 \times 2 \text{ feet}^1$$
$$= 12 \text{ feet}^{1+1+1} = 12 \text{ feet}^3$$
$$3 \text{ feet} \times 4 \text{ feet}^2 = 3 \text{ feet}^1 \times 4 \text{ feet}^2 = 12 \text{ feet}^{1+2} = 12 \text{ feet}^3$$

When dividing denominate numbers raised to a power, the powers are subtracted. For example:

$$12 \text{ feet}^3 \div 4 \text{ feet}^2 = 3 \text{ feet}^{3-2} = 3 \text{ feet}^1 = 3 \text{ feet}$$

Sample Problem 7-13.

Multiply: 8 inches $\times$ 6 inches.

Multiply the numbers together. Multiply the units together. Units will be of the same type but will be raised to a power. The new denominate number formed will fall in a new category (length $\times$ length = area).

$$
\begin{array}{r}
8 \text{ in.} \quad \text{(length)} \\
\times \quad 6 \text{ in.} \quad \text{(length)} \\
\hline
48 \text{ in.}^2 \quad \text{(area)}
\end{array}
$$

Sample Problem 7-14.

Multiply: 6 meters $\times$ 4 meters $\times$ 10 meters.

Multiply the numbers together. Multiply the units together. Raise the unit to the proper power.

$$
\begin{array}{r}
6 \text{ m} \quad \text{(length)} \\
\times \quad 4 \text{ m} \quad \text{(length)} \\
\hline
24 \text{ m}^2 \quad \text{(area)} \\
\times \quad 10 \text{ m} \quad \text{(length)} \\
\hline
240 \text{ m}^3 \quad \text{(volume)}
\end{array}
$$

Sample Problem 7-15.

Divide 16 gallons by 4.
Divide the numbers. Units stay the same.
> 16 gal. $\div$ 4 = 4 gal.

Sample Problem 7-16.

Divide 16 gallons by 4 gallons.
Divide the numbers. Divide the units.
> 16 gal. $\div$ 4 gal. = 4

Sample Problem 7-17.

Divide 8 feet² by 2 feet.
Divide the numbers. Divide the units.

$$8 \text{ ft.}^2 \div 2 \text{ ft.} = 8 \text{ ft.}^2 \div 2 \text{ ft.}^1 = 4 \text{ ft.}^{2-1} = 4 \text{ ft.}^1 = 4 \text{ ft.}$$

Note: It can be helpful to set these problems up in the form of a "conversion bar" as previously discussed.

Practice Problems

Perform the indicated arithmetic operations on a separate sheet of paper. If units are not the same, make conversions first. Leave your answer in the unit used for the arithmetic rather than changing the answer to a compound denominate number. Include units of measure (inches, feet, minutes, dollars, etc.).

1. 6 1/2 inches + 7 3/4 inches + 18 3/8 inches
2. 256 centimeters + 592 centimeters + 325 centimeters
3. 15 feet + 4 yards + 10 inches
4. 15 meters + 120 centimeters + 380 millimeters
5. 28 quarts − 3 gallons
6. 2 hours − 90 minutes
7. 4 pounds − 40 ounces
8. 6 kilograms − 120 grams
9. 40 feet + 6 meters
10. 30 kilograms − 25 pounds

Perform the indicated arithmetic operations on a separate sheet of paper. Leave your answer in the unit used for the arithmetic rather than changing the answer to a compound denominate number.

11. 4 feet × 8
12. 6 × 32 kilometers
13. 48 inches ÷ 6
14. 256 grams ÷ 16
15. 4 gallons × 10
16. 200 centimeters × 350 centimeters
17. 5 feet × 6 feet × 10 feet
18. 3 meters × 8 meters × 2 meters
19. 48 square inches ÷ 6
20. 128 square centimeters ÷ 8

ARITHMETIC WITH COMPOUND DENOMINATE NUMBERS

The U.S. Conventional system contains several units under the same quantity, such as inches, feet, yards, and miles. A value has **compound units** when combinations of such units exist. A measure given as 3 yards, 2 feet, 9 inches is in compound units. Numbers given in compound units are *compound denominate numbers.*

RULES FOR ADDITION WITH COMPOUND UNITS

1. Align in columns with like units. Add like units.
2. Units in the answer should be reduced to the lowest terms.

Sample Problem 7-18.

Add the following time periods:
15 hours, 52 minutes, 34 seconds
14 hours, 30 minutes, 45 seconds

Step 1. Align in columns. Add like units.

    15 hr., 52 min., 34 sec.
  + 14 hr., 30 min., 45 sec.
    29 hr., 82 min., 79 sec.

Step 2. In a manner similar to that of fractions, reduce the units to lowest terms. Start with the smallest unit (seconds). If it is large enough, convert it to the next larger unit (minutes). If you do not know how many smaller units make up a larger unit, consult the conversion tables.

(conversion: 60 seconds = 1 minute)

      79 sec.
    − 60 sec. (1 min.)
      19 sec.

79 sec. = 1 min., 19 sec.

(Another way to change 79 seconds into a compound unit would be to divide 79 by 60. The quotient would give minutes; a remainder would give seconds. This method applies anytime you want to convert single units to compound units.)

Step 3. Remove the reduced unit from the original answer. Then add the results of reducing to the modified original answer.

        29 hr., 82 min.
    +   1 min., 19 sec.
        29 hr., 83 min., 19 sec.

Step 4. Repeat the process of reducing with the next larger unit (minutes).

(conversion: 60 minutes = 1 hour)

      83 min.
    − 60 min. (1 hour)
      23 min.

83 min. = 1 hr., 23 min.

Step 5. Add the results of reducing to the proper column. Remember to exclude the reduced unit (83 minutes) from the original answer before adding.

      29 hr.,   19 sec.
    + 1 hr., 23 min.
      30 hr., 23 min., 19 sec.

Step 6. The largest unit (hours) can be converted to a larger unit (days).

(conversion: 24 hours = 1 day)

      30 hr.
    − 24 hr. (1 day)
       6 hr.

30 hr. = 1 day, 6 hr.

Step 7. Final answer.
1 day, 6 hours, 23 minutes, 19 seconds

RULES FOR SUBTRACTION WITH COMPOUND UNITS

1. Start with smallest units.
2. To borrow, follow these steps:
   a. Subtract one unit from the next larger unit.
   b. Convert the borrowed unit into the smaller unit. Use conversion tables, if necessary.
   c. Add the borrowed units to the smaller unit.
3. Perform the arithmetic.

Sample Problem 7-19.

Subtract the following lengths:
5 yards, 1 foot
3 yards, 2 feet

Step 1. Align in columns. See if you can subtract as is. If so, do so.

$$\begin{array}{r} 5 \text{ yd., } 1 \text{ ft.} \\ - \ 3 \text{ yd., } 2 \text{ ft.} \\ \hline \end{array}$$

Step 2. Since direct subtraction was not possible, you will have to "borrow." Subtract 1 unit from the larger unit (yards). Convert the borrowed unit into smaller units. Rewrite.

(conversion: 1 yard = 3 feet)

$$\begin{array}{r} 4 \text{ yd., } 4 \text{ ft.} \\ - \ 3 \text{ yd., } 2 \text{ ft.} \\ \hline \end{array}$$

Step 3. Perform the arithmetic.

$$\begin{array}{r} 4 \text{ yd., } 4 \text{ ft.} \\ - \ 3 \text{ yd., } 2 \text{ ft.} \\ \hline 1 \text{ yd., } 2 \text{ ft.} \end{array}$$

RULES FOR MULTIPLICATION WITH COMPOUND UNITS

1. Multiply each term of the value. All rules of multiplication with denominate numbers apply.
2. Reduce the answer to lowest terms.

Sample Problem 7-20.

Multiply: 5 × (5 gallon, 3 quart, 1 pint).
Note: The compound denominate number is shown in parentheses here to show that it is a single value.

Step 1. Multiply each unit individually.

$$\begin{array}{r} 5 \text{ gal., } \ 3 \text{ qt., } 1 \text{ pt.} \\ \times \phantom{25 \text{ gal., } 15 \text{ qt., } 5} 5 \\ \hline 25 \text{ gal., } 15 \text{ qt., } 5 \text{ pt.} \end{array}$$

Step 2. Reduce to lowest terms. Start with the smallest unit.
a. Convert pints to quarts.
   (conversion: 2 pints = 1 quart)
   5 pt. = 2 qt., 1 pt.
b. Add to larger unit.

$$\begin{array}{r} 25 \text{ gal., } 15 \text{ qt.} \\ + \phantom{25 \text{ gal., } 15} 2 \text{ qt., } 1 \text{ pt.} \\ \hline 25 \text{ gal., } 17 \text{ qt., } 1 \text{ pt.} \end{array}$$

  c. Convert quarts to gallons.
   (conversion: 4 quarts = 1 gallon)
   17 qt. = 4 gal., 1 qt.
  d. Add to larger unit for final answer.

$$
\begin{array}{r}
25 \text{ gal.,} \qquad 1 \text{ pt.} \\
+ \quad 4 \text{ gal., 1 qt.} \\
\hline
29 \text{ gal., 1 qt., 1 pt.}
\end{array}
$$

RULES FOR DIVISION WITH COMPOUND UNITS

1. Start dividing with the largest unit. If the division has a remainder, it will be of the same units. All rules of division with denominate numbers apply.
2. Convert the remainder, if any, and add to the next lower unit.
3. Continue dividing until all units of the problem have been divided. If, at this point, the remainder is zero, you are finished. Record the final answer.
4. If possible, continue dividing into smaller subunits until a remainder of zero is obtained or until the smallest possible subunit is divided.
5. Any remainder of the smallest subunit is written as a fractional part of that unit.

Sample Problem 7-21.

Divide (5 yards, 2 feet, 7 inches) by 3.

Step 1. Divide each unit individually, starting with the largest.

$$
\begin{array}{r}
1 \text{ yd. remainder 2 yd.} \\
3 \overline{\smash{)}5 \text{ yd.}}
\end{array}
$$

Step 2. Convert the remainder and add to the next lower unit.
   (conversion: 1 yard = 3 feet)
   2 yd. = 6 ft.
   2 ft. + 6 ft. = 8 ft.

Step 3. Divide the adjusted unit.

$$
\begin{array}{r}
2 \text{ ft. remainder 2 ft.} \\
3 \overline{\smash{)}8 \text{ ft.}}
\end{array}
$$

Step 4. Convert the remainder and add to the next lower unit.
   (conversion: 1 foot = 12 inches)
   2 ft. = 24 in.
   7 in. + 24 in. = 31 in.

Step 5. Divide the adjusted unit.

$$
\begin{array}{r}
10 \text{ in. remainder 1 in.} \\
3 \overline{\smash{)}31 \text{ in.}}
\end{array}
$$

Step 6. The final remainder is written as a fraction.

$$
\frac{1}{3} \text{ in.}
$$

Step 7. Final answer is the combination of all division answers (quotients).
   1 yard, 2 feet, 10 1/3 inches

Sample Problem 7-22.

Divide (2 miles, 1 yard) by 3.

Step 1. Divide each unit individually, starting with the largest.

$$
\begin{array}{r}
0 \text{ mi. remainder 2 mi.} \\
3 \overline{\smash{)}2 \text{ mi.}}
\end{array}
$$

Step 2. Convert the remainder and add to the next lower unit.
(conversion: 1 mile = 1760 yards)
2 mi. = 3520 yd.
1 yd. + 3520 yd. = 3521 yd.

Step 3. Divide the adjusted unit.

$$\frac{1173 \text{ yd.}}{3 \overline{)3521 \text{ yd.}}} \quad \text{remainder 2 yd.}$$

Step 4. Convert the remainder and add to the next lower unit.
(conversion: 1 yard = 3 feet)
2 yd. = 6 ft.
0 ft. + 6 ft. = 6 ft.

Step 5. Divide the adjusted unit.

$$\frac{2 \text{ ft.}}{3 \overline{)6 \text{ ft.}}} \quad \text{remainder 0}$$

Step 6. With all units divided and a remainder of zero, no further division is needed.
Answer: (1173 yards, 2 feet)

Practice Problems

Perform the indicated arithmetic operations on a separate sheet of paper. The answers should be reduced to the lowest form of compound denominate numbers. Include units of measure (inches, feet, minutes, etc.).

1. 12 yards, 1 foot, 9 inches
    + 2 yards, 5 feet, 15 inches

2. 6 gallons, 5 quarts
    +3 gallons, 3 quarts, 9 pints

3. 3 days, 32 hours, 15 minutes
    2 days, 16 hours, 65 minutes
    8 days, 21 hours, 30 minutes
    +2 days, 18 hours, 50 minutes

4. 15 yards, 4 inches
    − 7 feet, 8 inches

5. 5 weeks, 8 days, 4 hours, 32 minutes
    +2 weeks, 3 days, 28 hours, 45 minutes

6. 60 hours, 25 minutes
    − 1 day, 45 minutes

7. 4 × (12 feet, 9 inches)
8. 3 × (16 gallons, 7 quarts, 3 pints)
9. (3 days, 16 hours) ÷ 2
10. (32 feet, 12 inches) ÷ 6

CONVERTING COMPOUND AND SINGLE UNITS

The value of a compound denominate number has single-unit equivalents. Changing to single units requires addition and use of conversion bars.

Single units can, of course, be changed into compound units. Again, addition and use of the convesion bar are required.

Sample Problem 7-23.

Change (8 yards, 2 feet, 11 inches) to single units of yards.

Step 1. Convert feet to yards.

(conversion: 3 feet = 1 yard)

$$\left(\frac{2 \text{ ft.}}{}\right) \left(\frac{1 \text{ yd.}}{3 \text{ ft.}}\right) = \frac{2 \text{ yd.}}{3} = 0.667 \text{ yd.}$$

Step 2. Convert inches to yards.

(conversions: 12 inches = 1 foot; 3 feet = 1 yard)

$$\left(\frac{11 \text{ in.}}{}\right) \left(\frac{1 \text{ ft.}}{12 \text{ in.}}\right) \left(\frac{1 \text{ yd.}}{3 \text{ ft.}}\right) = \frac{11 \text{ yd.}}{36} = 0.306 \text{ yd.}$$

Step 3. Add all units for final answer.

$$
\begin{array}{r}
8. \quad \text{yards} \\
0.667 \text{ yards} \\
+ \ 0.306 \text{ yards} \\
\hline
8.973 \text{ yards}
\end{array}
$$

Note: An easier solution is to change feet into inches. Then, total inches and convert to yards (conversion: 36 inches = 1 yard).

Sample Problem 7-24.

Change 1.8356 tons into a compound denominate number.

Step 1. Determine the largest unit and number. Largest unit is tons. Number of whole tons is 1. Amount remaining is 0.8356 tons.

Step 2. Determine the next smaller unit. Pounds is the next smaller unit.

Step 3. Find the number of pounds.

(conversion: 1 ton = 2000 pounds)

$$\left(\frac{0.8356 \text{ tons}}{}\right) \left(\frac{2000 \text{ lb.}}{1 \text{ ton}}\right) = 1671.2 \text{ lb.}$$

Number of whole pounds is 1671. Amount remaining is 0.2 lb.

Step 4. Determine the next smaller unit. Ounces is the next smaller (and smallest) unit.

Step 5. Find the number of ounces.

(conversion: 1 pound = 16 ounces)

$$\frac{0.2 \text{ lb.}}{} \ \frac{16 \text{ oz.}}{1 \text{ lb.}} = 3.2 \text{ oz.}$$

Step 6. Add together for final answer.

(1 ton, 1671 pounds, 3.2 ounces)

Practice Problems

On a separate sheet of paper, change the number with compound units on the left to the single units shown on the right. If the answer has fractional parts of a larger unit, express it as a decimal. Round the answers to 3 decimal places. Include units of measure.

Example: 2.333 yards.

1. 5 yards, 2 feet, 6 inches = _____ inches = _____ yards
2. 4 pounds, 12 ounces = _____ ounces = _____ pounds
3. 1 gallon, 3 quarts, 5 pints = _____ pints = _____ quarts
4. 2 weeks, 12 days, 45 minutes = _____ hours = _____ days
5. 4 yards, 14 inches = _____ feet = _____ inches

Change the single unit on the left to the compound units on the right.

6. 100 inches = _____ yards, _____ feet, _____ inches
7. 38 ounces = _____ pounds, _____ ounces
8. 56 quarts = _____ gallons, _____ quarts, _____ pints
9. 200 minutes = _____ days, _____ hours, _____ minutes
10. 58 feet = _____ yards, _____ feet, _____ inches

TIME ZONES AND TRAVEL TIME

The world is divided into 24 time zones — one for each hour of the day. Each calendar day is designated to begin and end at the international date line. This imaginary line lies somewhere between Alaska and Russia. The time zones represent each hour of the day, going from east to west around the globe. Moving west, each time zone is 1 hour behind its neighbor to the east. The exception to this is at the time zone just east of the international date line. Moving west from here, time will advance 24 hours.

The four time zones of the continental United States are shown in Technical Section B. Traveling west, time is subtracted. Traveling east, time is added. *Local time* is the time at a particular location.

The concept of time zones can be quite confusing. To further complicate matters, the clock we are generally familiar with is based on 12 hours, rather than 24 hours. The day runs in 2 periods of 12 hours; from midnight to noon and back to midnight. The military uses the 24-hour clock. Military time makes calculating time periods much easier.

With the 24-hour clock, time starts at midnight (0000). It goes up to noon (1200). Past noon, hours continue up to 2400 (0000 midnight). Military time is given as a 4-place number with no punctuation. Hours of the day are given by the 2 leftmost numbers. Minutes are given by the 2 rightmost numbers. On a 24-hour clock, hours are given by numbers 1 through 24 (or 00 through 23). The unit of measure is hours. A comparison of the standard clock to the 24-hour clock is available in Technical Section B.

To convert from standard clock time to 24-hour time, add 12 hours to hours from 1 p.m. to midnight. To convert from 24-hour time to standard clock time, subtract 12 hours from hours of 1300 or greater. Except for punctuation, time up until 1 p.m. is the same. If minutes are given in standard clock time, they are written with a colon (:). The colon is dropped in 24-hour time. For example, 9:25 a.m. in standard clock time is written 0925 in 24-hour time.

Sample Problem 7-25.

A plane leaves New York at 11:15 a.m. and travels to Miami in 3 hours and 10 minutes. What is the arrival time in Miami?

Step 1. Convert the times given to 24-hour time.

11:15 a.m. = 1115 hours
3 hours and 10 minutes = 0310 hours

Step 2. Add the numbers in 24-hour time.

$$\begin{array}{r} 1115 \text{ hours} \\ + \ 0310 \text{ hours} \\ \hline 1425 \text{ hours} \end{array}$$

Step 3. Convert back to standard clock time by subtracting 12 hours (for 1300 hours or greater). Next, place a colon in the answer. The answer is the arrival time of the plane.

$$\begin{array}{r} 14 \ (25) \text{ hours} \\ - \ 12 \ (00) \text{ hours} \\ \hline 2 \ (25) \text{ hours} = 2.25 \text{ p.m.} \end{array}$$

Sample Problem 7-26.

A phone call is made from Seattle, Washington, to Washington, D.C., at 5:45 p.m. Seattle time. What is the time in Washington, D.C.?

Step 1. Use the time zone map in Technical Section B. Determine which zone each city is located in.

Washington, D.C., is in the eastern time zone.
Seattle is in the Pacific time zone.

Step 2. Find the time in the city in question, Washington. When going east, add one hour for each time zone. When going west, subtract one hour for each time zone.

> Pacific   = 5:45 p.m.
> mountain = 6:45 p.m.
> central   = 7:45 p.m.
> eastern   = 8:45 p.m. (answer)

Sample Problem 7-27.

A plane leaves St. Louis, Missouri, at 10:25 a.m. local time. If the travel time to Los Angeles, California, is 4 hours and 40 minutes, what is the time in Los Angeles when the plane arrives?

Step 1. Convert times to the 24-hour system.

> 10:25 a.m. = 1025 hours
> 4 hours and 40 minutes = 0440 hours

Step 2. Add the times. The time after addition will be the time in St. Louis when the plane lands in Los Angeles.

> 1025 hours (departure time)
> + 0440 hours (travel time)
> 1465 hours = 1505 (St. Louis time)

Note: Minutes should not exceed 60 in the answer. If they do, they should be changed to hours. It might help to think of 24-hour time as a compound denominate number, in units of minutes and seconds, and the same rules apply. (Often, if subtracting, 60 minutes will need to be "borrowed" from the hours column before the operation can be performed. For example [unrelated to sample problem], subtract: 1820 hr. − 1635 hr. Borrowing 60 minutes, 1820 becomes 1780 hr. Now, subtract as usual.)

Step 3. Change the St. Louis time to the standard clock time (subtract 12 hours; add colon).

> 1505 hours = 3:05 p.m.

Step 4. Locate the cities on the time-zone map.

> St. Louis = central time
> Los Angeles = Pacific time

Step 5. Change to Los Angeles time by counting backwards since travel is to the west.

> 3:05 p.m. central time
> 2:05 p.m. mountain time
> 1:05 p.m. Pacific time (answer)

Practice Problems

Complete the following problems on a separate sheet of paper. Show your work and your final answer.

Change the following from clock time to 24-hour time.

1. 8:35 a.m.
2. 9:06 a.m.
3. 10:14 a.m.
4. 11:55 a.m.
5. 2:40 p.m.
6. 5:30 p.m.
7. 10:20 p.m.
8. 11:00 p.m.
9. noon
10. midnight

Change the following from 24-hour time to clock time.
11. 0630
12. 0320
13. 1240

14. 1545
15. 1110
16. 1200
17. 0905
18. 0108
19. 2250
20. 1835

Change the time given on the left to the time zone on the right.
21. 1:30 p.m. eastern to Pacific
22. 4:00 a.m. central to eastern
23. 10:20 p.m. mountain to eastern
24. 9:20 a.m. mountain to Pacific
25. 5:10 p.m. central to mountain
26. 2:30 a.m. Pacific to central
27. 4:00 a.m. Pacific to central
28. 3:30 p.m. eastern to mountain
29. 1050 Atlantic to Pacific
30. 0715 mountain to Atlantic

31. A plane leaves Miami, Florida, at 7:30 a.m. local time. It arrives in Denver, Colorado, after 2 hours and 45 minutes travel time. What is the arrival time in Denver?
32. A plane leaves Seattle, Washington, at 10:20 p.m. local time. What is the arrival time in Boston, Massachusetts, after a 5 hour flight?

TEST YOUR SKILLS

Do *not* write in this book. Use a separate sheet of paper to complete the following problems. Show your work and your final answer. (Obtain conversion factors from Technical Section B.)

CONVERTING UNITS

Convert the measurement given to the new unit shown. If necessary, use decimal places in the answer rather than compound units.

U.S. Conventional:
1. 3 miles = _____ feet
2. 588 sq. in. = _____ ft.^{2}
3. 235 ft.^{3} = _____ cu. in.
4. 96 ounces = _____ pounds
5. 32 cups = _____ quarts

SI Metric:
6. 785 meters = _____ centimeters
7. 65 000 dm^2 = _____ m^2
8. 8500 m^3 = _____ dam^3
9. 20 kilograms = _____ grams
10. 250 ml = _____ liters

Time: (U.S. Conventional and SI Metric)
11. 52 weeks = _____ hours
12. 2080 hours = _____ days
13. month of May = _____ hours
14. 1 week = _____ minutes
15. 12,900 seconds = _____ hours

Combined units:
16. 40 miles per hour = _____ ft./sec.
17. 80 words per minute = _____ words/hour

18. 75 liters per kilometer = _____ l/m
19. 60 grams per carton = _____ kg/carton
20. 125 pounds per ft.² = _____ lb./in.²

Converting between U.S. Conventional and SI Metric:
21. 60 miles = _____ kilometers
22. 32 sq. meters = _____ ft.²
23. 180 cu. ft. = _____ m³
24. 86 kilograms = _____ pounds
25. 90 liters = _____ gallons

ARITHMETIC WITH DENOMINATE NUMBERS

Perform the indicated arithmetic operations. Answers in compound denominate numbers should be reduced to the lowest terms.

26.   6 yards, 8 feet, 15 inches
  +   27 feet, 18 inches

27.   12 feet², 280 inches²
  + 4 feet², 130 inches²

28.   6 gallons, 9 quarts, 5 pints, 2 cups
  + 2 gallons, 7 quarts, 1 pint , 6 cups

29.   5 tons, 1200 pounds
  − 3 tons, 1500 pounds

30.   50 kilograms
  − 36 grams

31.   3 weeks, 18 days, 16 hours, 33 minutes
  − 1 week,   9 days, 21 hours, 42 minutes

32. 2 × (14 feet, 9 inches)
33. 8 × (6 gallons, 3 quarts, 1 pint)
34. 14 feet × 10 feet
35. (4 days, 12 hours, 8 minutes) ÷ 3

WORD PROBLEMS WITH UNITS OF MEASURE

36. A communication satellite weighs 4125 pounds prior to launching. How many tons is this?
37. Cruising altitude of a certain airplane is 32,000 feet. What is its altitude in miles?
38. The Hoover Dam on the Colorado River is the highest dam in the United States at 726 feet. How high is this in meters?
39. By volume, the largest dam in North America is the Fort Peck Dam in Montana; it has a volume of 125.6 million cubic yards of concrete (rounded to nearest 100,000 cubic yards). Convert this amount to cubic meters. Round answer to nearest 100,000 cubic meters.
40. The scale model shown in Fig. 7-2 is to be made to full scale. The scale is 1 inch equals 1 foot. A metric rule is the only measurement tool readily available. What will dimensions a, b, and c be, in centimeters, when the part is full size?

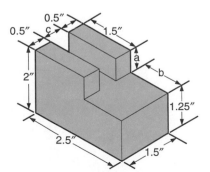

Fig. 7-2. Dimensions of a scale model of a part.

41. Speed limit on a certain highway in Canada is 100 kilometers per hour. If traveling the speed limit, what should a speedometer, calibrated in miles per hour, read?
42. The acceleration of gravity is 32 feet per second². What is this in meters per second²?
43. Atmospheric pressure at sea level is about 14.7 pounds per inch². What is this in grams per centimeter²?
44. Waiting for a train at a railroad crossing, a motorist estimates each car to be about 40 feet long. The motorist counts 40 cars crossing the road in one minute. If the train moves at this same rate and lasts for 10 minutes, (a) what is the estimated length of the train, in miles, and (b) how fast is it moving?
45. The unit rate for a concrete installation, including formwork, rebar, placing concrete, etc., is given at 3 worker-hours per cubic yard. At a labor rate of $17 per hour, what is the labor cost for installing 200 cubic yards of concrete?

TIME ZONES AND TRAVEL TIME

46. Change 2:15 a.m. eastern time to mountain time.
47. Change 3:08 p.m. Pacific time to central time.
48. Change 5:31 p.m. to 24-hour time.
49. A plane leaves San Francisco, California, at 9:25 a.m. local time. It arrives in Boston, Massachusetts, at 5:45 p.m. local time. What is the flight time?
50. The flight time from Bangor, Maine, to Denver, Colorado, is 4 3/4 hours. What time would the plane arrive in Denver if it left Bangor at 9:35 p.m.?

PROBLEM-SOLVING ACTIVITIES
Activity 7-1.
Units of liquid measure
Objective: To convert units by pouring water between pint, quart, and gallon containers.
Instructions:
1. Fill several pint containers with water.
2. Using the conversion factor, convert the pints to quarts.
3. Pour the water from the pint containers to the quart containers. Compare the number of quart containers filled to your calculations.
4. Using the conversion factors, convert the quarts to gallons.
5. Pour the water from the quart containers to the gallon containers. Compare the number of gallon containers filled to your calculations.

Activity 7-2.
Comparing meters and yards
Objective: To use a meter stick and yardstick to measure the same objects.
Instructions:
1. Select 3 different objects to measure, such as a desk, a book, and a chair.
2. Measure each object with a yardstick.
3. Use conversion factors to convert your measurements from U.S. Conventional to SI Metric.
4. Measure the objects with a meter stick.
5. Compare these measurements to your calculations.

Activity 7-3.
Flash cards: 24 hour and clock time
Objective: To use flash cards with 24-hour time written on one side and clock time written on the other side.
Instructions:
1. Using 3″ × 5″ index cards, make at least 10 different flash cards with 24-hour time on one side and the equivalent clock time on the opposite side.
2. Use the cards to quiz another student.

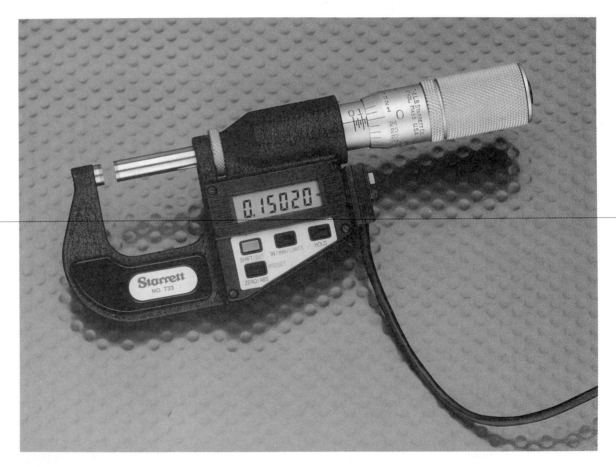

A micrometer can be used to take linear measurements. This particular model has both a conventional scale and a digital readout. The digital readout can be converted from inches to millimeters by simply pushing a button. (Starrett)

Chapter 8

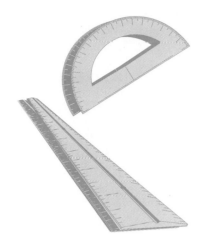

Reading Tools of Measurement

OBJECTIVES

After studying this chapter, you will be able to:
- Explain basic terminology of some common tools of measurement.
- Read and use a variety of tape measures.
- Read and use architects', engineers', and metric scales.
- Read and use standard-inch, metric, and vernier micrometers.
- Read and use standard-inch and metric vernier calipers.
- Read a multimeter.

Tools of measurement are available for all kinds of measurable quantities. It is important to know how to read them properly. The accuracy with which a scale is read helps to determine the accuracy of a measurement. This chapter presents a variety of measurement tools and discusses how to interpret markings on their scales.

U.S. CONVENTIONAL RULES

In regard to linear measure, spacing between whole numbers can be divided into as many parts as desired. These *fractional parts* are divisions of a whole. The denominator of the fraction gives the number of parts the whole was divided into. (Refer to Chapter 2 for an in-depth explanation of fractional parts.) Although any number can be used, rules that measure inches use denominators that can be easily reduced to lower terms. Standard divisions include fractions of:

$$\frac{1}{2} \quad \frac{1}{4} \quad \frac{1}{8} \quad \frac{1}{16} \quad \frac{1}{32} \quad \frac{1}{64}$$

To determine the value of any particular division, if not otherwise marked, count how many spaces there are between the whole numbers. The number of spaces is equal to the denominator of the fraction. To determine the value of a particular division, count the spaces to the measurement and place this value over the denominator.

A measurement will often contain a whole number and a fraction. The fractional part of the measurement should always be reduced to lowest terms. Fig. 8-1 shows the space between the whole numbers 0 and 1 divided into 16 parts. Each of the 16 parts is numbered in consecutive order; then, below, is the value as it is reduced to lower terms.

The following sample problems explain how to accurately read rulers, tape measures, and the like. The figures shown are segments of rulers and tape measures. Naturally, due to space limitations, the whole scale could not be used. Assume, however, that the measurements are extended from zero.

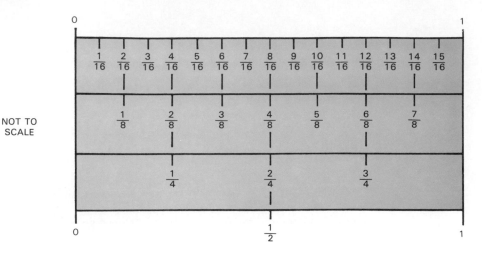

Fig. 8-1. Fractional parts typical of a ruler or tape measure.

Sample Problem 8-1.

Read measurements given at *a* and *b* in Fig. 8-2.
Characteristics:
- The large numbers mark inches. The lines to the right of each large number are the inch marks.
- The "1F" markings represent 1 foot. This segment of tape is taken from a long tape measure marked in feet. Between each foot are inches marked 1 through 11. The mark where 12 inches would fall is marked instead with the next unit of feet.
- Fractional inches are obtained by counting over from the inch mark. Give measurement in units of feet and inches.

Measurement given at a: 1 ft., $3\frac{3}{8}$ in.

Measurement given at b: 1 ft., $4\frac{3}{4}$ in.

Note: Reduce answer to lowest terms.

NOT TO
SCALE

Fig. 8-2. Tape measure marked in feet, inches, and fractional inches. The fractions on this tape are marked as small as 1/8 inch. (Cooper Tools)

Sample Problem 8-2.

Read measurements given at *a*, *b*, *c*, and *d* in Fig. 8-3.
Characteristics:
- Inches are marked continuously, without markings for feet.
- Give measurement in units of inches. Do not give feet since feet are not marked.

Measurement given at a: $13\frac{5}{16}$ in.

Measurement given at b: $14 \frac{1}{8}$ in.

Measurement given at c: $14 \frac{15}{16}$ in.

Measurement given at d: $15 \frac{3}{4}$ in.

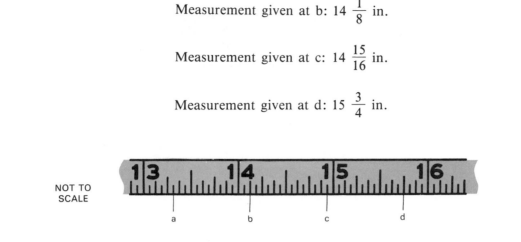

Fig. 8-3. Tape measure marked in inches and fractional inches. The fractions on this tape are marked as small as 1/16 inch. (Cooper Tools)

Sample Problem 8-3. Read measurements given at *a, b, c,* and *d* in Fig. 8-4.
Characteristics:
- Tape measure is a combination of tapes in Fig. 8-2 and Fig. 8-3.
- Give measurements in units of feet and inches or just inches.
- If you need to measure inches to 1/32, use scale between 0 and 12 inches.

Measurement given at a: $11 \frac{5}{32}$ in.

Measurement given at b: $11 \frac{11}{16}$ in.

Measurement given at c: $12 \frac{7}{16}$ in. or 1 ft., $\frac{7}{16}$ in.

Measurement given at d: $13 \frac{1}{2}$ in. or 1 ft., $1 \frac{1}{2}$ in.

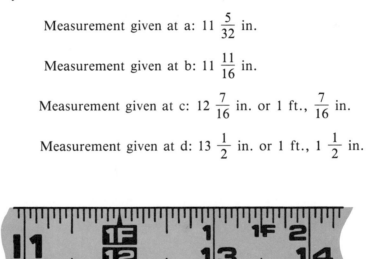

Fig. 8-4. Tape measure marked in either feet and inches (top scale) or continuous inches (bottom scale). Fractions are marked to 1/16 inch; however, fractions between 0 and 12 inches are marked to 1/32 inch. (Cooper Tools)

Sample Problem 8-4.
Read measurements given at *a, b, c,* and *d* in Fig. 8-5.
Characteristics:
- Rather than the usual divisions of 1/8, 1/16, or 1/32, this scale is divided into 1/10 inches. Fractions between 0 and 6 inches are marked as small as 1/50 inch. Notice, here, that between each 1/10 division are 5 divisions (count the spaces). Remember: 1/10 = 5/50.

- The decimal 0.1 can be used rather than the fraction 1/10. If desired, 1/50 can be changed to the decimal 0.02 inch. The value of each small line starting right of zero is: 0.02, 0.04, 0.06, 0.08 inch.
- It is best to give these measurements in decimal, rather than in fractional form. Decimal numbers are easier to work with.

    Measurement given at a: 5.14 in.
    Measurement given at b: 5.7 in.
    Measurement given at c: 6.4 in.
    Measurement given at d: 7.8 in.

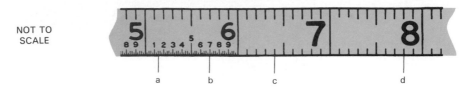

Fig. 8-5. Tape measure marked in inches and fractional inches. Fractions of this tape measure are given in 1/50 and 1/10 inches. These are best read as decimal fractions (i.e., 0.02 and 0.1 inches). (Cooper Tools)

Sample Problem 8-5.

Read measurements given at *a, b, c,* and *d* in Fig. 8-6.
Characteristics:
- These **graduations**, or ruled markings, are better read as decimals than as fractions. Read 1/10 foot as 0.1 foot. Read 1/100 foot as 0.01 foot.
- The large "1F" marks the start of 1 foot. The small "1F" is marked periodically following 1 foot as a reminder. The 2-foot section is marked periodically with a "2F," etc.
- The numbers marked are incremented by 0.1 foot (1/10 foot). Notice the numbers go to 9 (0.9 foot), then to the 1-foot marking, then repeat from 1 (0.1 foot).
- There are 10 divisions (spaces) between each number, making each of these divisions 1/100 of a foot (0.01 foot).
- Remember: 0.1 = 0.10.
- Starting right of zero, smallest lines have values as follows: 0.01 foot, 0.02 foot, 0.03 foot, etc.

    Measurement given at a: 0.9 feet
    Measurement given at b: 0.96 feet
    Measurement given at c: 1.03 feet
    Measurement given at d: 1.15 feet

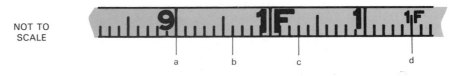

Fig. 8-6. Tape measure marked in feet and fractional feet. The fractions are marked to 0.1 and 0.01 feet. (Cooper Tools)

Sample Problem 8-6.

Read measurements given at *a, b,* and *c* in Fig. 8-7.
Characteristics:
- Feet are marked in large numbers.
- Inches are marked along the bottom from 1 to 11, repeating with each foot. Inches are divided as small as 1/16 of an inch.

- Markings for each 0.1 foot are along the top. Each 0.1 foot increment has 10 divisions of 0.01 foot.
- Notice that since the foot is divided into twelfths (inches) and tenths (decimal feet), a 1 inch increment is smaller than a tenth of a foot increment.
  Measurement given along top and bottom scales for a:

  0.9 ft. or $10\frac{13}{16}$ in.

  Measurement given along top and bottom scales for b:

  1.02 ft. or 1 ft., $\frac{1}{4}$ in.

  Measurement given along top and bottom scales for c:

  1.14 ft. or 1 ft., $1\frac{11}{16}$ in.

  Note: Measurements given for top and bottom scales are not exact equivalents.

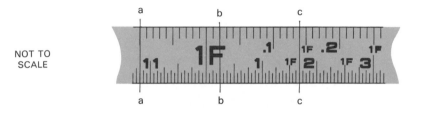

NOT TO
SCALE

Fig. 8-7. Tape measure marked in feet and fractional feet (top scale) and feet and inches (bottom scale). (Cooper Tools)

SI METRIC RULES

The SI Metric System is based on powers of 10. The decimal system is based on powers of 10. Therefore, all metric measurements use decimals, rather than fractional parts. As learned in an earlier chapter, the base unit of length measure is the meter. The meter, however, can be a fairly large unit. (A meter is slightly more than a yard.) Since meters are fairly large units, rulers and tape measures are graduated into smaller units of decimeters, centimeters, and millimeters.
- 1 meter = 10 decimeters = 100 centimeters = 1000 millimeters.
- 1 decimeter = 10 centimeters = 100 millimeters.
- 10 millimeters = 1 centimeter = 0.1 decimeter.

Sample Problem 8-7.

Read measurements given at *a* through *f* in Fig. 8-8.
Characteristics:
- Graduated in decimeters, centimeters, and millimeters, the size of the ruler in Fig. 8-8 is about 12 inches (not actual size).
- Divisions along top scale are marked in one direction. Divisions on the bottom scale are marked in the opposite direction.
- The top edge has every centimeter marked with a number. The bottom has every 10 millimeters marked with a number. (Notice that 10 millimeters indeed equal 1 centimeter.)
- Each decimeter is marked, with this ruler being 3 decimeters long. (Notice that each line equals 1 millimeter, 10 millimeters equal 1 centimeter, 10 centimeters equal 1 decimeter, and 10 decimeters equal 1 meter.)
  Measurement given at a: 3 cm = 0.3 dm
  Measurement given at b: 10 cm = 1 dm
  Measurement given at c: 22.8 cm = 2.28 dm

Measurement given at d: 52 mm = 5.2 cm = 0.52 dm
Measurement given at e: 200 mm = 20 cm = 2 dm
Measurement given at f: 255 mm = 25.5 cm = 2.55 dm

Fig. 8-8. Ruler marked in decimeters, centimeters, and millimeters. (Hearlihy & Co.)

Sample Problem 8-8.

Read measurements given at *a* through *e* in Fig. 8-9.
Characteristics:
- Every 10 millimeters are marked up to the meter marks and then repeat. Remember: 1 meter = 1000 millimeters.
- Numbers are marked at each centimeter. To change from millimeters to centimeters, move the decimal one place to the left (for example, 990 millimeters = 99 centimeters). Remember: 1 meter = 100 centimeters.
- To change millimeters to decimeters, move the decimal two places to the left (for example, 970 millimeters = 9.7 decimeters).

Measurement given at a: 980 mm = 98.0 cm = 9.80 dm = 0.980 m
Measurement given at b: 996 mm = 99.6 cm = 9.96 dm = 0.996 m
Measurement given at c: 1009 mm = 100.9 cm = 10.09 dm
    = 1.009 m
Measurement given at d: 1025 mm = 102.5 cm = 10.25 dm
    = 1.025 m

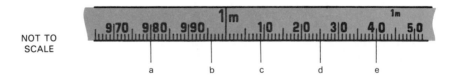

Fig. 8-9. Tape measure marked in meters, centimeters, and millimeters. (Cooper Tools)

Measurement given at e: 1040 mm = 104.0 cm
    = 10.40 dm = 1.040 m

Sample Problem 8-9.

Read measurements given at *a* through *e* in Fig. 8-10.
Characteristics:
- Across the top is inches, graduated in increments as small as 1/16 of an inch.
- Across the bottom are decimeters, centimeters, and millimeters.

Measurements given along top and bottom scales for a:
$2 \frac{9}{16}$ in. or 65 mm

Measurements given along top and bottom scales for b:
$3 \frac{7}{16}$ in. or 87 mm

Measurements given along top and bottom scales for c:

$4\frac{3}{4}$ in. or 121 mm

Measurements given along top and bottom scales for d:

$5\frac{7}{8}$ in. or 149 mm

Measurements given along top and bottom scales for e:

$7\frac{3}{16}$ in. or 182 mm

Note: Measurements given for top and bottom scales are not exact equivalents.

NOT TO
SCALE

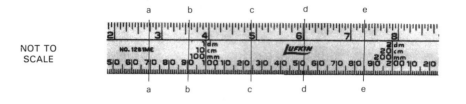

Fig. 8-10. Combination U.S. Conventional and SI Metric rule. (Cooper Tools)

TRIANGULAR DRAFTING SCALES

Triangular scales are used by architects, engineers, designers, and drafters. The scales are graduated so that a drawing can be made at a reduced or enlarged size. The word *scale* is used in two ways. A scale can be a tool used to make measurements, or a ratio between a dimension on paper and actual physical size. The previous sections of this chapter dealt with rulers and tape measures used to measure the actual size. These measurements have a *scale ratio* of 1:1. This is read 1 to 1 and, in this context, means 1 inch measured is 1 inch actual. This is stated in other words as 1 inch = 1 inch, 1 foot = 1 foot, etc.

Different scales are needed for different drawings. For example, a regional map might have a scale of 1 inch = 6 miles. The map reader could use a ruler, measure inches, and multiply by 6 to get miles. An easier way to get miles, however, would be to use a scale that gives 6 graduations to the inch. This way, miles could be read directly off the scale. Another map might use a different scale. The triangular scale has a variety of scales (ratios) to allow reduction or enlargement of measurements to best fit the size of paper. For example, given the same size of paper, the scale used for the architectural drawing of a school would be different from the scale used for the drawing of a birdhouse.

Architect's scale

The **triangular architect's scale** is divided into feet, inches, and fractional parts of an inch, as you will see in the sample problems to follow. Each figure shown has a zigzag line in the center. This tells you that the portion of the scale between the zigzag is missing. Doing this allows the scale to be as close to actual size as possible and still fit on the page. General instructions for reading the architect's scale are given in the rules of reading architect's scales.

RULES OF READING THE ARCHITECT'S SCALE

- If the desired scale is marked with 0 on the left, read the graduations from left to right. See Fig. 8-11.
- If the desired scale is marked with 0 on the right, read the graduations from right to left.
- Read feet at whole units from 0 to each of the marked numbers. Do not try to estimate fractional amounts by reading between numbers.
- Inches are read before the 0, using the smaller graduations. This area represents 12 inches. If smaller graduations are provided, they represent fractional parts of an inch.
- Units of measure are in feet and inches (for example, 7 feet, 6 inches). Do not state units as a fraction of a foot (for example, 7 1/2 feet).
- One way to use the scale is to first find the number of whole feet in a dimension by lining up one side of the dimension with 0 and reading over to get feet. Then, move the scale and line to the other side of the dimension with the division that marks the whole feet. Now, read back to the other side of the dimension to get inches.

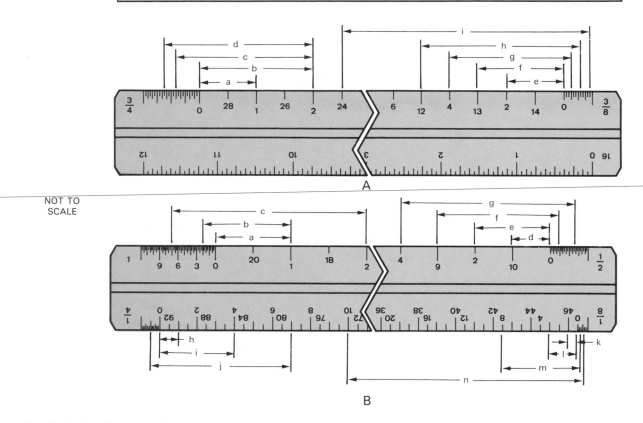

Fig. 8-11. A—One side of an architect's scale showing 3/4'', 3/8'', and full-size scales. B—Architect's scale showing 1'', 1/2'', 1/4'', and 1/8'' scales. (Hearlihy & Co.)

Sample Problem 8-10.

A drafter makes a drawing using a scale ratio of 3/4" = 1'-0". What are the drafter's measurements of lengths a, b, c, and d in Fig. 8-11A.

Since scale ratio is 3/4" = 1'-0", use the scale marked 3/4.

length a   = 1 ft.
length b   = 2 ft.
length c   = 2 ft., 5 in.
length d   = 2 ft., 7 1/2 in.

Sample Problem 8-11.

Given a drawing with a scale ratio of 3/8" = 1'-0", determine the dimensions of lengths e through i in Fig. 8-11A.

Since scale ratio is 3/8" = 1'-0", use the scale marked 3/8.

length e = 2 ft.
length f = 3 ft.
    Note: On this scale, each line is 1 ft.; so the line with 13 is 3 ft.
    The 13 is really feet on the 3/4 scale.
length g = 4 ft., 3 in.
length h = 5 ft., 7 in.
length i = 24 ft., 11 in.

Sample Problem 8-12.

Give dimensions *a* through *c* (1" scale), *d* through *g* (1/2" scale), *h* through *j* (1/4" scale), and *k* through *n* (1/8" scale). Use Fig. 8-11B.

length a = 1 ft.
length b = 1 ft., 2 in.
length c = 2 ft., 7 in.
length d = 1 ft.
length e = 2 ft.
length f = 3 ft., 3 in.
length g = 4 ft., 8 in.
length h = 1 ft.
length i = 4 ft.
length j = 7 ft., 6 in.
length k = 1 ft.
length l = 3 ft.
length m = 8 ft., 4 in.
length n = 73 ft., 8 in.

Engineer's scale

Civil engineer's scales are sometimes called decimal scales or, simply, engineer's scales. They are sometimes called decimal scales because there are 10 minor divisions to a major division. (Decimal numbers are based on powers of 10.) General instructions for reading engineer's scales are set forth in the rules of reading engineer's scale.

RULES OF READING THE ENGINEER'S SCALE

- The number representing the particular scale (the number set off to the side) states into how many parts the inch is divided. Fig. 8-12 has a scale of 10 across the top and 20 across the bottom. The "10" gives 10 minor divisions (or 1 major division) per inch. The "20" gives 20 minor divisions (or 2 major divisions) per inch.
- Actual size of each minor division on the 10 scale is 1/10 inch (0.1 inch); on the 20 scale is 1/20 inch (0.05 inch), etc.
- The six scales are graduated by 10, 20, 30, 40, 50, and 60 minor divisions per inch.
- Each major division represented is subdivided into 10 parts.
- The scale can be used to represent inches, feet, yards, miles, kilometers, or any other unit of measure.
- All measurements are in decimal parts of the unit, rather than combined units; for example, 2.5 feet, not 2 feet, 6 inches.
- A scale can represent any power of 10, by moving the decimal point. For example, the same point on a scale could be: 1.7 feet, 17 feet, 170 feet, 1.7 miles, etc.
- The unit of measure must be assigned to the dimension.

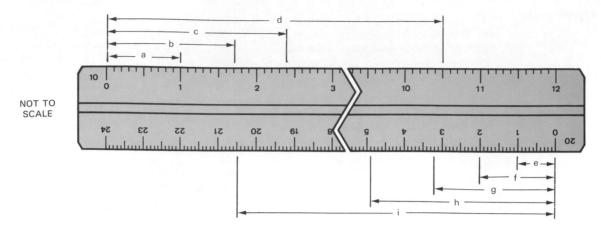

Fig. 8-12. One side of a triangular, civil engineer's scale. (Hearlihy & Co.)

Sample Problem 8-13.

Make up a table giving measurements *a, b, c,* and *d* in Fig. 8-12 in terms of each of the following scales:

1" = 1'
1" = 10'
1" = 100'
1" = 1000'

		Scale			
	Minor Units	1" = 1'	1" = 10'	1" = 100'	1" = 1000'
a	10	1'	10'	100'	1000'
b	17	1.7'	17'	170'	1700'
c	24	2.4'	24'	240'	2400'
d	105	10.5'	105'	1050'	10,500'

Sample Problem 8-14.

Make up a table giving measurements *e* through *i* in Fig. 8-13 in terms of each of the following scales:

1" = 2'
1" = 20'
1" = 200'
1" = 2000'

		Scale			
	Minor Units	1" = 2'	1" = 20'	1" = 200'	1" = 2000'
e	10	1'	10'	100'	1000'
f	20	2'	20'	200'	2000'
g	32	3.2'	32'	320'	3200'
h	49	4.9'	49'	490'	4900'
i	205	20.5'	205'	2050'	20,500'

Metric scale

The **metric scale** is used to make drawings with dimensions in metric units. For engineering drawings, basic units are millimeters. For architectural drawings, basic units are meters. Scales are shown as ratios, such as 1:1 for full size. A 1:2 reduction scale means that 1 mm on the drawing equals 2 mm on the object being drawn. In other words, with this scale ratio, the drawing is half the size of the object. A 2:1 scale means that 2 mm on the drawing equals 1 mm on the object; the drawing is twice the size of the object at this scale ratio. A variety of scale ratios are available. One triangular scale will have six different ratios. Typical ratios are 1:1, 1:2, 2:20, 1:50, 1:100—more than 20 types in all.

One metric scale can be used for various reductions and enlargements. For example, a 1:100 scale (Fig. 8-13) can be used for ratios of 1:1, 1:10, 1:1000, etc., in addition to 1:100. Multiply the measurement on the 1:100 trangular scale by 10 for each zero to be added to the scale ratio.

Suppose, for example, a 4 m line is to be drawn using a 1:100 metric scale. Using the numbers directly from the 1:100 scale, a line representing 4 m (one-hundredth actual size) may be drawn. If, instead, a 1:1000 scale ratio is desired, the same 1:100 scale can be used, but the values of the divisions are now multiplied by 10. Therefore, what appears on the scale as 4 m, now has a value of 40 m (4 × 10 = 40). On the 1:100 scale, the 0.4 m increment is now interpreted as 4 m (0.4 × 10 = 4). This would be the length representing the 4 m line (now, one-thousandth actual size).

In order to change from a smaller to larger ratio (i.e., 1:100 to 1:1), divide the measurement on the scale by 10 for each zero to be removed. For example, 4 m (4000 mm) represented on a 1:100 triangular scale would represent 40 mm if used for a 1:1 scale ([4000 ÷ 10] ÷ 10 = 40), etc.

Before using a metric scale, always be sure you know what the primary units are. Larger scale ratios, such as 1:1 or 1:5, are usually in millimeters. Smaller scale ratios, such as 1:100 or 1:500, are usually in meters.

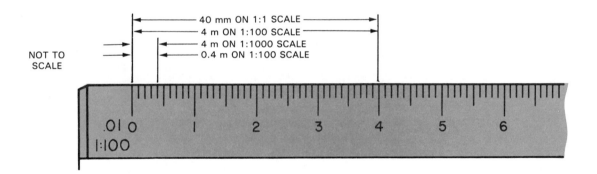

Fig. 8-13. Measuring with a metric scale.

MECHANICAL ENGINEER'S SCALE

You have learned about the *civil* engineer's scale. The *mechanical* **engineer's scale** is another type. They typically have scales of full size, half size, quarter size, and eighth size. Mechanical engineer's scales are divided into inches and fractional inches, rather than into feet and inches. Fractions of inches are in increments of 1/2", 1/4", 1/8", etc.

Divisions may be either *fully divided* or *open divided*. Fully divided means that the subdivided units appear along the entire length of the scale. Only the first major unit of an open divided scale is subdivided (i.e., an architect's scale).

Mechanical engineer's scales are useful when dimensions are in inches, and when fractional inches are expressed as fractions. To read them, apply the principles just learned for reading triangular scales.

MICROMETERS

Next to the ruler or tape measure, the **micrometer** might be the most commonly used tool for taking linear measurements. Its purpose is to measure objects with extreme accuracy. Standard-inch micrometers measure to thousandths of an inch (0.001"). Vernier micrometers measure to ten-thousandths of an inch (0.0001"). Standard-metric micrometers measure to hundredths of a millimeter (0.01 mm) and vernier metric, to 2-thousandths of a millimeter (0.002 mm).

Fig. 8-14 is a picture of a micrometer. In using the micrometer, the spindle is opened, and the object to be measured is held in place at this location. Measurements are performed by turning the thimble, which closes the spindle until it touches the object. The value of the measurement is read as a combination of the markings on the sleeve and thimble.

A micrometer allows for very accurate measurements. The accuracy is limited, mostly, by the accuracy of the reading. Follow applicable rules of reading micrometers to ensure proper readings.

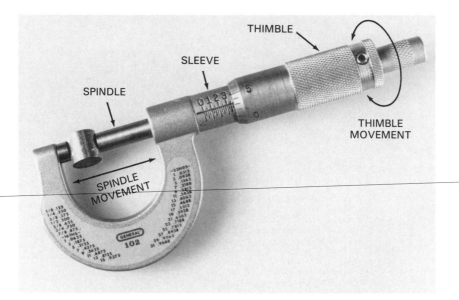

Fig. 8-14. Photograph of a micrometer. Notice the fraction/decimal equivalents provided to help convert the decimal readings to fractions if required. (Hearlihy & Co.)

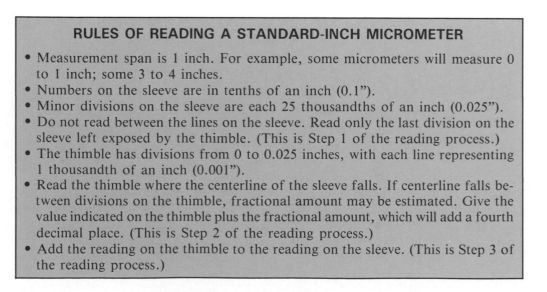

RULES OF READING A STANDARD-INCH MICROMETER

- Measurement span is 1 inch. For example, some micrometers will measure 0 to 1 inch; some 3 to 4 inches.
- Numbers on the sleeve are in tenths of an inch (0.1").
- Minor divisions on the sleeve are each 25 thousandths of an inch (0.025").
- Do not read between the lines on the sleeve. Read only the last division on the sleeve left exposed by the thimble. (This is Step 1 of the reading process.)
- The thimble has divisions from 0 to 0.025 inches, with each line representing 1 thousandth of an inch (0.001").
- Read the thimble where the centerline of the sleeve falls. If centerline falls between divisions on the thimble, fractional amount may be estimated. Give the value indicated on the thimble plus the fractional amount, which will add a fourth decimal place. (This is Step 2 of the reading process.)
- Add the reading on the thimble to the reading on the sleeve. (This is Step 3 of the reading process.)

Sample Problem 8-15.

What is the reading indicated on the micrometer scale of Fig. 8-15A?

Step 1. Read sleeve to the last exposed division.

Sleeve = 0.525"

Step 2. Read thimble where it aligns with centerline.

Thimble = 0.018"

Step 3. Add sleeve and thimble.

$$
\begin{array}{r}
0.525" \text{ (sleeve)} \\
+\ 0.018" \text{ (thimble)} \\
\hline
0.543" \text{ (measurement)}
\end{array}
$$

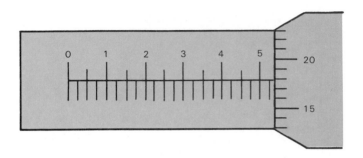

Fig. 8-15A.

Sample Problem 8-16.

What is the reading indicated on the micrometer scale of Fig. 8-15B?

Step 1. Read sleeve to the last exposed division.

Sleeve = 0.000"

Step 2. Read thimble where it aligns with centerline.

Thimble = 0.020"

Step 3. Add sleeve and thimble.

$$
\begin{array}{r}
0.000" \text{ (sleeve)} \\
+\ 0.020" \text{ (thimble)} \\
\hline
0.020" \text{ (measurement)}
\end{array}
$$

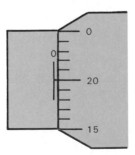

Fig. 8-15B

Sample Problem 8-17.

What is the reading indicated on the micrometer scale of Fig. 8-15C?

Step 1. Read sleeve to the last exposed division.

Sleeve = 0.075"

Step 2. Read thimble where it aligns with centerline.

Thimble = 0.001"

Step 3. Add sleeve and thimble.

$$
\begin{array}{r}
0.075\text{" (sleeve)} \\
+\ 0.001\text{" (thimble)} \\
\hline
0.076\text{" (measurement)}
\end{array}
$$

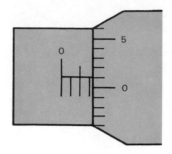

Fig. 8-15C

Sample Problem 8-18.

What is the reading indicated on the micrometer scale of Fig. 8-15D?

Step 1. Read sleeve to the last exposed division.

Sleeve = 0.250"

Step 2. Read thimble where it aligns with centerline.

Thimble = 0.017"

Step 3. Add sleeve and thimble.

$$
\begin{array}{r}
0.250\text{" (sleeve)} \\
+\ 0.017\text{" (thimble)} \\
\hline
0.267\text{" (measurement)}
\end{array}
$$

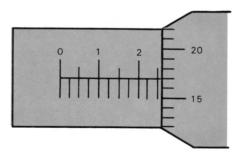

Fig. 8-15D

Sample Problem 8-19.

What is the reading indicated on the micrometer scale of Fig. 8-15E?

Step 1. Read sleeve to the last exposed division.

Sleeve = 0.975"

Step 2. Read thimble where it aligns with centerline.

Thimble = 0.010"

Step 3. Add sleeve and thimble.

$$
\begin{array}{r}
0.975\text{" (sleeve)} \\
+\ 0.010\text{" (thimble)} \\
\hline
0.985\text{" (measurement)}
\end{array}
$$

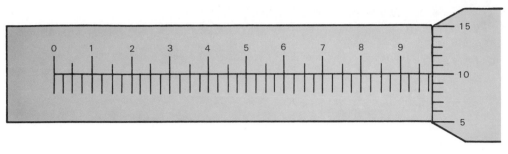

Fig. 8-15E

RULES OF READING A VERNIER MICROMETER

- Measures to ten-thousandths of an inch.
- The sleeve and thimble are exactly the same as a standard-inch micrometer, giving the first 3 decimal places.
- A fourth decimal place in ten-thousandths of an inch (0.0001") is obtained by reading the vernier lines on the sleeve. (A *vernier* is a short auxiliary scale that further subdivides lines permitting a very fine measurement.) One of the lines on the thimble will align with one of the vernier lines on the upper portion of the sleeve.
- Reading the vernier portion involves some judgment on the part of the reader. The first three decimal places are accurate readings, with the fourth place being more difficult to read accurately.

Sample Problem 8-20.

What are the readings indicated on the micrometer scales of Fig. 8-16?

Fig. 8-16A: 0.8750" (sleeve)
                0.0040" (thimble)
           + 0.0009" (vernier)
              _____
                0.8799" (measurement)

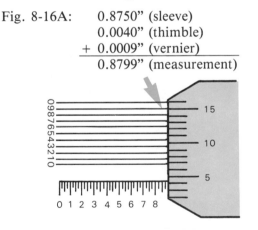

Fig. 8-16A

Fig. 8-16B: 0.6250" (sleeve)
                0.0090" (thimble)
           + 0.0002" (vernier)
              _____
                0.6342" (measurement)

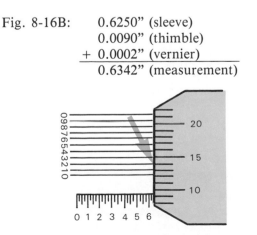

Fig. 8-16B

Fig. 8-16C: 0.1750" (sleeve)
0.0010" (thimble)
+ 0.0003" (vernier)
0.1763" (measurement)

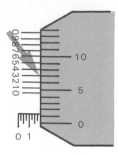

Fig. 8-16C

Fig. 8-16D: 0.3500" (sleeve)
0.0190" (thimble)
+ 0.0000" (vernier)
0.3690" (measurement)

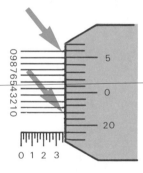

Fig. 8-16D

RULES OF READING A METRIC MICROMETER

- Standard measures to hundredths of a millimeter; vernier to 2-thousandths.
- The sleeve is in units of millimeters.
- Between each numbered line are 10 divisions—5 above the centerline and 5 below the centerline. Below the centerline are whole millimeters; above are half millimeters (0.5 mm).
- The thimble has 50 divisions, from 0 to 0.50 mm.
- Determine measurement by adding the thimble and the sleeve. For vernier metric, add thimble, sleeve, and vernier readings. Each line on vernier is a 0.002 mm increment.

Sample Problem 8-21.

What are the readings indicated on the micrometer scales of Fig. 8-17?

Fig. 8-17A: 12.00 mm (sleeve)
+   .24 mm (thimble)
12.24 mm (measurement)

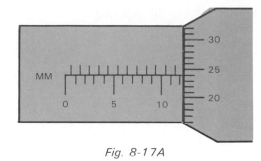

Fig. 8-17A

Fig. 8-17B: 6.50 mm (sleeve)
+   .06 mm (thimble)
6.56 mm (measurement)

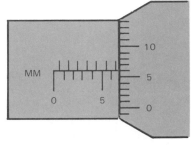

Fig. 8-17B

Fig. 8-17C: 24.50 mm (sleeve)
+   .45 mm (thimble)
24.95 mm (measurement)

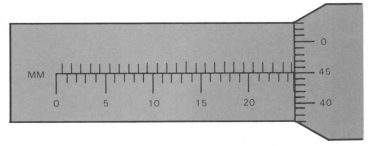

Fig. 8-17C

Fig. 8-17D: 16.50 mm (sleeve)
+   .26 mm (thimble)
16.76 mm (measurement)

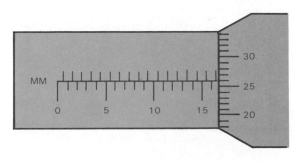

Fig. 8-17D

Fig. 8-17E: 3.00 mm (sleeve)
              +   .28 mm (thimble)
                 3.28 mm (measurement)

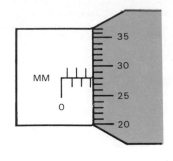

Fig. 8-17E

VERNIER CALIPERS

Unlike the micrometer, the vernier caliper can be used to measure both inside and outside dimensions. Standard vernier calipers are capable of making accurate measurements to 0.001" and are available in lengths of 6", 12", 24", 36", and 48". Metric vernier calipers can be used to take measurements to 0.02 mm and are available in lengths of 150 mm, 300 mm, and 600 mm. Vernier calipers are available with either a 25-division vernier plate or a 50-division vernier plate. A typical vernier caliper is shown in Fig. 8-18.

To use a vernier caliper, slide the movable jaw assembly until the jaws almost touch the object being measured. Then, use the adjusting nut to move the jaws into contact with the object. The jaws should contact the object firmly, but they must not be too tight. Overtightening may damage the caliper, causing inaccurate readings.

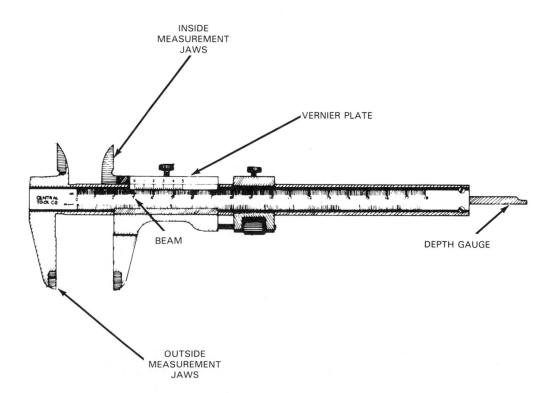

Fig. 8-18. Typical vernier caliper. (Central Tools)

RULES FOR READING A STANDARD 25-DIVISION VERNIER CALIPER

- There are 25 equal divisions on the vernier plate. Each line represents 0.001". Every fifth line is numbered.
- Each inch section on the beam is graduated into forty equal divisions. Each division equals .025". Every fourth division, which is equal to 0.1", is numbered.
- When reading the caliper:
    1. Determine the number of inches, tenths of an inch, and fortieths of an inch between the "0" line on the beam and the "0" line on the vernier plate. Add these quantities together.
    2. Count the number of graduations that lie between the "0" line on the vernier plate and a line on the plate that coincides with a line on the beam. Each of these graduations is equal to 0.001". Multiply the number of graduations by 0.001". It is important to note that only one line on the plate will coincide with a line on the beam.
    3. Add the quantities from Steps 1 and 2 to get the total reading.

Sample Problem 8-22.

What is the reading indicated on the vernier caliper shown in Fig. 8-19?

Step 1. Determine the number of inches, tenths, and fortieths between the "0" line on the beam and the "0" line on the vernier plate. Add these quantities together.

$$\begin{array}{r} 5" \\ 0.30" \\ +\,0.050" \\ \hline 5.350" \end{array}$$

Step 2. Count the number of graduations that lie between the "0" line on the vernier plate and a line on the plate that coincides with a line on the beam. Each graduation equals 0.001". Multiply the number of graduations by 0.001".
18 graduations × 0.001" = 0.018".

Step 3. Add the numbers from Steps 1 and 2 to get the total reading.

$$\begin{array}{r} 5.350" \text{ (quantity from Step 1)} \\ +\,0.018" \text{ (quantity from Step 2)} \\ \hline 5.368" \text{ (total reading)} \end{array}$$

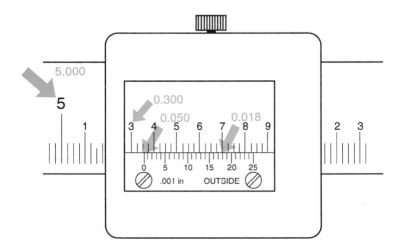

Fig. 8-19. Reading a 25-division vernier scale.

RULES FOR READING A STANDARD 50-DIVISION VERNIER CALIPER

- There are 50 equal divisions on the vernier plate. Every division is equal to 0.001". Every fifth or tenth line is numbered.
- Each inch section on the beam is graduated into 20 equal parts. Each part is equal to 0.050". Every other graduation line on the beam is numbered and represents 0.100".
- When reading the caliper:
  1. Determine the number of inches, tenths, and twentieths between the "0" line on the beam and the "0" line on the vernier plate. Add these quantities together.
  2. Count the number of graduations between the "0" line on the vernier plate and a line on the plate that coincides with a line on the beam. Each of these graduations is equal to 0.001". Multiply the number of graduations by 0.001". It is important to note that only one line on the plate will coincide with a line on the beam.
  3. Add the quantities from Steps 1 and 2 to get the total reading.

Sample Problem 8-23.

What is the reading indicated on the vernier caliper shown in Fig. 8-20?

Step 1. Determine the number of inches, tenths, and twentieths between the "0" line on the beam and the "0" line on the vernier plate. Add these quantities together.

$$
\begin{array}{r}
6" \\
0.20" \\
+\,0.050" \\
\hline
6.250"
\end{array}
$$

Step 2. Count the number of graduations between the "0" line on the vernier plate and a line on the plate the coincides with a line on the beam. Each graduation equals 0.001". Multiply the number of graduations by 0.001".

$$15 \text{ graduations} \times 0.001" = 0.015".$$

Step 3. Add the quantities from Steps 1 and 2 to get the total reading.

$$
\begin{array}{l}
6.250" \text{ (quantity from Step 1)} \\
+\,0.015" \text{ (quantity from Step 2)} \\
\hline
6.265" \text{ (total reading)}
\end{array}
$$

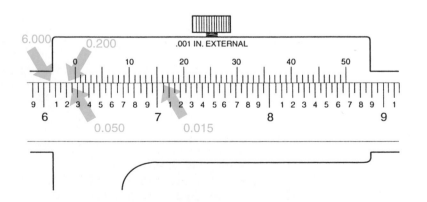

Fig. 8-20. Reading a 50-division vernier scale.

RULES FOR READING A METRIC 25-DIVISION VERNIER CALIPER

- There are 25 equal divisions on the vernier plate. Each division is equal to 0.02 mm. Every fifth line is numbered (0.10 mm, 0.20 mm, etc.).
- The beam is graduated in 0.5 mm divisions. Every twentieth division is numbered (10 mm, 20 mm, etc.).
- When reading the metric vernier caliper:
    1. Determine the number of millimeters between the "0" line on the beam and the "0" line on the vernier plate.
    2. Locate the line on the vernier plate that coincides with one of the lines on the beam. Note the value of this line as indicated on the vernier plate.
    3. Add the quantities from Steps 1 and 2 to arrive at the total reading.

Sample Problem 8-24.

What is the reading indicated on the vernier caliper shown in Fig. 8-21?

Step 1. Determine the number of millimeters between the "0" line on the beam and the "0" line on the vernier plate. In Fig. 8-21, there are 29 millimeters between the "0" line on the beam and the "0" line on the plate.

Step 2. Locate the line on the vernier plate that coincides with one of the lines on the bar. In Fig. 8-21, the line represents 0.28 mm.

Step 3. Add the quantities from Steps 1 and 2 to arrive at the total reading.

29 mm (quantity from Step 1)
+0.28 mm (quantity from Step 2)
29.28 mm (total reading)

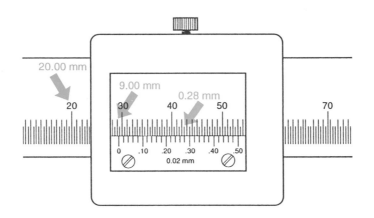

Fig. 8-21. Reading a metric 25-division vernier scale.

RULES FOR READING A METRIC 50-DIVISION VERNIER CALIPER

- There are 50 equal divisions on the vernier plate. Each division is equal to 0.02 mm. Every fifth line is numbered (.10 mm, .20 mm, etc.).
- The beam is graduated in 0.1 mm divisions. Every tenth division is numbered (10 mm, 20 mm, etc.).
- When reading the metric vernier caliper,
    1. Determine the number of millimeters between the "0" line on the beam and the "0" line on the vernier plate.
    2. Locate the line on the vernier plate that coincides with one of the lines on the beam. Note the value of this line as indicated on the vernier plate.
    3. Add the quantities from Steps 1 and 2 to arrive at the total reading.

Sample Problem 8-25.

What is the reading indicated on the vernier caliper shown in Fig. 8-22?

Step 1. Determine the number of millimeters between the "0" line on the beam and the "0" line on the vernier plate. In Fig. 8-22, there are 19 millimeters between the "0" line on the beam and the "0" line on the plate.

Step 2. Locate the line on the vernier plate that coincides with one of the lines on the bar. In Fig. 8-24, the line represents 0.28 mm.

Step 3. Add the quantities from Steps 1 and 2 to arrive at the total reading.

$$19 \text{ mm (quantity from Step 1)}$$
$$\underline{+0.28 \text{ mm} } \text{ (quantity from Step 2)}$$
$$19.28 \text{ mm (total reading)}$$

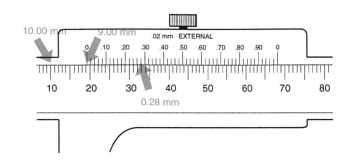

Fig. 8-22. Reading a metric 50-division vernier scale.

ELECTRICAL MULTIMETERS: VOLT-OHM-MILLIAMMETERS

Multimeters are used to measure electrical quantities such as voltage (volts), resistance (ohms), and current (amperes, or amps). Meters are available in two types— analog, with a needle, and digital, with digits. The difference between the two types is like the difference between the two types of wristwatches. The watch with two hands and a face is an analog type. The watch with just numbers is a digital type. In this chapter, the analog-type multimeter will be examined.

Fig. 8-23 shows a photograph of a multimeter. The meter face is the large white portion with the needle and the various sets of numbers. The various sets represent different scales. On the left side of the lower portion is the range switch. The range switch is used to select which scale is to be used. On the right side of the lower portion is a switch used to select the measurement to be made.

Fig. 8-23. Photograph of a multimeter. (B & K Precision)

Fig. 8-24 is a drawing of a meter showing the scales and range switch. This figure is to accompany the rules of reading a multimeter. These rules should help you understand how to read a multimeter.

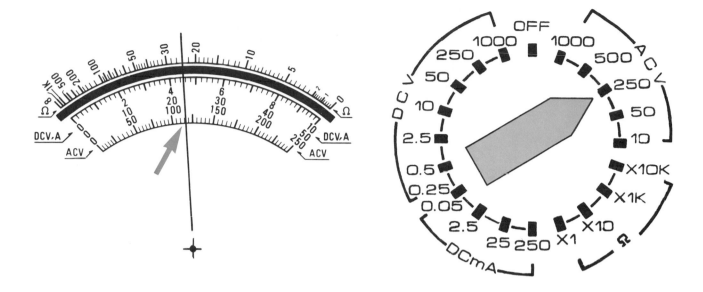

Fig. 8-24. Give the reading on this sample multimeter.

RULES OF READING A MULTIMETER

- The range is indicated by the range switch pointer. The range is the *maximum* for that scale. For example, the 250 volt range means the scale used has a maximum reading of 250 volts.
- The scale on the meter face to use is determined by using the numbers on the right side of the scale that correspond to the range setting. However, if the decimal place on the scale does not correspond, it can be adjusted as needed. For example, the 500 range would use the scale marked with a maximum of 50. The marked numbers would be mentally adjusted to read 0, 100, 200, 300, 400, and 500.
- Once the range is selected, the scale is determined, and the decimal place is mentally adjusted, use the set of numbers for that scale and ignore all others.
- The values of the smaller lines change with each range and scale. There are 10 lines (spaces) between 0 and the first marked number. After the scale's decimal has been mentally adjusted, divide the adjusted number, closest to zero, by 10. For example: Range = 1000; first number = 200; divide 200 by 10; answer is 20. Therefore, each line = 20. This can be verified by counting off each line by 20.
- DC volts (V) and milliamps (mA) are read using the set of lines above the scales, as indicated by the arrows at the ends of the scale. AC volts are read using the lines below the scales, as indicated by the arrows.
- The symbol for ohms is Ω. Ohms, as stated before, is the unit of measure of resistance. The ohms scale is numbered backwards, with zero on the right and infinity (∞) on the left. (The infinity sign looks like an 8 on its side.) The value of each line on this scale must be determined between each number. The reading from the scale will be *multiplied* by the multiplier on the range switch. For example, in Fig. 8-24, suppose the range has been selected as "x1k." With the needle in the position shown, the measurement would be 24 × 1k = 24 kΩ. The "k" stands for "kilo," which means 1000. Thus, 24 kΩ = 24,000Ω.

Sample Problem 8-26.

What is the reading on the meter in Fig. 8-24?

Characteristics:

• Range switch setting = 250 ACV.
• Set of numbers on the scale:

    0   50   100   150   200   250

    Note: No decimal adjustment is necessary.

• Value of each line is found by dividing first number by 10:

$$\frac{50}{10} = 5 \text{ volts per line}$$

• Read ACV on lower set of lines.
    Measurement = 116 ACV (ACV = volts AC [alternating current])

Sample Problem 8-27.

What is the reading on the meter at *a*, then *b*, in Fig. 8-25?

Characteristics:

• Range switch setting = 500 ACV.
• Set of numbers on the scale with the decimal adjusted.

    0   100   200   300   400   500

• Value of each line is found by dividing first number by 10:

$$\frac{100}{10} = 10 \text{ volts per line}$$

• Read ACV on lower set of lines.
    Measurement given at a: 150 ACV
    Measurement given at b: 380 ACV

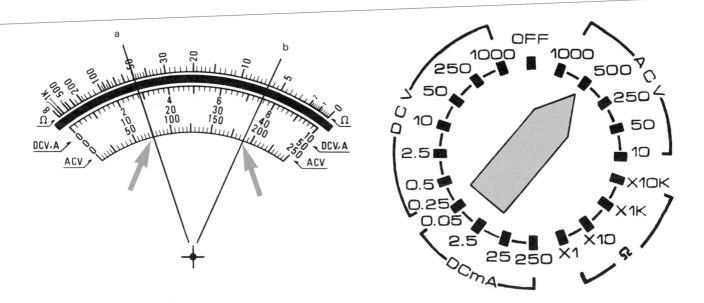

Fig. 8-25. Give the reading at a, then b, on this sample multimeter.

Sample Problem 8-28.

What is the reading on the meter at *a*, then *b*, in Fig. 8-26?

Characteristics:

• Range switch setting = 10 DCV.
• Set of numbers on the scale.

    0   2   4   6   8   10

• Value of each line is found by dividing first number by 10:

$$\frac{2}{10} = 0.2 \text{ volts per line}$$

• Read DCV on upper set of lines.
  Measurement given at a: 3.5 DCV
  Measurement given at b: 8.2 DCV

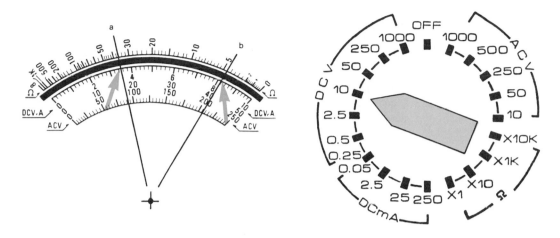

Fig. 8-26. Give the reading at a, then b, on this sample multimeter.

Sample Problem 8-29.

What is the reading on the meter at *a*, then *b*, then *c*, in Fig. 8-27?
Characteristics:

• Range switch setting = × 10Ω. All readings on the ohms scale must be multiplied by the range setting.
• Use the set of numbers across the top of the meter.
• Value of each smaller line is found on an individual basis between each marked number.
• Measurements given:

Needle	Small Line Value	Reading	Range	Measurement
a	20Ω	140Ω	× 10	1400Ω
b	2Ω	25Ω	× 10	250Ω
c	0.5Ω	7.5Ω	× 10	75Ω

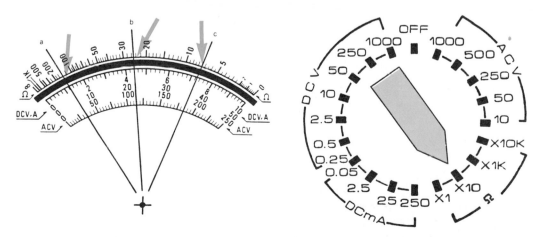

Fig. 8-27. Give the reading at a, then b, then c, on this sample multimeter.

TEST YOUR SKILLS

Do *not* write in this book. Use a separate sheet of paper to complete the following problems. Use the drawings shown with each problem.

U.S. CONVENTIONAL RULES

1. In Fig. 8-28, what are the measurements at points *a, b,* and *c*? (feet/inches from zero)
2. In Fig. 8-29, what are the measurements at points *a, b,* and *c*? (feet/inches from zero)
3. In Fig. 8-30, what are the measurements at points *a, b,* and *c*? (inches from zero)
4. In Fig. 8-31, what are the measurements at points *a, b,* and *c*? (decimal inches from zero)
5. In Fig. 8-32, what are the measurements at points *a, b,* and *c*? (decimal feet from zero)

Fig. 8-28. Tape measure in feet, inches, and fractional inches. (Cooper Tools)

Fig. 8-29. Tape measure in feet, inches, and fractional inches. (Cooper Tools)

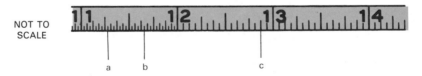

Fig. 8-30. Tape measure in inches and fractional inches. (Cooper Tools)

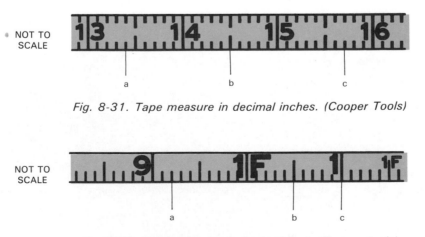

Fig. 8-31. Tape measure in decimal inches. (Cooper Tools)

Fig. 8-32. Tape measure in decimal feet. (Cooper tools)

METRIC RULES

6. In Fig. 8-33, what are the measurements of *a*, *b*, and *c*? (centimeters)
7. In Fig. 8-34, what are the measurements of *a*, *b*, and *c*? (a, b in centimeters, c in millimeters)

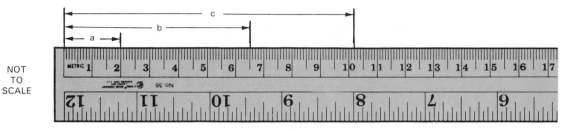

Fig. 8-33. Ruler in U.S. Conventional and SI Metric. (Hearlihy & Co.)

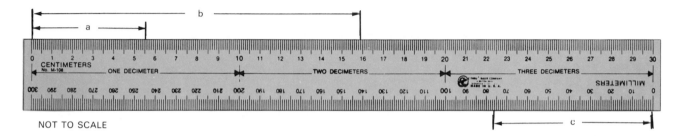

Fig. 8-34. Metric ruler. (Hearlihy & Co.)

TRIANGULAR DRAFTING SCALES

8. In Fig. 8-35, what are the measurements of *a*, *b*, and *c*? (feet/inches)
9. In Fig. 8-36, what are the measurements of *a*, *b*, and *c*? (feet inches)
10. In Fig. 8-37, what are the measurements of *a*, *b*, and *c*? (decimal feet)
11. In Fig. 8-38, what are the measurements of *a*, *b*, and *c*? (feet/inches)

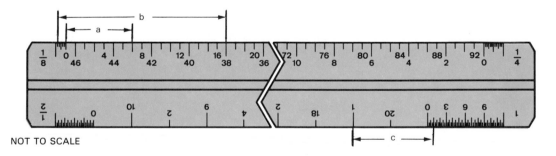

Fig. 8-35. An architect's scale. (Hearlihy & Co.)

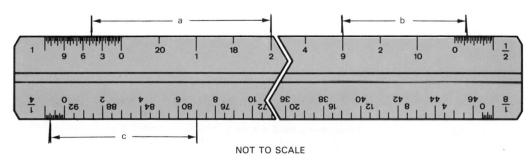

Fig. 8-36. An architect's scale. (Hearlihy & Co.)

20 SCALE: 1 INCH = 2 FEET

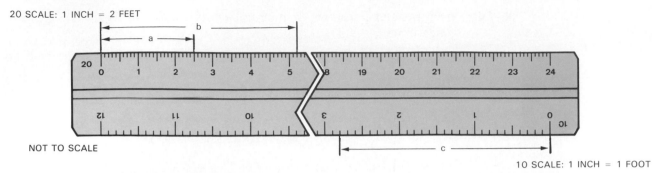

NOT TO SCALE

10 SCALE: 1 INCH = 1 FOOT

Fig. 8-37. A civil engineer's scale. (Hearlihy & Co.)

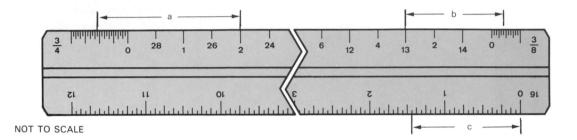

NOT TO SCALE

Fig. 8-38. An architect's scale. (Hearlihy & Co.)

MICROMETERS

12. What measurement is read on the micrometer scale shown in Fig. 8-39? (inches)

Sleeve = _____
Thimble = _____
Measurement = _____

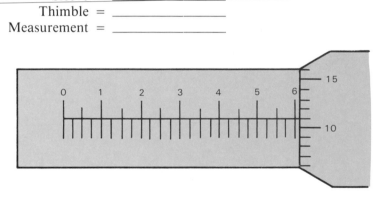

Fig. 8-39. A micrometer scale giving thousandths of inches.

13. What measurement is read on the micrometer scale shown in Fig. 8-40? (inches)

Sleeve = _____
Thimble = _____
Measurement = _____

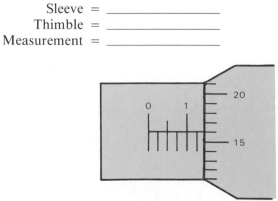

Fig. 8-40. A micrometer scale giving thousandths of inches.

14. What measurement is read on the micrometer scale shown in Fig. 8-41? (inches)

    Sleeve = _____

    Thimble = _____

    Measurement = _____

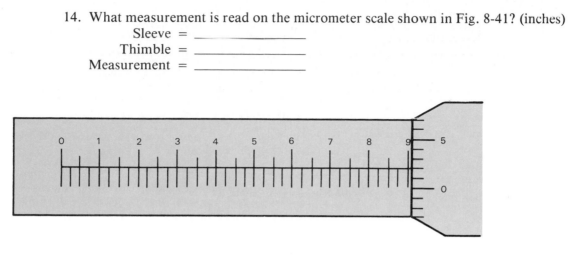

Fig. 8-41. A micrometer scale giving thousandths of inches.

15. What measurement is read on the micrometer scale shown in Fig. 8-42? (inches)

    Sleeve = _____

    Thimble = _____

    Measurement = _____

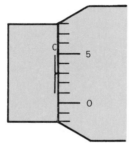

Fig. 8-42. A micrometer scale giving thousandths of inches.

16. What measurement is read on the micrometer scale shown in Fig. 8-43? (inches)

    Sleeve = _____

    Thimble = _____

    Measurement = _____

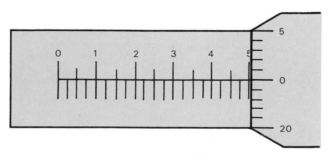

Fig. 8-43. A micrometer scale giving thousandths of inches.

17. What measurement is read on the micrometer scale shown in Fig. 8-44? (milli-meters)

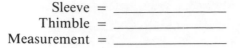

    Sleeve = _____
    Thimble = _____
    Measurement = _____

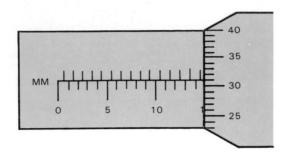

Fig. 8-44. A micrometer scale giving millimeters.

18. What measurement is read on the micrometer scale shown in Fig. 8-45? (milli-meters)

    Sleeve = _____
    Thimble = _____
    Measurement = _____

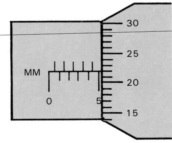

Fig. 8-45. A micrometer scale giving millimeters.

19. What measurement is read on the micrometer scale shown in Fig. 8-46? (millimeters)

    Sleeve = _____
    Thimble = _____
    Measurement = _____

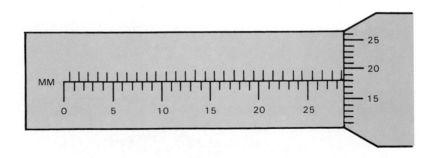

Fig. 8-46. A micrometer scale giving millimeters.

20. What measurement is read on the micrometer scale shown in Fig. 8-47? (millimeters)

        Sleeve = _____

       Thimble = _____

Measurement = _____

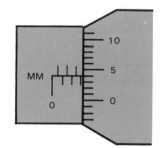

Fig. 8-47. A micrometer scale giving millimeters.

21. What measurement is read on the micrometer scale shown in Fig. 8-48? (millimeters)

        Sleeve = _____

       Thimble = _____

Measurement = _____

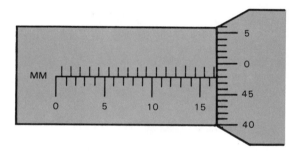

Fig. 8-48. A micrometer scale giving millimeters.

22. What measurement is read on the micrometer scale shown in Fig. 8-49? (inches)

        Sleeve = _____

       Thimble = _____

        Vernier = _____

Measurement = _____

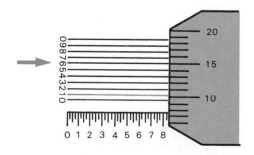

Fig. 8-49. A micrometer scale with vernier.

23. What measurement is read on the micrometer scale shown in Fig. 8-50? (inches)

Sleeve = _____

Thimble = _____

Vernier = _____

Measurement = _____

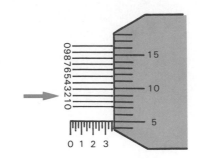

Fig. 8-50. A micrometer scale with vernier.

24. What measurement is read on the micrometer scale shown in Fig. 8-51? (inches)

Sleeve = _____

Thimble = _____

Vernier = _____

Measurement = _____

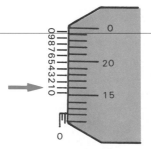

Fig. 8-51. A micrometer scale with vernier.

25. What measurement is read on the micrometer scale shown in Fig. 8-52? (inches)

Sleeve = _____

Thimble = _____

Vernier = _____

Measurement = _____

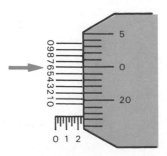

Fig. 8-52. A micrometer scale with vernier.

VERNIER CALIPERS

26. What is the reading indicated on the vernier caliper shown in Fig. 8-53?

$$\text{Quantity indicated on beam} = \text{_____}$$
$$\text{Quantity indicated on vernier plate} = \text{_____}$$
$$\text{Total reading} = \text{_____}$$

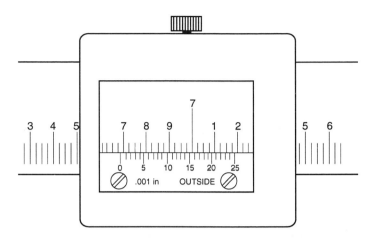

Fig. 8-53. Standard 25-division vernier caliper scale.

27. What is the reading indicated on the vernier caliper shown in Fig. 8-54?

$$\text{Quantity indicated on beam} = \text{_____}$$
$$\text{Quantity indicated on vernier plate} = \text{_____}$$
$$\text{Total reading} = \text{_____}$$

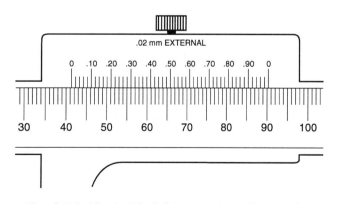

Fig. 8-54. Metric 50-division vernier caliper scale.

ELECTRICAL MULTIMETERS: VOLT-OHM-MILLIAMMETERS

(For Questions 28, 31, and 33, "range" is the position of the range switch; "scale" is the numbers across the scale, with the decimal adjusted for the range; "line" is the value of each minor line of the adjusted scale.)

28. With the multimeter shown in Fig. 8-55:

    range = _____
    scale = 0, _____, _____, _____, _____, _____
    line = _____

29. Read the meter needles at *a,* then *b,* and then *c.*

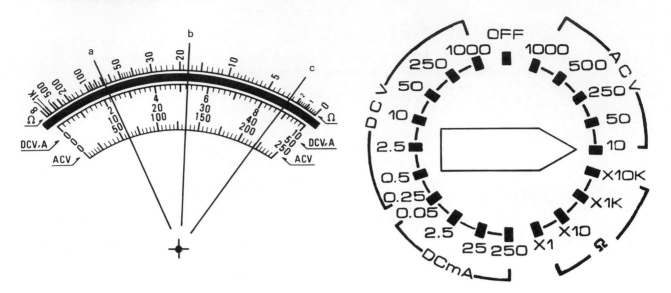

Fig. 8-55. Sample multimeter indications.

30. With the multimeter shown in Fig. 8-56, set up a table as shown and fill in values.

Needle	Small Line Value	Reading	Range	Measurement
a				
b				
c				

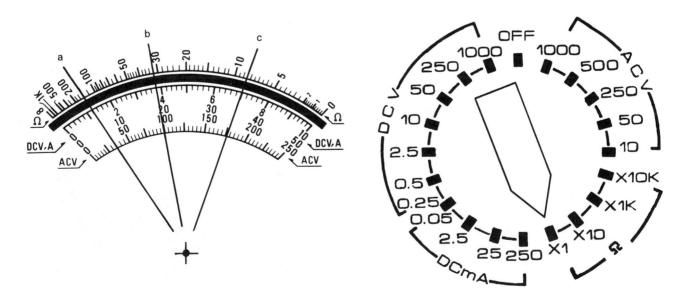

Fig. 8-56. Sample multimeter indications.

31. With the multimeter shown in Fig. 8-57:
    range = _____
    scale = 0, _____, _____, _____, _____, _____
    line = _____
32. Read the meter needles at: *a,* then *b,* and then *c.*

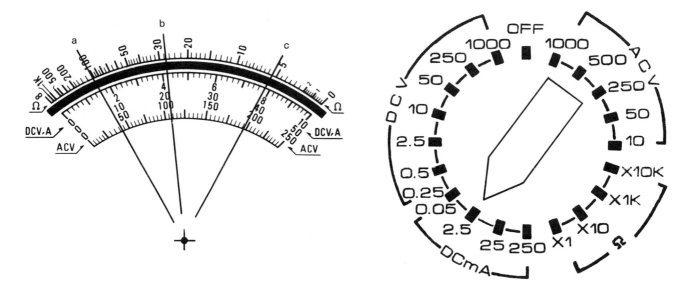

Fig. 8-57. Sample multimeter indications.

33. With the multimeter shown in Fig. 8-58:
    range = _____
    scale = 0, _____, _____, _____, _____, _____
    line = _____
34. Read the meter needles at *a,* then *b,* and then *c.*

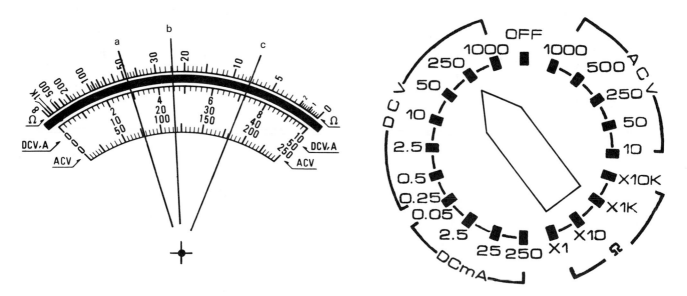

Fig. 8-58. Sample multimeter indications.

PROBLEM-SOLVING ACTIVITIES

Activity 8-1.
Tape measure flash cards
> **Objective:** To use flashcards to practice reading a tape measure.
> **Instructions:**
> 1. Make photocopies of different sections of various tape measures and rulers.
> 2. Mark each photocopy with the point of the measurement.
> 3. Write the answer on the back of the photocopy.
> 4. Use the photocopies to quiz your classmates.

Activity 8-2.
Poster of a tape measure
> **Objective:** To make a poster-size drawing of a tape measure.
> **Instructions:**
> 1. Make a poster of a tape measure. The poster should be large enough to be used as a classroom teaching aid.
> 2. The markings on the poster should duplicate the markings on the tape measure, with numbers marked only where they actually appear on the tape measure.
> 3. Whenever a measurement is made as part of an activity, it can be shown to the entire class. A paper clip or some other temporary marker can be used to indicate the measurement.
> 4. A sheet of clear plastic showing the fractional values of each of the markings can be placed over the poster.

Activity 8-3.
Using a micrometer
> **Objective:** To measure a set of drill bits using a micrometer.
> **Instructions:**
> 1. The drill bits used in this activity should be marked in decimal inches. If only bits with fractional markings are available, change the fractions to decimals. Metric sizes may also be used, if available.
> 2. Tighten the micrometer until it fully touches the drill bit to be measured. Do not overtighten.
> 3. Read the micrometer as instructed earlier in this chapter. Compare your reading to the dimension marked on the drill bit.

Chapter 9

Perimeter, Area, and Volume

OBJECTIVES

After studying this chapter, you will be able to:
- Explain the basic terminology of geometry.
- Compute the perimeter and area of plane shapes.
- Compute the volume of solids.

Perimeter, area, and volume are the bases for calculating required materials for many tasks. For example, if fencing in a section of land, it is necessary to know perimeter in order to know how much fencing to buy. The area of a room must be figured to determine how much carpeting will be needed to cover a floor. If concrete is needed to make a new sidewalk, the required volume must be determined.

Perimeter is defined as distance around. A fence is an application of perimeter; wood trim at the base of a wall is another. **Area** is often considered the surface, such as the surface of a floor. The label on a can of paint gives how much area the paint will cover. Measurements must be given with a roll of wallpaper to know the area it will cover. You must determine the area of your ceiling to know how much ceiling tile to purchase. **Volume** is the calculation of the area combined with the height, or thickness. One example is the concrete needed for a sidewalk; another is gallons of water, figured from cubic feet, needed to fill a swimming pool.

SHAPES IN GENERAL

There are two different kinds of lines—straight and curved. Outlines of shapes have either, or both, types of lines. Formulas exist that give perimeter, area, and volume. They are based on the basic shapes of the rectangle, the triangle, and the circle. You may come across a shape that is not a circle, but is formed of curves. For most such shapes, you would, at this point, have to approximate perimeter, area, and volume. This is because a formula to give an exact answer is not readily available. Finding an exact answer for odd, curved shapes requires use of calculus.

BASIC TERMINOLOGY

Before going further, a few terms should be explained. Knowing their meaning should aid your understanding of this chapter.
- **Planes**—are defined by 2-dimensional, flat surfaces.
- **Intersect**—means to cross.
- **Parallel**—nonintersecting lines or planes. The distance between straight parallel lines is the same at all locations. The same is true for parallel planes. Therefore, by measuring any two locations, it can be determined if two straight lines are parallel. To determine if two planes are parallel, measure any two locations *plus* a third location that is not in line with the other two. If the distances are equal, the planes

are parallel. Railroad tracks are *parallel*. Opposite walls in a room are generally *parallel*.

- **Degrees**—A circle is divided so that 360 equal-size angles result. Degrees are an angular measure. Each of the 360 angles formed is defined as one degree (1°). This is the basis for any angular measurement. A quarter of a circle has an angular measurement of 90 degrees.
- **Right angles**—are angles that measure 90 degrees. A right angle is often shown with a small square drawn in the corner of the angle. Lines or planes that together form a right angle are said to be **perpendicular**, or **square**.

PLANE SHAPES

Plane shapes, or figures, as the name implies, are shapes that are on plane surfaces. They are flat. Four common plane shapes are the rectangle, the square, the triangle, and the circle. (The ellipse [oval] is another fairly common plane shape. Beyond this, nothing more will be said about the ellipse.)

The rectangle

A **rectangle** is a four-sided figure with opposite sides equal in length and parallel to each other. A rectangle has four, 90-degree angles. Windows are commonly in the shape of rectangles; so are pictures and notebook paper. All rectangles conform to the rules of rectangles.

RULES OF RECTANGLES

- See Fig. 9-1 for an example.
- A rectangle is a four-sided figure.
- Opposite sides are parallel.
- Opposite sides are equal in length. It is not necessary for all four sides to be equal, only opposite sides. Therefore, rectangles can be short and wide, tall and narrow, or have all sides equal.
- The four inside angles each measure 90 degrees. Notice the squares in the corners indicating right angles.
- Lines drawn to opposite corners, forming an "X" inside of the rectangle, will be exactly the same length. The point where the lines cross will be the exact center of the rectangle.

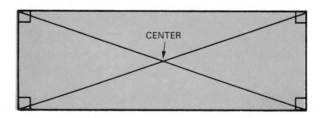

Fig. 9-1. A rectangle showing markings for center and right angles.

The square

A **square** is a special type of rectangle. It is a rectangle with four equal sides. All squares are rectangles; however, not all rectangles are squares. Floor and ceiling tiles are commonly in the shape of squares. Squares always conform to the rules of squares.

RULES OF SQUARES

- See Fig. 9-2 for an example.
- All sides are of equal length. Otherwise, all rules of rectangles apply.

Fig. 9-2. A square is a rectangle with four equal sides.

The triangle

A **triangle** is a three-sided figure. A triangle has three angles. The inside measurements of these angles add up to 180 degrees. Triangles take the forms of equilateral, isosceles, and scalene. **Equilateral triangles** have three equal sides and, therefore, three equal angles. Each angle of an equilateral triangle measures 60 degrees. **Isosceles triangles** have two equal sides, hence, two equal angles. **Scalene triangles** have no equal sides; no equal angles.

Triangles can also be classified as right or oblique. A **right triangle** is a triangle with a 90-degree angle. The side opposite this angle is called the hypotenuse of the triangle. Right triangles can be isosceles or scalene, but never equilateral. An **oblique triangle** is a triangle that does not contain a right angle. Therefore, any triangle that is not a right triangle is an oblique triangle. One particular oblique triangle is the **obtuse triangle.** It has one angle greater than 90 degrees. The other oblique triangle is the **acute triangle.** All of its angles are less than 90 degrees. Examples of triangular shapes include roof trusses and drafting triangles. All triangles conform to the rules of triangles. (More about triangles is presented in Chapter 15 of this book.)

RULES OF TRIANGLES

- See Fig. 9-3 for an example.
- A triangle is a three-sided figure. Sides may be any length.
- The three inside angles add up to 180 degrees.
- A right triangle may be thought of as one-half of a rectangle.

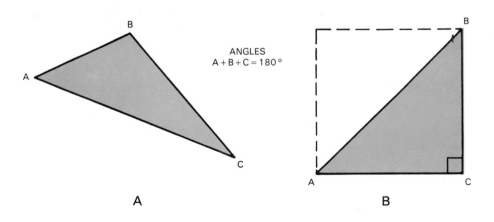

ANGLES
$A + B + C = 180°$

A

B

Fig. 9-3. A—An oblique triangle. B—A right triangle. Do you see why a right triangle is "1/2 of a rectangle"?

The circle

A **circle** is a continuous line, or it is a plane surface bounded by the line. Every point on the line, or circumference, is always the same distance from a center point that lies in the plane. Certain terms relating to circles were discussed previously in Chapter 6. These terms were sector, radius, arc, and circumference. It is important to be familiar with these terms and with circles in general because so many objects are in the shape of a circle. Some common examples of circular shapes include a coin, a tire, and a pie graph. All circles conform to the rules of circles.

RULES OF CIRCLES

- See Fig. 9-4 for an example.
- All points along the circle are the same distance from the center.
- By definition, there are 360 degrees in a circle.
- The radius, r, is the distance from the center to the circumference.
- The **diameter, d,** is the distance from one side of the circle to the other, measured through the center.
- The diameter is twice as long as the radius.
- Circumference, c, is both the boundary of a circle and the measurement of the distance around a circle.
- The ratio of circumference to diameter (c/d) of *any* circle is 3.1416. This ratio is denoted by the Greek letter π (pi).
- A **radian** is an angle, the measure of which is defined by two radii of a circle and an arc joining them, all of the same length. The number of radians in any circle is *always* 2π ($2 \times \pi$), or 6.28.
- Conversion factor: 2π radians $= 360°$.

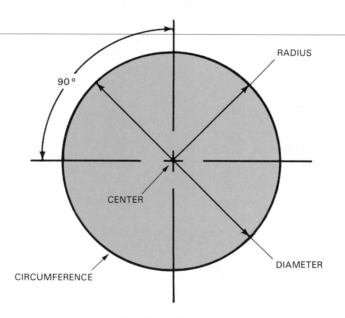

Fig. 9-4. The circle.

PERIMETER

Perimeter is both the boundary of a plane shape and the distance around it. Before fencing in a yard, before installing gutters or wood trim, before building a picture frame, you need to figure perimeter so you know how much material to buy. Figuring perimeter is a very common application of arithmetic.

How to calculate perimeter (p)

In general, perimeter of a plane figure bounded by straight lines (a **polygon**) is found by adding together the length of all sides. Curved figures are different. Specifics are given below and in the formulas to calculate perimeter.

Perimeter of polygons in general. To calculate perimeter of polygons, add the lengths of all sides. Refer to Formula 9-1 of the formulas to calculate perimeter. Imagine taking a piece of string and wrapping it around the outside of the object. If the string is then measured, the total length of it will be equal to the perimeter. (This is a practical way of measuring perimeter of any smaller shape if you have string and a yard-stick, but do not have a tape measure.)

Perimeter of rectangles. Rectangles are a special-case polygon. Since the length of opposite sides of a rectangle are equal, the general formula (Formula 9-1) can be written in a simpler form. Refer to Formula 9-2. (One side of a rectangle is the length; the other is the width. It does not especially make any difference which is which. Furthermore, you may come across a dimension referred to as breadth, height, altitude, thickness, or depth. As long as you have the two dimensions, regardless of what they are referred to as, you can use the Formula.) The rectangle contains two sides equal to the length and two sides equal to the width. Therefore, to find perimeter, double the length and double the width. Then, add them together.

Perimeter of squares. Squares are also a special-case polygon. With the length of the sides of a square all equal, an even simpler formula can be written. Refer to Formula 9-3. Find the perimeter of a square by multiplying the length of one side by 4.

Perimeter of triangles in general. To find the perimeter of a triangle, in general, add together the lengths of the three sides. Refer to Formula 9-4.

Perimeter of equilateral triangles. This is a special-case triangle. Since all three sides are equal, a simpler formula can be written. Refer to Formula 9-5. Find the perimeter by multiplying the length of one side by 3.

Perimeter of circles. When making calculations concerning circles the Greek letter π is involved. Perimeter of a circle is the same as circumference. Use either Formula 9-6 or Formula 9-7 depending on which is given, radius or diameter. The two formulas are equal because diameter is equal to twice the radius.

FORMULAS TO CALCULATE PERIMETER

Formula 9-1. Perimeter of plane figures with straight lines (polygons). See Fig. 9-5A.
$$p = \text{side a} + \text{side b} + \text{side c} + \text{side d} + \text{side e} + \ldots, \text{ where}$$
side a = length of side a; etc.

Formula 9-2. Perimeter of rectangles. See Fig. 9-5B.
$$p = 2l + 2w, \text{ where } l = \text{length}; w = \text{width.}$$
Note: Remember, when a letter is written alongside a number or another letter in an equation or formula, it signifies multiplication.
(i.e., $p = 2 \times l + 2 \times w$). Perform multiplication first, then, addition.

Formula 9-3. Perimeter of squares. See Fig. 9-5C.
$$p = 4s, \text{ where } s = \text{length of one side.}$$

Formula 9-4. Perimeter of triangles. See Fig. 9-5D.
$$p = \text{side a} + \text{side b} + \text{side c}, \text{ where side a} = \text{length of side a; etc.}$$

Formula 9-5. Perimeter of equilateral triangles. See Fig. 9-5E.
$$p = 3s, \text{ where } s = \text{length of one side.}$$

Formula 9-6. Perimeter of circles, using radius.
$$p = 2\pi r$$

Formula 9-7. Perimeter of circles, using diameter.
$$p = \pi d$$

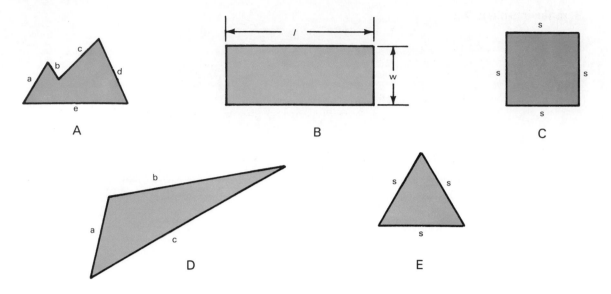

Fig. 9-5. A—A polygon. B—A rectangle. C—A square. D—A scalene triangle. E—An equilateral triangle.

Sample Problem 9-1.

Find the perimeter of the rectangle in Fig. 9-6.

Solution A:

Step 1. Select a suitable formula. Use Formula 9-1.

$$p = \text{side a} + \text{side b} + \text{side c} + \text{side d}$$

Step 2. Substitute in values and solve.

$$p = 4 \text{ in.} + 6 \text{ in.} + 4 \text{ in.} + 6 \text{ in.}$$
$$= 20 \text{ in.}$$

Solution B:

Step 1. Select a suitable formula. Use Formula 9-2.

$$p = 2l + 2w$$

Step 2. Substitute in values and solve.

$$p = (2 \times 6 \text{ in.}) + (2 \times 4 \text{ in.})$$
$$= 12 \text{ in.} + 8 \text{ in.}$$
$$= 20 \text{ in.}$$

Note: Multiplication is performed first. Also, it does not matter which side is used as l and which is used as w.

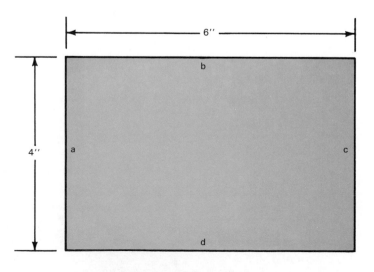

Fig. 9-6. Find the perimeter.

Sample Problem 9-2.

Find the perimeter of the square in Fig. 9-7.

Solution A:

Step 1. Select a suitable formula. Use Formula 9-1.

$$p = \text{side } a + \text{side } b + \text{side } c + \text{side } d$$

Step 2. Substitute in values and solve.

$$p = 25 \text{ ft.} + 25 \text{ ft.} + 25 \text{ ft.} + 25 \text{ ft.}$$
$$= 100 \text{ ft.}$$

Solution B:

Step 1. Select a suitable formula. Use Formula 9-2.

$$p = 4s$$

Step 2. Substitute in values and solve.

$$p = 4 \times 25 \text{ ft.}$$
$$= 100 \text{ ft.}$$

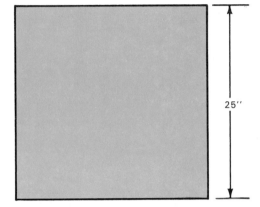

Fig. 9-7. Find the perimeter.

Sample Problem 9-3.

Find the perimeter of the triangle in Fig. 9-8.

Step 1. Select the suitable formula. Use Formula 9-4.

$$p = \text{side } a + \text{side } b + \text{side } c$$

Step 2. Substitute in values and solve.

$$p = 15 \text{ m} + 10 \text{ m} + 20 \text{ m}$$
$$= 45 \text{ m}$$

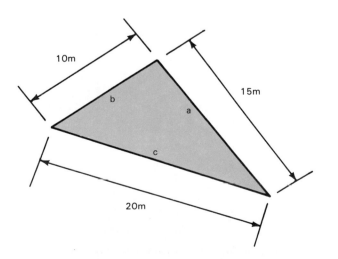

Fig. 9-8. Find the perimeter.

Sample Problem 9-4.

Find the amount of wire mesh fencing needed to enclose a yard having the dimensions shown in Fig. 9-9.

Step 1. Select the suitable formula. Use Formula 9-1.

p = side a + . . . + side f

Step 2. Substitute in values and solve.

p = 250 ft. + 150 ft. + 200 ft. + 400 ft. + 300 ft. + 200 ft.
= 1500 ft.

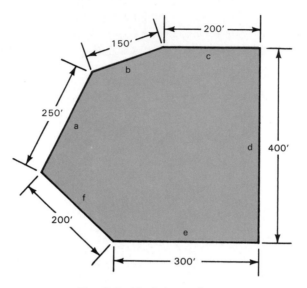

Fig. 9-9. Find the perimeter.

Sample Problem 9-5.

Find the perimeter of the circle in Fig. 9-10.

Solution A:

Step 1. Select a suitable formula. Use Formula 9-6.

p = 2πr

Step 2. Substitute in values and solve.

p = 2 × π × 15 yd.
= 94.2 yd.

Solution B:

Step 1. Select a suitable formula. Use formula 9-7.

p = πd

Step 2. Substitute in values and solve.

p = π × 30 yd. (d = 2 × r)
= 94.2 yd.

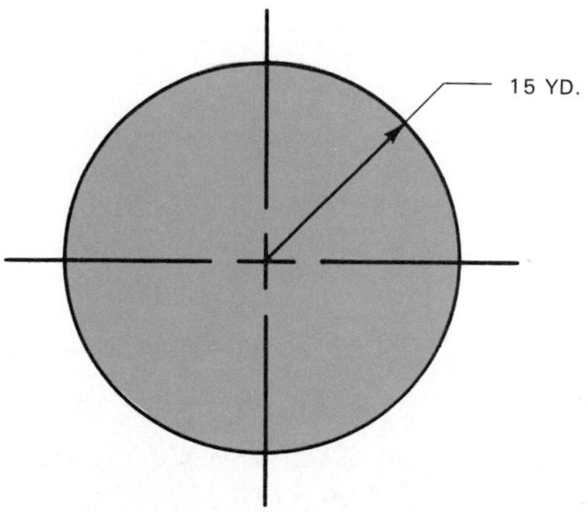

Fig. 9-10. Find the perimeter.

Sample Problem 9-6.

The rectangle in Fig. 9-11 has a perimeter of 1100 feet. One side has a length of 150 feet. Find the other three sides.

Solution A:

Step 1. Select a suitable formula. Use Formula 9-2.

$$p = 2l + 2w$$

Step 2. Substitute values into formula.

$$1100 \text{ ft.} = (2 \times 150 \text{ ft.}) + 2w$$

Step 3. Perform the multiplication.

$$1100 \text{ ft.} = 300 \text{ ft.} + 2w$$

Step 4. Solve for w. To do this, you must "rearrange" the formula so that w is by itself. Whatever operation is performed on the "w side" must be reversed. Also, whatever reversal is done on the w side, must be done on the other side of the equal sign. To reverse, since 300 feet was to be added to the w side, subtract 300 feet from both sides. This process of rearranging is called *transposing the equation.* It is covered in Chapter 12 in some detail.

$$1100 \text{ ft.} - 300 \text{ ft.} = (300 \text{ ft.} - 300 \text{ ft.}) + 2w$$
$$800 \text{ ft.} = 2w$$

Next, since w was to be multiplied by 2, reverse this and divide both sides by 2.

$$800 \text{ ft.}/2 = 2w/2$$
$$400 \text{ ft.} = w$$

Answer: side a = side c = 150 ft.
side b = side d = 400 ft.

Solution B:

Step 1. Use the relationship in rectangles that opposite sides are equal. Given one side equals 150 feet, then its opposite side also equals 150 feet.

Side a = side c = 150 ft.

Step 2. Add these two sides together:

150 ft. + 150 ft. = 300 ft.

Step 3. Subtract the known sides from the perimeter to find the combined value of the other two sides.

1100 ft. - 300 ft. = 800 ft.

Step 4. Divide by 2 to get the value of one side.

800 ft./2 = 400 ft.

Therefore:

Side b = side d = 400 ft.

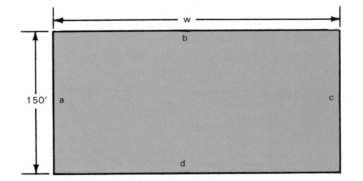

Fig. 9-11. Find the sides.

Practice Problems

Complete the following problems on a separate sheet of paper. Include units of measure.

1. Find the length of string needed to wrap around the perimeter of a cardboard box that is 24" long and 18" wide.
2. Find the perimeter of the object shown in Fig. 9-12A.

3. Find the perimeter of the object shown in Fig. 9-12B.
4. Determine how much wood must be purchased for the trim on the base of the walls of a 9' × 14' room. The only opening in the room is a 36" wide door.
5. Find the perimeter of the rectangle shown in Fig. 9-12C.
6. Find the perimeter of the square shown in Fig. 9-12D.
7. Determine how much fence material is needed to enclose a swimming pool in a 50' × 50' area.
8. Find the perimeter of the triangle shown in Fig. 9-12E.
9. Find the perimeter of the circle shown in Fig. 9-12F.
10. A flower garden is to be formed with bricks in the shape of an 8' diameter circle. How many 2" wide bricks are needed to form the circle?
11. Find the perimeter of the triangle shown in Fig. 9-12G.
12. Find the perimeter of the circle shown in Fig. 9-12H.
13. A string attached to each of the four boundary markers of a rectangular-shaped property measures 1000 feet long. If the side along the road is 150', how deep is the property?
14. The perimeter of the rectangle shown in Fig. 9-12I is 700'. How long is the unknown side?
15. The perimeter of the triangle shown in Fig. 9-12J is 150 cm. How long is the unknown side?

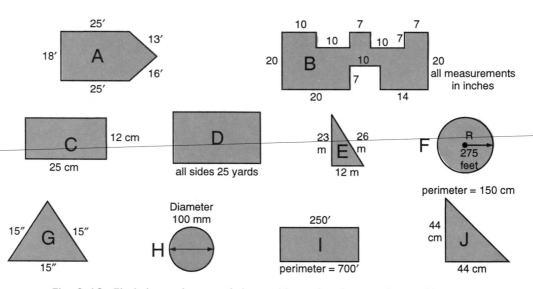

Fig. 9-12. Find the perimeter of these objects for the practice problems.

AREA

Area is a measure of the surface of a plane figure. It is a term used everyday. People talk about area of a room, area of a wall, land area, etc. The area of a room is the floor space. You need to know area of a floor before buying tile or carpeting. You need to know the area of a wall if you plan to paint or wallpaper. The size of a county, a state, or a country is given in terms of land area.

Area is measured in square units such as square feet, square inches, square meters, square centimeters, etc. Fig. 9-13 shows a rectangle subdivided into squares the size of 1 square unit of measure. If the unit of measure were 1 inch, each small square would measure 1 inch wide and 1 inch high (1 square inch). Abbreviations commonly used for measures of area include:

- square inches = sq. in. = in.²
- square feet = sq. ft. = ft.²
- square yards = sq. yd. = yd.²
- square miles = sq. mi. = mi.²
- square centimeters = cm²
- square meters = m²

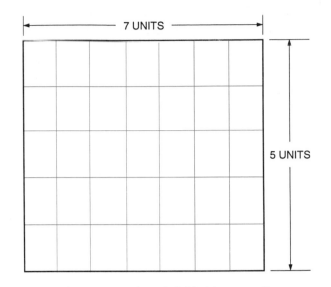

Fig. 9-13. A rectangle subdivided into small squares.

How to calculate area

Specifics for calculating area of various plane figures are given below. Corresponding formulas are found in the formulas to calculate area.

Area of rectangles. Refer again to Fig. 9-13. The large rectangle is subdivided into small squares. Along one side are 7 square units. Along the adjacent (next to) side are 5 square units. If you count all of the subdivided squares, you will find a total of 35 small squares. This is the same answer as would be obtained by multiplying the length by the width. Refer to Formula 9-8.

Area of squares. A square is a special-case rectangle. Since length and width have the same dimension, the length of one side can be squared. If you recall, squaring a number is the same as multiplying a number by itself. Refer to Formula 9-9.

Area of triangles. A triangle can be subdivided into small squares in much the same manner as a rectangle. Remember that a right triangle is one-half of a rectangle. Therefore, area of a right triangle is one-half the area of a rectangle, or 1/2 times length times width. In terms of triangles, you will find length and width expressed as base and height, or altitude. Base is one side of the triangle. Height is the highest point perpendicular to the base. See Fig. 9-14. Area of a triangle is one-half of base times height no matter what the type of triangle. Refer to Formula 9-10.

Area of circles. In the same manner as the rectangle and the triangle, a circle can be subdivided into small squares. See Fig. 9-15. The factor of π is involved as before with perimeter. Use either Formula 9-11 or Formula 9-12, depending on whether radius or diameter is given.

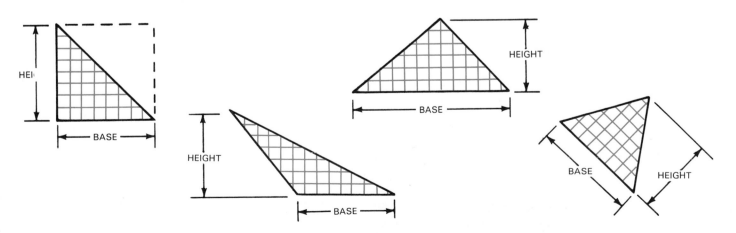

Fig. 9-14. Triangles subdivided into small squares.

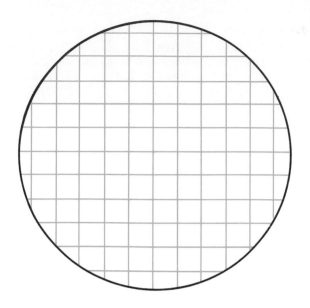

Fig. 9-15. A circle subdivided into small squares.

FORMULAS TO CALCULATE AREA

Formula 9-8. Area of rectangles.
   $A = l \times w$, where l = length; w = width.
Formula 9-9. Area of squares.
   $A = s^2$, where s = length of one side.
Formula 9-10. Area of triangles. See Fig. 9-14.
   $A = 1/2 (b \times h)$, where b = base; h = height.
   Note: In equations or formulas, when numbers or letters are next to parentheses without a sign of operation, it signifies multiplication. (i.e., $A = 1/2 \times b \times h$)
Formula 9-11. Area of circles using radius.
   $A = \pi r^2$
Formula 9-12. Area of circles using diameter.
   $A = \pi d^2/4$

Sample Problem 9-7.

Find the area of a rectangular floor measuring 12 feet by 16 feet.
Step 1. Find the suitable formula. Use Formula 9-8.
   $A = l \times w$
Step 2. Substitute in values and solve.
   $A = 12 \text{ ft.} \times 16 \text{ ft.}$
   $= 192 \text{ ft.}^2$

Sample Problem 9-8.

Find the area of a triangular window, measuring 32 inches at its base and 36 inches in height.
Step 1. Find the suitable formula. Use Formula 9-10.
   $A = 1/2 (b \times h)$
Step 2. Substitute in values and solve.
   $A = 1/2 \times 32 \text{ in.} \times 36 \text{ in.}$
   $= 576 \text{ in.}^2$

Sample Problem 9-9.

Find the area covered by a circular swimming pool having a diameter of 24 feet.

Step 1. Find the suitable formula. Use Formula 9-12.

$$A = \pi d^2/4$$

Step 2. Substitute in values and solve.

$$A = \pi \times (24 \text{ ft.})^2/4$$
$$= \pi \times 24 \text{ ft.} \times 24 \text{ ft.}/4$$
$$= \pi \times 576 \text{ ft.}^2/4$$
$$= 452 \text{ ft.}^2$$

Sample Problem 9-10.

Find the area of a room whose floor plan is shown in Fig. 9-16.

Step 1. Divide the floor plan into 2 rectangles. (See dashed line.)

Step 2. Find area of Section 1 (A_1).

Use Formula 9-8.

$$A = l \times w$$

Find *l;* then substitute.

$$l = 20 \text{ ft.} - 10 \text{ ft.} = 10 \text{ ft.}$$
$$A_1 = 10 \text{ ft.} \times 10 \text{ ft.}$$
$$A_1 = 100 \text{ ft.}^2$$

Step 2. Find area of Section 2 (A_2).

Use Formula 9-8.

$$A = l \times w$$

Substitution:

$$A_2 = 10 \text{ ft.} \times 30 \text{ ft.}$$
$$A_2 = 300 \text{ ft.}^2$$

Step 3. Add together area of each section.

$$A_{total} = 100 \text{ ft.}^2 + 300 \text{ ft.}^2$$

Answer: $A_{total} = 400 \text{ ft.}^2$

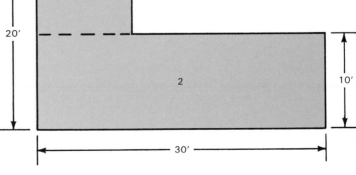

Fig. 9-16. A floor plan.

Practice Problems

Complete the following problems on a separate sheet of paper. Include units of measure (inches, feet, minutes, dollars, etc.).

1. How many 1' square floor tiles are needed to cover the 36' × 48' floor in the basement of a house?
2. How many 18" × 24" boxes can be stored in a 96" × 120" closet (without stacking the boxes)?

3. Find the area of the rectangle shown in Fig. 9-17A.
4. A room with an 8' high ceiling has two walls that are 18' long and two walls that are 24' long. What is the combined surface area of the walls? If a gallon of paint covers 360 sq. ft., how many gallons are needed to paint the walls?
5. A roll of wallpaper is 100' long and 24" wide. How many rolls would be needed to cover the walls in problem 4?
6. Find the area of the square shown in Fig. 9-17B.
7. How many 12" square ceiling tiles are used in a 14' × 16' room?
8. How many 24" square ceiling tiles are used in a 16' × 18' room?
9. Find the area of the triangle shown in Fig. 9-17C.
10. What is the area of a 4' × 8' sheet of plywood?
11. How many 4' × 8' sheets of plywood are needed to cover the first floor of house that is 48' long and 32' wide?
12. Find the area of the circle shown in Fig. 9-17D.
13. How much floor space (area) is taken up by a 10' diameter circular staircase?
14. How many square yards of carpeting are needed to cover the floor of a 12' × 16' room?
15. Find the total area of the building shown in Fig. 9-17E.

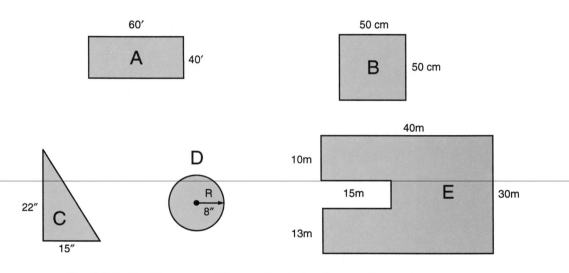

Fig. 9-17. Find the area of these objects for the practice problems.

SOLID SHAPES

Solids are objects with three dimensions. These are the dimensions of length, width, and height. Again, these terms may be interchanged with breadth, thickness, depth, etc. In the context used here, the term "solids" relates to *hollow* as well as solid, three-dimensional shapes. Common solid shapes are cubes and other prisms, cones, cylinders, and spheres. These shapes are extensions of plane figures including squares, triangles, and circles.

The cube and other prisms

The **cube** is a solid—the faces (sides) of which are all squares. A cube has six faces. Each is at a right angle to each adjacent face. See Fig. 9-18. Dice are common examples of cubes. Cubes are members of a broader class of solids called **prisms**. Prisms are solids whose bases (ends) are parallel and the same shape. Most boxes are prisms. A house attic is often prism-shaped. Some tents are prism-shaped.

The cone

The **cone** is a solid—the base of which is a circle; the other surface tapers to a point. See Fig. 9-18. Looking at it from the side, a cone looks like a triangle. A funnel is cone-shaped. A hopper, used for loading and unloading bulk materials, is very often cone-shaped. A cyclone particle separator is cone-shaped.

The cylinder

The **right circular cylinder** is a solid—the bases of which are equal-sized circles. The side of a right circular cylinder (hereafter **cylinder**) is at right angles to the bases. See Fig. 9-18. From a side view, a cylinder looks like a rectangle. Cans, pipes, storage tanks, and pistons are all examples of cylinders.

The sphere

A **sphere** is a solid bounded by a curved surface of which any point on it is equally distant from a center point within. The distance from the surface to the center is given by the radius. See Fig. 9-18. From any side view, a sphere looks like a circle. A ball bearing is a sphere, as are some storage, pressure, and reactor vessels.

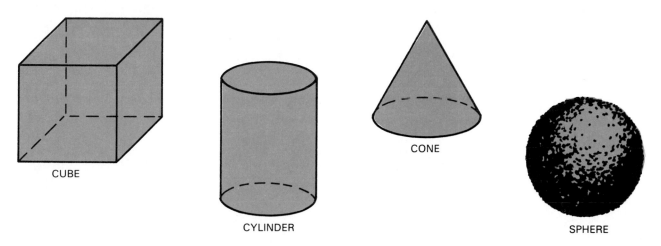

CUBE   CYLINDER   CONE   SPHERE

Fig. 9-18. Some common solid shapes.

VOLUME

Volume is the capacity of an object, or the amount of space it occupies. The volume of an empty room is the space occupied by air. The interior volume of a box is the space inside. The volume of a concrete pillar is the space it occupies.

Fig. 9-19 is a cube that is subdivided into smaller cubes each of one cubic unit of measure. A cube is the primary unit of volume measure. Cubic units of volume include cubic inches, cubic feet, cubic centimeters, etc. A cubed unit is a unit that is repeated as a factor, three times. If the unit of measure in Fig. 9-19 were 1 inch, each small cube would measure 1 inch wide, 1 inch high, and 1 inch deep (1 cubic inch). Since there would be a total of eight 1-cubic-inch cubes, the total volume would be 8 cubic inches. Abbreviations commonly used for measures of volume include:

- cubic inches = cu. in. = in.³
- cubic feet = cu. ft. = ft.³
- cubic yards = cu. yd. = yd.³
- cubic centimeters = cm³
- cubic meters = m³

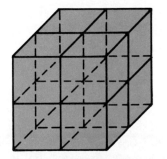

Fig. 9-19. A cube subdivided into small cubes.

How to calculate volume

Specifics for calculating volume of various solid shapes are given below. Corresponding formulas are found in the formulas to calculate volume.

Volume of cubes and other prisms. The volume of cubes and other rectangular prisms can be found by multiplying length times width times height. Refer again to Fig. 9-19. If the unit of measure was 1 inch as before, the volume would be 2 inches times 2 inches times 2 inches, or 8 cubic inches. (Remember, 8 cubic inches was the total volume previously determined by counting the number of cubes.) Notice that what you are doing, in effect, is cubing the dimension of one side. This is because all three dimensions of a cube are the same; therefore, you are repeating a factor, three times. Refer to Formula 9-13 for the volume of rectangular prisms and Formula 9-14, specifically for volume of cubes. To find the volume of a triangular prism, multiply 1/2 times the triangle base times its height times the depth of the prism. See Fig. 9-20 and refer to Formula 9-15.

There is a general formula that will work for any prism. Multiply area of one of the bases by the depth. Refer to Formula 9-16. To calculate the volume in an attic space, for example, you could use Formula 9-15, or you could use the general formula. Find the area of one of the bases of the prism. In this case, the bases are triangles. Multiply area of one of the triangles times the depth. The answer is the volume of the attic space.

Volume of cones, cylinders, and spheres. The volumes of cones and cylinders depend on area of the base, or πr^2, times height of the side. With cones, a factor is applied. Volume of spheres depends on radius cubed times a factor. Refer to Formulas 9-17 through 9-19.

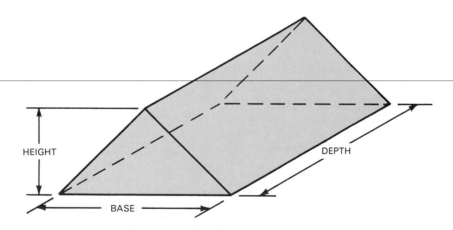

Fig. 9-20. A triangular prism.

HEIGHT   DEPTH   BASE

FORMULAS TO CALCULATE VOLUME

Formula 9-13. Volume of rectangular prisms.
  $V = l \times w \times h$, where l = length; w = width; h = height.
Formula 9-14. Volume of cubes.
  $V = s^3$, where s = length of one side.
Formula 9-15. Volume of triangular prisms.
  $V = 1/2 (b \times h \times d)$, where b = base of triangle; h = height of triangle;
    d = depth (or length) of prism.
Formula 9-16. Volume of prisms in general.
  $V = A_b \times d$, where A_b = area of base; d = depth (or length).
Formula 9-17. Volume of cones.
  $V = 1/3 (\pi r^2 \times h)$, where r = radius of circular base; h = height.
Formula 9-18. Volume of cylinders.
  $V = \pi r^2 \times h$, where r = radius of circular base; h = height (or depth).
Formula 9-19. Volume of spheres.
  $V = 4/3 (\pi r^3)$, where r = radius.

Sample Problem 9-11.

Find the volume of the room shown in Fig. 9-21.

Step 1. Find the suitable formula. Use Formula 9-13.

$$V = l \times w \times h$$

Step 2. Substitute in values and solve.

$$V = 12 \text{ ft.} \times 8 \text{ ft.} \times 24 \text{ ft.}$$
$$= 2304 \text{ ft.}^3$$

Note: It is not important which dimension is chosen for length, width, or height.

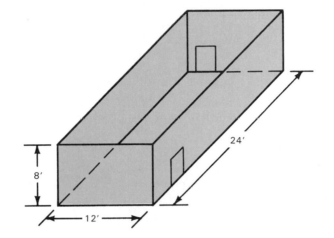

Fig. 9-21. This basic shape is an example of a rectangular prism.

Sample Problem 9-12.

Find the volume of the box shown in Fig. 9-22.

Step 1. Find the suitable formula. Use Formula 9-14.

$$V = s^3$$

Step 2. Substitute in values and solve.

$$V = (2 \text{ ft.})^3 = 2 \text{ ft.} \times 2 \text{ ft.} \times 2 \text{ ft.}$$
$$= 8 \text{ ft.}^3$$

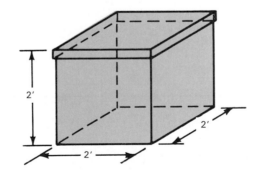

Fig. 9-22. A cube has equal dimensions of length, width, and height.

Sample Problem 9-13.

Find the volume of the triangular prism shown in Fig. 9-23.

Step 1. Find the suitable formula. Use Formula 9-15.

$$V = 1/2 (b \times h \times d)$$

Step 2. Substitute in values and solve.

$$V = 1/2 \times 16 \text{ in.} \times 12 \text{ in.} \times 32 \text{ in.}$$
$$= 3072 \text{ in.}^3$$

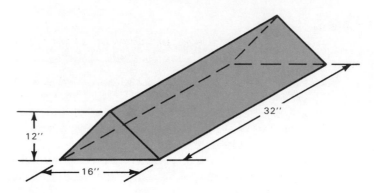

Fig. 9-23. Find the volume.

Sample Problem 9-14.

Find the volume of the pipe shown in Fig. 9-24.

Step 1. Find the suitable formula. Use Formula 9-18.

$$V = \pi r^2 \times h$$

Step 2. Substitute in values and solve.

$$V = \pi \times (2 \text{ in.})^2 \times 120 \text{ in.}$$
$$= 1508 \text{ in.}^3$$

Note: Feet were first converted to inches. Otherwise, inches would first be converted to feet.

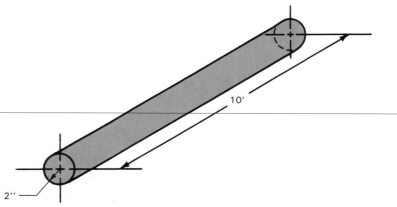

Fig. 9-24. This section of pipe is a good example of a cylinder.

Sample Problem 9-15.

Find the cement needed for a sidewalk 30 feet long, 4 feet wide, and 4 inches deep.

Step 1. Cement is sold by the cubic yard. Therefore, all measurements should first be converted to yards.

Unit conversion: 1 yard = 3 feet
1 yard = 36 inches

Use the conversion bar:

$$\left(\frac{4 \text{ ft.}}{}\right) \left(\frac{1 \text{ yard}}{3 \text{ ft.}}\right) = 1.33 \text{ yd.}$$

$$\left(\frac{30 \text{ ft.}}{}\right) \left(\frac{1 \text{ yard}}{3 \text{ ft.}}\right) = 10 \text{ yd.}$$

$$\left(\frac{4 \text{ in.}}{}\right) \left(\frac{1 \text{ yard}}{36 \text{ in.}}\right) = 0.111 \text{ yd.}$$

Step 2. Find the suitable formula. Use Formula 9-13.

$$V = l \times w \times h$$

Step 3. Substitute in values and solve.

$$V = 10 \text{ yd.} \times 1.33 \text{ yd.} \times 0.111 \text{ yd.}$$
$$= 1.48 \text{ yd.}^3$$

Practice Problems

Complete the following problems on a separate sheet of paper. Include units of measure (inches, feet, minutes, dollars, etc.).

1. Find the volume of the box shown in Fig. 9-25A.
2. How many cubic feet of compressed sawdust can be packed into a 36" tall cardboard box with an 18" × 24" bottom (and top)?
3. Find the volume of the cylinder shown in Fig. 9-25B.
4. Find the volume of the triangular prism shown in Fig. 9-25C.
5. Calculate the volume of air contained in the attic of a house to assist in planning for air conditioning. The attic is shaped like a triangular prism — 6' high, 32' wide, and 52' long.
6. Calculate the volume of air contained in the room shown in Fig. 9-25D.
7. Calculate the volume that would be displaced if the sphere shown in Fig. 9-25E were submersed in water.
8. Calculate the cubic yards of gravel needed to fill a rectangular-shaped hole that is 3' deep, 9' wide, and 24' long.
9. Calculate the cubic yards of concrete needed to create the sidewalk shown in Fig. 9-25F.
10. How many ounces of liquid can be contained in a glass jar that is 5" tall and 3" in diameter? (1 gallon = 231 cubic inches)

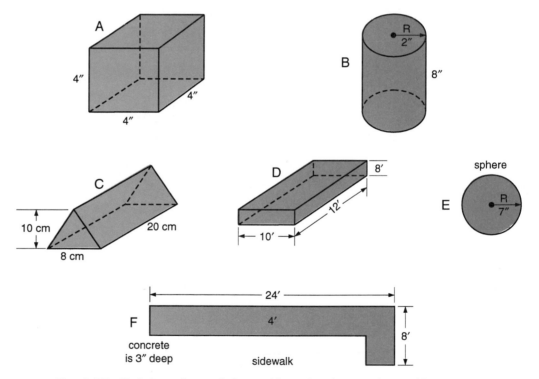

Fig. 9-25. Find the volume of these objects for the practice problems.

TEST YOUR SKILLS

Do *not* write in this book. Use a separate sheet of paper to complete the following problems. Show your work and your final answer. In so doing, write the formula used and substitute the values into the formula. Include any other steps necessary to complete the problem. Include units of measure (for example, inches, feet, etc.). Round decimal fractions to 2 decimal places.

PERIMETER

1. Find the perimeter of a rectangle with sides measuring 30 feet and 60 feet.
2. Find the perimeter of the figure shown in Fig. 8-26.

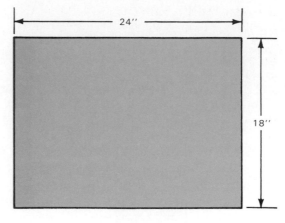

Fig. 9-26. Find perimeter.

3. Find the perimeter of a rectangle with sides of 25 feet and 120 inches.
4. Find the length of fence required to enclose a pasture having the dimensions shown in Fig. 9-27.

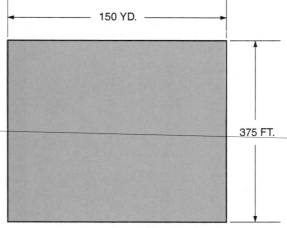

Fig. 9-27. Dimensions of pasture.

5. Fig. 9-28 depicts an 8" x 10" picture. Find the length of the wood necessary to make a frame using wood 2 inches wide.

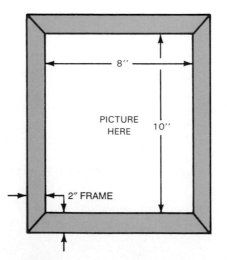

Fig. 9-28. A picture frame showing dimensions.

6. A rectangle has a perimeter of 2400 feet. One side equals 900 feet, find the lengths of the other three sides.
7. Find the perimeter of a square with sides of 15 inches.
8. Find the perimeter of the square shown in Fig. 9-29.

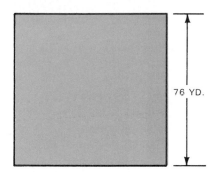

Fig. 9-29. A square.

9. A square has a perimeter of 80 meters. What is the length of each side?
10. Find the perimeter of a square with sides of 6 feet, 9 inches.
11. Find the perimeter, in *feet,* of a square with sides of 56 inches.
12. Find the perimeter, in *meters,* of a square with sides of 75 centimeters.
13. Find the perimeter of a triangle with sides of 14 inches, 20 inches, and 18 inches.
14. A triangle has a perimeter of 86 feet. One side measures 34 feet; another, 22 feet. Find the length of the third side.
15. Find the perimeter of a tire with a radius of 12 inches.
16. Find the perimeter of a circle with a radius of 75 centimeters.
17. Find the perimeter of a circle with a diameter of 18 inches.
18. A circle has a perimeter of 18 inches. Find the radius.
19. A circle has a perimeter of 30 centimeters. Find the diameter.
20. Find the perimeter of Fig. 9-30.

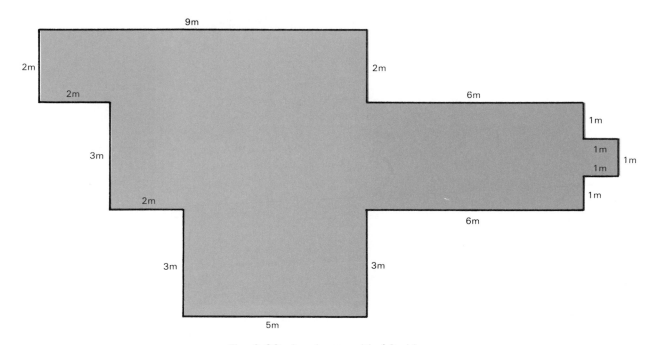

Fig. 9-30. A polygon with 16 sides.

AREA

21. Find the area of a rectangle with sides of 18 inches and 20 inches.
22. Square, 12-inch tiles are to be used to cover a floor 15 feet wide by 24 feet long. How many tiles will be needed?

23. Square, 2-foot tiles are to be used to cover a ceiling 14 feet by 20 feet. How many tiles will be needed?

24. Carpeting sold by the square yard is to be used to cover a floor. The room is 16 feet by 12 feet. How much carpeting will be needed?

25. A roll of wallpaper is 100 feet long and 24 inches wide. It will be used to cover four walls. Each wall is 8 feet high and 12 feet long. How many rolls of wallpaper must be bought?

26. A roll of wallpaper is 100 feet long and 18 inches wide. It will be used to cover two 8-foot-high and 10-foot-long walls and two 8-foot-high and 16-foot-long walls. How many rolls of wallpaper must be bought?

27. A gallon of paint covers 400 square feet. How many gallon-sized cans must be bought to paint the walls of a large room, 60 feet long and 32 feet wide? Wall height is 8 feet. (Ignore area lost due to windows, doors, etc.)

28. Find the area of a square with a side of 15 inches.

29. If a square has an area of 128 square centimeters, what is the length of its side?

30. Find the area of a triangle with a base of 25 feet and height of 14 feet.

31. Fig. 9-31 depicts the peak of a house. Find the area.

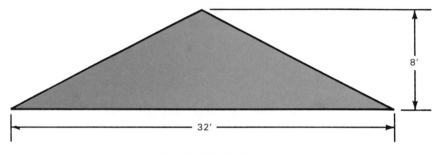

Fig. 9-31. Find area.

32. Fig. 9-32 depicts the side of a house. Find the area.

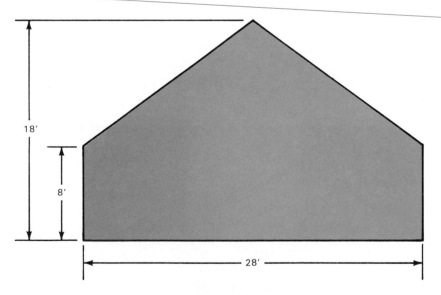

Fig. 9-32. Side view of a house.

33. Find the area of a circle with a radius of 15 inches.

34. Find the area of a circle with a diameter of 28 meters.

VOLUME

35. Find the volume of a box 12 inches wide, 14 inches long, and 8 inches deep.

36. Find the volume of a box 2 feet wide, 1 1/2 feet long, and 6 inches deep.

37. What is the volume of air that is contained in a room 14 feet by 16 feet with an 8-foot-high ceiling?
38. Find the volume of a cube with a side of 36 centimeters.
39. Find the volume of a cylinder with a radius of 32 inches and a length of 96 inches.
40. Find the maximum gallons of water a holding pond with the following dimensions can contain: 36 feet long, 12 feet wide, 9 feet deep.

PROBLEM-SOLVING ACTIVITIES

Activity 9-1.
Perimeter of a room

Objective: To measure the perimeter of a room.
Instructions:
1. Using a tape measure, determine the length of each wall of a room.
2. Add the lengths of the walls to determine the perimeter of the room.

Activity 9-2.
Circumference and diameter of a ball

Objective: To measure the circumference and diameter of a basketball.
Instructions:
1. Wrap a string around the center of the ball.
2. Measure the length of the string to determine the ball's circumference.
3. Place the ball against a wall. Push a cardboard box against the ball. Refer to Fig. 9-33.

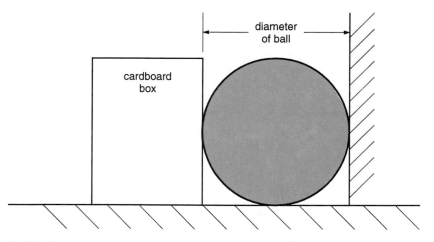

Fig. 9-33. Measure between the box and the wall to find the diameter of the ball.

4. Measure the distance between the box and the wall to determine the diameter of the ball.
5. Use the diameter to calculate the circumference of the ball. Compare your calculations to the measured value.

Activity 9-3.
Floor tiles for a room

Objective: To measure the area of a room and calculate the number floor tiles in the room.
Instructions:
1. Locate a room with 12" square floor tiles.
2. Measure the length and width of the room.
3. Calculate the area of the floor. Because the floor tiles are 12" square, the number of tiles is equal the square footage of the room.
4. Count the tiles to see how the actual number compares to your calculations.

Part III

Basic and Applied Algebra

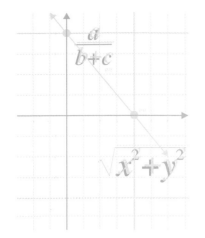

Chapter 10

Signed Numbers and Operations

OBJECTIVES

After studying this chapter, you will be able to:
- Explain basic terminology concerning signed numbers, powers, and roots.
- Determine if "signed" fractions are positive or negative.
- Perform operations of addition, subtraction, multiplication, and division using signed numbers.
- Find the solutions of numbers raised to powers.
- Find square and cube roots of numbers.
- Evaluate expressions with more than one operation by following the correct order of operations.

Signed numbers are numbers that have either a positive or negative position relative to zero. Sometimes signed numbers also indicate positive or negative *changes* in values; sometimes they give direction. The value of a signed number, without regard to its sign, is known as the **magnitude** of a number. Magnitude, in other words, is the size of the number itself. (When a signed number shows direction, it is called a **vector**. A vector is a quantity with both magnitude and direction. For example: They drove 4 miles *west*. The vector quantity is −4 miles. Magnitude is 4 miles. Direction is indicated by the negative sign. In this case, the direction of west has been assigned a negative value.)

Working with signed numbers requires a basic knowledge of the operations of addition, subtraction, multiplication, and division. It further requires a basic knowledge of powers and roots and the proper sequence of performing these operations. These will be explained in this chapter.

THE NUMBER LINE

Fig. 10-1 is a number line. **Negative numbers** are shown to the left of zero. Negative numbers must have a minus sign (−) written before them. **Positive numbers** are shown to the right of zero. They may or may not have a plus sign (+) written before them. Zero has no sign. It is neither positive nor negative. Examples of signed numbers include:
- A football team gains 20 yards: +20.
- A football team loses 20 yards: −20.
- The temperature is 70 degrees: +70.
- The temperature is 15 degrees below zero: −15.
- You lose a $3.50 bet: −3.50.
- Death Valley is 280 feet below sea level: −280.
- You gain 14 1/2 pounds: +14 1/2.

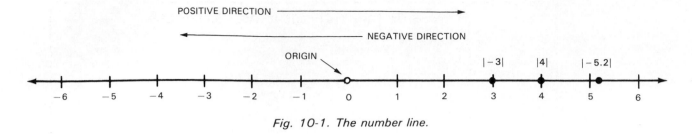

Fig. 10-1. The number line.

Absolute value

The **absolute value** is the magnitude of a number. It is the number with the sign removed. The symbol for absolute value is two straight lines (| |). Examples of absolute value, also shown in Fig. 10-1, include:

$$|-3| = 3$$
$$|+4| = 4$$
$$|-5.2| = 5.2$$

Inequalities

An **inequality** is a mathematical statement. It is similar to an equation in that two or more numbers or mathematical expressions are compared in value. It is not an equation, however, because one side of the sign does not equal the other side. On the number line, numbers to the right of a given number are *greater than* the given number. Numbers to the left are *less than* the given number. The "greater than" sign is a caret that points to the right (>). The "less than" sign is a caret that points to the left (<). An inequality is read as a sentence, from left to right.

Inequality symbols:
≠ Not equal to
> Greater than
< Less than
≥ Greater than or equal to
≤ Less than or equal to

Examples of inequalities include:
- 5 > 3:   Five is *greater than* three.
- 0 > −2:   Zero is *greater than* negative two.
- −4 < −1: Negative four is *less than* negative one.
- 3 + 2 > 0: Three plus two is *greater than* zero.

Note that the caret always points to the smaller number. On a number line, numbers to the left are less than numbers to the right, regardless of the magnitude of the number.

Positive and negative fractions

All numbers have a sign, either positive or negative, regardless of the fact that positive signs are not always written. Fractions may have a sign in front, a sign in the numerator, a sign in the denominator, or a sign with both the numerator and denominator. The combination of signs determines if the fraction is positive or negative. As examples:

$$-\frac{4}{7} \quad +\frac{5}{9} \quad \frac{-2}{5} \quad \frac{+3}{4} \quad \frac{1}{-2} \quad \frac{6}{+3} \quad \frac{-6}{+3}$$

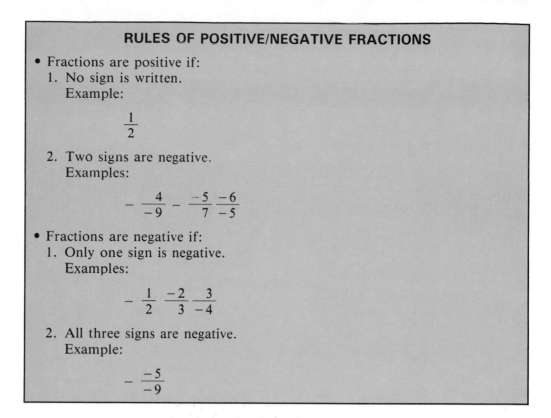

RULES OF POSITIVE/NEGATIVE FRACTIONS

- Fractions are positive if:
  1. No sign is written.
     Example:

$$\frac{1}{2}$$

  2. Two signs are negative.
     Examples:

$$-\frac{4}{-9} \quad -\frac{-5}{7} \quad \frac{-6}{-5}$$

- Fractions are negative if:
  1. Only one sign is negative.
     Examples:

$$-\frac{1}{2} \quad \frac{-2}{3} \quad \frac{3}{-4}$$

  2. All three signs are negative.
     Example:

$$-\frac{-5}{-9}$$

Practice Problems

Complete the following problems on a separate sheet of paper.

Show each of the following as a signed number. It is not necessary to include units.
1. Interest of $10 is posted to a savings account.
2. A $600 check is written from a check book to pay the rent.
3. The outside temperature, including wind chill factor, is 35 degrees below zero.
4. The inventory is short 3 parts.
5. Air pressure in a car's tires is set at 32 pounds per square inch (psi).

Give the absolute value of each of the following numbers.
6. $|32|$
7. $|-15|$
8. $|-1|$
9. $|0|$
10. $|-5|$

Give the correct inequality sign for the following inequalities.
11. 4 _____ 9
12. -5 _____ -8
13. 0 _____ -6
14. 9 _____ 0
15. 6 _____ -6
16. $6 - 3$ _____ $4 - 9$
17. On a 25 question test: 3 wrong answers _____ a grade of 75%
18. Area of a floor: 14' × 10' _____ 12' × 12'
19. Liquid measure: 1 gallon _____ 5 quarts
20. Weight: 125 pounds _____ 60 kilograms

Determine if the following fractions are positive or negative.

21. $-\dfrac{1}{3}$

22. $\dfrac{-2}{5}$

23. $-\dfrac{3}{-5}$

24. $-\dfrac{-5}{7}$

25. $\dfrac{1}{2}$

26. $\dfrac{-4}{5}$

27. $\dfrac{6}{-9}$

28. $\dfrac{-7}{-8}$

29. $-\dfrac{-1}{2}$

30. $-\dfrac{8}{-8}$

ADDITION AND SUBTRACTION OF SIGNED NUMBERS

Until now, you have seen that a number always increased in size with addition and decreased in size with subtraction. This is not always the case, as will be seen from working with signed numbers. In performing addition or subtraction of signed numbers, follow the applicable rules set forth below.

RULES OF ADDITION OF SIGNED NUMBERS

- If the signs of the two numbers being added are the *same*, *add* the numbers and *keep the sign*.
  Examples:
  $$7 + 3 = 10$$
  $$-7 + (-3) = -10$$
- If the signs of the numbers being added are *different, subtract* the smaller number from the larger number and *take the sign of the larger number*.
  Examples:
  $$-7 + 3 = -4$$
  $$7 + (-3) = 4$$
- No sign if answer is zero.
  Examples:
  $$-5 + 5 = 0$$
  $$5 + (-5) = 0$$

RULES OF SUBTRACTION OF SIGNED NUMBERS

When subtracting two numbers, change the sign of the number being subtracted, and change the subtraction sign to addition. Then, follow the rules of addition. Example:

Subtraction	Change signs	Answer
$4 - 7$	$4 + (-7)$	-3
$8 - (-3)$	$8 + (+3)$	11
$-9 - 5$	$-9 + (-5)$	-14
$-7 - (-3)$	$-7 + (+3)$	-4

After working with signed numbers for a while, you may become proficient at adding and subtracting without actually going through the complete process given in the rules of addition (or subtraction) of signed numbers. You may be taking a "shortcut approach" to working these problems on the basis that when adding or subtracting:
- Two adjacent positive signs simplify to one positive.
- Opposite and adjacent signs simplify to a negative.
- Two adjacent negatives simplify to one positive.
- Once simplified, subtraction yields an answer toward the negative direction along a number line; addition, toward the positive direction.

For example:

Problem	Simplifies to	Answer
$6 + (+2)$	$6 + 2$	8
$-6 + (-2)$	$-6 - 2$	-8
$6 - (+2)$	$6 - 2$	4
$-6 - (-2)$	$-6 + 2$	-4

Arriving at the answer becomes a one-step process through experience in working with these numbers and understanding addition and subtraction on a number line.

Note: Notice that parentheses have been used with addition and subtraction of signed numbers to separate the addition or subtraction sign from the positive or negative sign. In math, parentheses [( )] are used to give the order of performing operations, to show multiplication, or to separate numbers. Here, the parentheses do not change the order of working the problem nor do they mean to multiply.

Using the number line to add and subtract

Employing a number line is a good way of showing what happens when adding or subtracting signed numbers. Fig. 10-2 provides several examples. Notice that after the operations are simplified, a subtraction operation moves in the negative direction and an addition operation moves in the positive direction.

Examples:
- Fig. 10-2A. $-2 + (-3) = -2 - 3 = -5$
- Fig. 10-2B. $2 + 3 = 5$
- Fig. 10-2C. $5 + (-10) = 5 - 10 = -5$
- Fig. 10-2D. $-6 - (-4) = -6 + 4 = -2$
- Fig. 10-2E. $2 - 6 = -4$

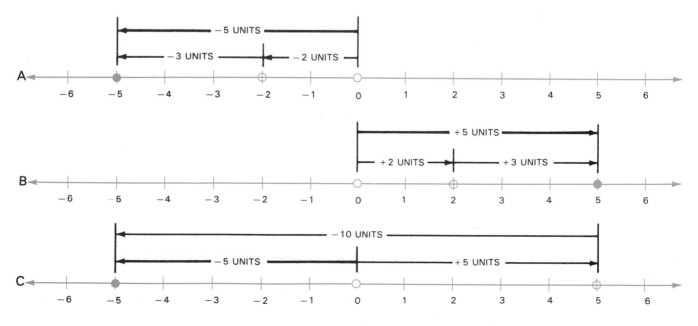

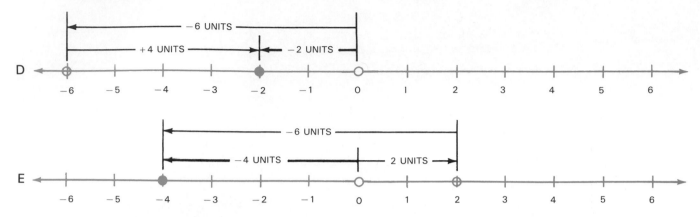

Fig. 10-2. Adding/subtracting signed numbers on a number line. Note that the resultant arrow (heavy line) always points in the direction of the longest arrow.

Arithmetic with several numbers

There are a couple of ways to approach a problem that contains more than two numbers. One way is to perform the arithmetic operation on two numbers at a time. For example, if several numbers are to be added together, the first two numbers are added, and that result is added to the next number. This process continues until all of the numbers have been added. For example, add $5 + (-3) - 6 + 7$. Proceed as follows:

$$5 + -3 = 2$$
$$2 - 6 = -4$$
$$-4 + 7 = 3$$

The answer is 3.

The other way to approach a problem that has more than two numbers is to simplify the problem by adding all of the positive numbers and all of the negative numbers (absolute values) in two separate steps. Then, subtract them and keep the sign of the larger number. For example, add $5 + (-3) - 6 + 7$. Proceed as follows:

$$5 + (-3) - 6 + 7 = (5 + 7) - (3 + 6) = 12 - 9 = 3$$

Note: Addition and subtraction can be performed in any order, provided there is no multiplication or division in the problem. However, the operations in parentheses are performed first. Order of performing operations will be discussed shortly.

Practice Problems

Complete the following problems on a separate sheet of paper.

1. $2 - 4$
2. $8 - 3$
3. $6 + -3$
4. $9 + -15$
5. $-5 + 2$
6. $-7 + 9$
7. $-3 - 7$
8. $-8 - 4$
9. $-21 - -3$
10. $-2 - -5$

11. $\dfrac{1}{2} - \dfrac{1}{3}$

12. $\dfrac{1}{4} - \dfrac{3}{8}$

13. $\dfrac{2}{3} - \dfrac{1-5}{6}$

14. $\dfrac{1}{6} - \dfrac{1}{-9}$

15. $\dfrac{4}{5} + \dfrac{-3}{10}$

16. $\dfrac{-6}{7} + \dfrac{3}{-7}$

17. $\dfrac{-1}{4} + \dfrac{-1}{-8}$

18. $\dfrac{-2}{-5} - \dfrac{-3}{-4}$

19. $\dfrac{3}{5} + \dfrac{2}{3}$

20. $\dfrac{-2}{6} + \dfrac{1}{-3}$

21. $0.4 + 1.2 - 2.4$
22. $-4.1 - -6.8 + -9 - 22$
23. $-36 - -7.2 + -12 - 6 - -5.8$
24. $-3 - 0 + 8.1 + -4.2 + 1 - 6.8 - -3 + -5$
25. $0.5 + 0.5 - 1.0 + 0.5 - 2.5 - -0.5 - 2.0$

MULTIPLICATION AND DIVISION OF SIGNED NUMBERS

The rules for multiplication/division are different from those for addition and subtraction. In performing multiplication/division of signed numbers, follow the applicable rules set forth below.

RULES OF MULTIPLICATION/DIVISION OF SIGNED NUMBERS

- Disregard the signs and multiply or divide the numbers as required.
- If the numbers being multiplied or divided have the same sign, the result is positive. (An even number of negative signs results in a positive answer.) Examples:

$$(5)(3) = 15$$
$$-6(-4) = 24$$
$$9 \div 3 = 3$$
$$-12 \div (-2) = 6$$

- If the numbers being multiplied or divided have opposite signs, the result is negative. (An odd number of negative signs results in a negative answer.) Examples:

$$(-2)4 = -8$$
$$(3)(-7) = -21$$
$$-18 \div 3 = -6$$
$$25 \div (-5) = -5$$

Note: Notice that multiplication in these examples is signified by adjacent numbers enclosed in one or two sets of parentheses. The signed numbers in the addition and subtraction problems were also enclosed in parentheses; however, their purpose was to separate positive and negative signs from addition and subtraction signs. When using parentheses with multiplication, enclosing the second signed number will make clear that the operation to be performed is multiplication and not addition or subtraction.

Practice Problems

Perform the indicated multiplication and division operations on a separate sheet of paper.

1. 4×-1
2. -6×2
3. -3×-2
4. 9×2
5. $-4 \times -2 \times -1 \times 2$
6. $-2 \times 3 \times -2 \times 2 \times -1$
7. $(3)(-3)$
8. $-2(-5)(2)$
9. $3.1(-1.2)$
10. $(-0.4)(-0.25)$
11. $\dfrac{1}{2} \times \dfrac{2}{3}$
12. $-\dfrac{3}{4} \times \dfrac{-1}{5}$
13. $\dfrac{-7}{9} \times \dfrac{-3}{-4}$
14. $\dfrac{-3}{-5} \times \dfrac{-10}{-9} \times \dfrac{-1}{-9} \times -1$

Perform the indicated division.

15. $\dfrac{-18}{6}$ 16. $\dfrac{60}{-10}$

17. $\dfrac{-12}{-2}$ 18. $-\dfrac{-16}{-32}$

19. $\dfrac{-81}{9}$ 20. $\dfrac{24}{-3}$

21. $\dfrac{-36}{-4}$ 22. $\dfrac{-8}{64}$

23. $6 \div 3$ 24. $-4 \div 2$

25. $-2 \div -8$ 26. $9 \div -27$

27. $-3\,\overline{)9}$ 28. $6\,\overline{)-12}$

29. $18\,\overline{)-6}$ 30. $-20\,\overline{)-5}$

APPLICATIONS OF SIGNED NUMBERS

Applications of mathematics are frequently found in the form of word problems. Almost any word problem using arithmetic can be modified to use positive and negative numbers. Rules presented in Chapter 1 apply here when solving problems with signed numbers. In addition to these, follow the rules for solving word problems of signed numbers.

RULES FOR SOLVING WORD PROBLEMS OF SIGNED NUMBERS

1. Determine if the numbers are positive or negative or designate *positive* and *negative*. Give designated sign to each number.
2. Determine what type of math will be needed.
3. Perform the math using signed numbers. Include the units.

Sample Problem 10-1.

A person drives a car north 8 miles to pick up a friend. From there, they ride south 13 miles on the same highway to pick up another friend. Together, they ride south another 4 miles to the shopping mall. How far is the mall from the driver's home?

Step 1. This is a vector application. Designate positive and negative. Assign "+" to either north or south. Assign "−" to the other direction. Give designated sign to each number.

        Assign: north = +
        Assign: south = −
        +8 miles (north)
        −13 miles (south)
        −4 miles (south)

Step 2. Determine what type of math will be needed. In this case, add the numbers. Use the rules of signed numbers.

        $+8 + (-13) + (-4)$

Step 3. Perform the math. Include the units.

        $+8 + (-13) + (-4) = -9$ miles

Answer: 9 miles south

Note: Notice that in the final answer, -9 miles was changed to 9 miles south because the "−" was assigned the direction of south. Unless the designation of negative as being "east" or "west" is given, a final answer of -9 miles would be somewhat meaningless.

Since the question did not ask for a specific direction, the answer would be correct if it only gave distance. Distance only is given by the absolute value.

        $|-9|$ miles = 9 miles

Sample Problem 10-2.

With a single check, a bank customer overdraws a checking account by three times the amount of the balance. The balance was $500. What was the amount of the check?

Step 1. Determine if positive or negative. Give sign to each number.

old balance $= +\$500$

new balance $= -(+\$500 \times 3) = -\$1500$ (overdraft)

Step 2. Determine what type of math will be needed. This problem is looking for a difference. Subtract old balance from new balance.

$-1500 - (+500)$

Step 3. Perform the math. Include the units.

$-1500 - (+500) = -1500 - 500 = -\$2000$

Answer: $2000 check

Note: In the final answer, the amount of the check was given as $2000, not $-\$2000$. In terms of bookkeeping, a check would be deducted from a balance; therefore, this check would represent $-\$2000$. The amount written on the check, however, would be $2000.

Sample Problem 10-3.

A student weighed 210 pounds at one time but lost 25 pounds. What does the student weigh?

Step 1. Determine if positive or negative. Give sign to each number.

old weight $= +210$ lb.

weight loss $= -25$ lb.

Step 2. Determine what type of math will be needed. This problem is looking for a difference. Subtract amount lost from old weight.

$210 - 25$

Step 3. Perform the math. Include the units.

$210 - 25 = 185$ lb.

Practice Problems

Complete the following problems on a separate sheet of paper. Include units of measure (inches, feet, minutes, dollars, etc.).

1. A credit card balance is $450 at the end of January. During February, the card holder charges $5.50; pays $250; and then charges $45.50, $125, and $375. What is the final balance?
2. An inventory clerk in a supermarket counted 250 boxes of cereal before the store opened for the day. Customers removed 52 boxes before the stock clerk placed 10 cases, each containing 24 boxes, on the shelf. At the end of the day, 375 boxes remained on the shelf. How many were purchased?
3. An individual weighed 200 pounds before going on a crash diet. Twenty-five pounds were lost during the first month, and 15 pounds were lost during the second month. Unfortunately, 10 pounds were gained back during the third month. What was the person's weight at the end of the third month?
4. The daily high temperatures for the first week in January were 35°F, 15°F, −6°F, −10°F, 0°F, 3°F, and −4°F. What was the average high temperature for the week?
5. A pre-hung 36" wide door comes with a 3/4" frame on the top and both sides. To frame the rough opening, the carpenter adds 3/4" to each side of the frame width. How wide is the rough opening?

POWERS AND ROOTS

In preceding chapters, powers and roots were introduced. In these chapters, however, numbers were always positive. Now that you know how to multiply signed numbers, you are prepared for a more in-depth look at powers and roots. Oftentimes, a number is multiplied by itself a number of times. In math, the term **power** refers to the number of times the *factor* (number being multiplied) is repeated. It also refers to the product of such a multiplication. The **root** of a number is given by the number, or factor, that is multiplied by itself to give the power. Just as division is the *inverse* (reverse operation) of multiplication, finding the root of a number is the inverse of finding a power.

In application, you used powers to find area and volume. In finding area of a square, for instance, the length of a side was repeated twice, or **squared**. In other words, the factor (length of one side, in this case) was repeated by a *power* of 2. In finding volume of a cube, the factor was repeated 3 times, or **cubed**. In other words, the factor was repeated by a *power* of 3. In application, you used roots to find the length of a side when the area of a square was known. To do this, you took the **square root** of the area. You would also use roots to find the length of a side of a cube if you knew its volume. To do this, you would take the **cube root** of the volume.

Powers of numbers

As just stated, the power of a number states how many times a factor is repeated. The power is written as a *superscript*, a small number written slightly above and to the right of the factor. The factor is referred to as the **base**. The superscript is referred to as the **exponent**. The exponent gives the power. See Fig. 10-3 for examples. When working with powers, follow the rules of powers.

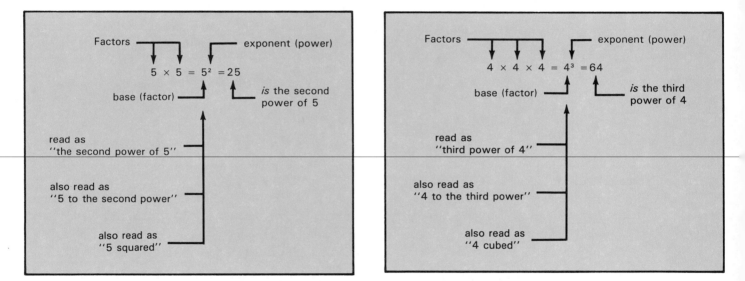

Fig. 10-3. Example of powers.

RULES OF POWERS

- Repeat the factor the number of times indicated by the exponent. Multiplication is the operation.
  Examples:
  $$6^2 = 6 \times 6 = 36$$
  $$1^3 = 1 \times 1 \times 1 = 1$$
- Negative numbers raised to a power follow the rules of multiplication with signed numbers.
- To determine if a negative number raised to a power will have a negative or positive answer:
  1. If the exponent is *odd,* the solution is *negative.*
  2. If the exponent is *even,* the solution is *positive.*
  Remember, multiplying an odd number of negative signs results in a negative answer; multiplying an even number of negative signs results in a positive answer.

Examples:
$$(-2)^3 = -2 \times -2 \times -2 = -8$$
$$(-3)^2 = -3 \times -3 = 9$$
$$(-2)^4 = -2 \times -2 \times -2 \times -2 = 16$$

Note: The exponent applies to everything within the parentheses. If not enclosed in parentheses, it only applies to the number (or symbol) immediately preceding it. A negative sign, therefore, would be excluded. This is especially important where even powers are concerned.

Examples:
$$-3^2 = -(3^2) = -(3 \times 3) = -9$$
$$-2^4 = -(2^4) = -(2 \times 2 \times 2 \times 2) = -16$$
$$-2^3 = -(2^3) = -(2 \times 2 \times 2) = -8$$

- Any number, except zero, raised to the zero power is 1.

Examples:
$$6^0 = 1$$
$$3.25^0 = 1$$
$$(-5)^0 = 1$$

- Zero cannot be raised to the zero power. The answer is undefined.
- Any number raised to the 1 power is the number itself.

Examples:
$$7^1 = 7$$
$$(-3)^1 = -3$$
$$0^1 = 0$$

- A number, or base, with a negative exponent can be written as a fraction. The numerator, in this case, is 1, and the denominator is the base with a *positive* exponent.

Examples:
$$5^{-2} = \frac{1}{5^2} = \frac{1}{25}$$
$$(-5)^{-2} = \frac{1}{(-5)^2} = \frac{1}{25}$$
$$(-2)^{-3} = \frac{1}{(-2)^3} = \frac{1}{-8}$$

- An exponent written as a fraction may be converted to a "root" form. See Rules of Roots.

Roots of numbers

As mentioned before, the root is the number that is multiplied by itself to arrive at the power. The square root and the cube root are the two roots that people are generally most familiar with. The symbol for the root of a number is a **radical sign** ($\sqrt{\ }$). Under the radical sign is the number or expression whose root is to be found. When in this form, they are called **radicals.** As examples:

$\sqrt{25}$ is read as "the square root of 25," "the second root of 25," or "radical 25."

$\sqrt[6]{128}$ is read as "the sixth root of 128."

$\sqrt[3]{8}$ is read as "the cube root of 8," or "the third root of 8."

When working with roots, follow the rules of roots.

RULES OF ROOTS

- The small number written outside of the radical sign states the root to be found. The number, called the index, must be a whole number, 2 or greater. The index is a positive number. No index means square root.

- Radicals may be written as powers, where the exponents are fractions. The index becomes the denominator. The power of the number under the radical sign becomes the numerator. The number under the radical sign becomes the base. Examples:

$$\sqrt{25} = (25)^{1/2}$$
$$\sqrt[3]{64} = (64)^{1/3}$$
$$\sqrt[3]{8^2} = (8)^{2/3}$$

- To find the square root of a number by inspection, determine what number can be multiplied by itself to equal the number inside the bracket. Examples:

$$\sqrt{25} = 5 \ (5 \times 5 = 25)$$
$$\sqrt{81} = 9 \ (9 \times 9 = 81)$$
$$\sqrt{144} = 12 \ (12 \times 12 = 144)$$
$$\sqrt{36^2} = 36 \ (36 \times 36 = 36^2)$$
$$\sqrt{0.16} = 0.4 \ (0.4 \times 0.4 = 0.16)$$

Note: The square root of a negative number is an *imaginary number* given by $i = \sqrt{-1}$. Imaginary numbers will not be discussed in this book except to give a brief example: $\sqrt{-4} = \sqrt{-1} \cdot \sqrt{4} = i\sqrt{4} = 2i$. In the system of *real numbers,* the system we deal with in this book, it is not possible to take the square root of a negative number.

- The square root of a number can be either positive or negative. (Keep in mind that two negative numbers multiplied together result in a positive answer.) To indicate a positive or negative answer, use this notation: ± (plus or minus). Examples:

$$\sqrt{64} = \pm 8 \qquad (8 \times 8 = 64) \quad (8 \times -8 = 64)$$
$$\sqrt{80^2} = \pm 80 \qquad (80 \times 80 = 80^2) \quad (-80 \times -80 = 80^2)$$
$$\sqrt{0.25} = \pm 0.5 \qquad (0.5 \times 0.5 = 0.25) \quad -0.5 \times (-0.5 = 0.25)$$

Note: The positive square root is called the *principal square root.* For the most part, we will be more concerned with principal square roots.

- To find the cube root of a number by inspection, determine what factor can be repeated three times to equal the number inside of the radical sign.
Note: It is possible to find the cube root of a negative number. Examples:

$$\sqrt[3]{216} = 6 \ (6 \times 6 \times 6 = 216)$$
$$\sqrt[3]{-125} = -5 \ (-5 \times -5 \times -5 = -125)$$
$$\sqrt[3]{7^3} = 7 \ (7 \times 7 \times 7 = 7^3)$$

Estimating square roots

With calculus or the use of a calculator, the square root of *any* number can be found. It is desirable, however, to be able to approximate the answer to difficult square roots, without using calculus or a calculator. In the previous examples, the numbers selected were simple and obvious. They were also **perfect squares**. The square root of a perfect square is an exact number. It is a number in which its square root leaves no remainder. Examples of perfect squares include 1, 4, 9, 4/9, and 0.25.

So far, you have found square roots by inspection. It is possible to approximate square roots by trial and error. Start with a number that you think might be close. Multiply the number by itself to determine if it is too small or too large. Try the next higher or lower number to find the numbers between which the square root lies. Repeat until you have reached the accuracy desired.

Sample Problem 10-4.

Find: $\sqrt{30}$. (Approximate to two significant figures.)

Step 1. Start with 5.

$5 \times 5 = 25$ (too low)

Step 2. Try 6.

$6 \times 6 = 36$ (too high)

Step 3. Answer is between 5 and 6. Try 5.5.
    $5.5 \times 5.5 = 30.25$ (too high)
Step 4. Try 5.4.
    $5.4 \times 5.4 = 29.16$ (too low)
Step 5. The approximate square root is between 5.4 and 5.5. The answer is the factor whose product is closest. Therefore, 5.5 is the answer.

Practice Problems

Complete the following problems on a separate sheet of paper. Show your work and your final answer.

1. 3^2	2. (-4^2)	3. -5^2	4. $(-2)^5$
5. 1^4	6. $(-3)^3$	7. -5^0	8. -4^1
9. 2^{-2}	10. -1^{-3}	11. 3^{-1}	12. -2^{-4}
13. $\sqrt{9}$	14. $\sqrt{81}$	15. $\sqrt[3]{27}$	16. $\sqrt[4]{16}$
17. $0.16^{1/2}$	18. $0.25^{1/2}$	19. $-8^{1/3}$	20. $25^{-1/2}$

21. A square room is measured for wood flooring. The area of the room is 196 square feet. What length should the boards be if they are to be installed from one end of the room to the other without making a joint?

In the following problems, approximate the square root to 2 significant figures.

22. $\sqrt{18}$ 23. $\sqrt{0.5}$ 24. $\sqrt{6}$ 25. $\sqrt{0.8}$

ORDER OF OPERATIONS

Math operations must be performed in the proper order. Failure to do so can lead to a wrong answer. Therefore, it is necessary to have rules that state the order in which to perform the operations.

Symbols of grouping

Symbols are used to group numbers and math operations. The primary purpose of grouping symbols is to determine the order in which to perform operations. Grouping symbols, listed by level of priority, include:
- Parentheses: ( )—innermost level
- Brackets: [ ]—second level
- Braces: { }—third level

Other grouping symbols include:
- Absolute value: | |
- Fraction bars: −
- Radical sign: $\sqrt{}$

Systematic removal of parentheses and brackets is performed to accomplish one step of what is referred to as simplification. (The other step is to combine like terms. You will learn more about this as you study algebra.) In simplifying, operations are continued until all of the math that can be, is performed. At this point, a problem is in its simplest form. Many problems get quite complicated. Simplification can ease the task of finding solutions. Once a problem is simplified, you are ready to find the answer. Sometimes, the process of simplifying will lead you to a direct answer, as you will see from the rules of removing grouping symbols.

RULES OF GROUPING SYMBOLS

- Remove the parentheses by performing the operation *inside* the symbols.
  Examples:
    $(-8 + 3) = -5$
    $(5 - 9) + (6 - 3) = -4 + 3 = -1$
    $(5 - 6 - 4) = -5$
    $(3 + 1) - (2 - 7) = 4 - (-5) = 4 + 5 = 9$

- If a problem contains more than one set of parentheses, remove the *innermost* first.
  Examples:
  $$2 - [5 - (3 + 1)] = 2 - [5 - 4] = 2 - 1 = 1$$
  $$[(5 + 2 - 1) + (9 - 4)] = [6 + 5] = 11$$
- A number written outside of and next to the parentheses indicates multiplication.
  Examples:
  $$4(-5) = -20$$
  $$4[(6 - 3) + 2(-3)] = 4[3 + (-6)] = 4[-3] = -12$$
- A negative sign written outside of the parentheses changes the solution inside to the opposite sign.
  Examples:
  $$-(5 + 8) = -13$$
  $$-(5 - 8) = 3$$
  $$-6 - (4 - 6) = -6 - (-2) = -6 + 2 = -4$$
- A negative sign written outside of parentheses can be removed by changing the signs inside. Note: If two signs are written together, only change the arithmetic sign.
  Examples:
  $$-[7 - 2 + (-6)] = -7 + 2 - (-6) = 1$$
  $$-(5 + 8) = -5 - 8 = -13$$
- Parentheses used to separate a negative sign from an arithmetic sign do not need to be removed until they are no longer needed for clarity.
  Examples:
  $$-[5 - 3 + (-4)] = -5 + 3 - (-4) = 2$$
- Fraction bars are removed after expressions above and below are simplified.
  Examples:
  $$2\left[\frac{4(-5)}{(8 + 2)} - 2\right] = 2\left[\frac{-20}{10} - 2\right] = 2[-2 - 2] = 2[-4] = -8$$
  $$\frac{2[5 + (2 + 3)]}{(8 + 2)} = \frac{2[5 + 5]}{10} = \frac{2[10]}{10} = \frac{20}{10} = 2$$
- Radical signs and absolute value symbols are removed after everything inside is simplified.
  Examples:
  $$3 + \sqrt{3 + (4 + 2)} = 3 + \sqrt{3 + 6} = 3 + \sqrt{9} = 3 + 3 = 6$$
  $$-[4 + |3 - 5|] = -[4 + |-2|] = -[4 + 2] = -[6] = -6$$

RULES OF ORDER OF OPERATIONS

Arithmetic operations *must* be performed in this order:
1. Within grouping symbols, starting with the innermost parentheses and working outward.
2. Powers and roots in any order.
3. Multiplication and division in order from left to right.
4. Addition and subtraction in order from left to right.

Sample Problem 10-5.

What is $3 + 4 \times 2$?

Step 1. Perform multiplication.
$$4 \times 2 = 8$$
Step 2. Perform addition.
$$3 + 8 = 11$$
The correct answer is 11.

Note: If addition was performed first, then, multiplication, the answer would be wrong:
$$3 + 4 = 7 \text{ and } 7 \times 2 = 14$$

Sample Problem 10-5.

What is $\dfrac{2 + 3 \times 5^2}{11}$?

Step 1. Determine power.
$$5^2 = 25$$
Step 2. Perform multiplication.
$$3 \times 25 = 75$$
Step 3. Perform addition.
$$2 + 75 = 77$$
Step 4. Expressions above and below are simplified. Therefore, remove the fraction bar.
$$\frac{77}{11} = 7$$

Practice Problems

Find the solution to the following problems by performing the indicated operations in the proper order.

1. $(3 + 6)2 - 6$
2. $10 - 3(-4 - 2) + 8$
3. $24 \div 6 + 2(5 - 8)$
4. $4 - 3^2 - (3 - 2^2 + 1)$
5. $|-25| + 4(-3) - [5(2 - 1)]$
6. $\dfrac{2(-3 + -5 - -7)}{3 + 4(3 - 9^{1/2})} \times \dfrac{6(21 + -17 - 5)}{4(2 - 3) + 2(-3)}$
7. $\dfrac{-1(9 + -6/3 + 1}{5 + 1[9 - 3(2 + -1) + 7]} \div \dfrac{(4 - 6)^2}{21 - (5)(-2)(-1)}$
8. $\dfrac{[4 + 1(3 + 5) - 1]^0}{2(6 + -5) + 10 \div -5} + \dfrac{-35 + 3(-4 - -5) \div -3}{12(3 + -3 - -6) \div 5 + 1}$
9. $\dfrac{\sqrt{36} - 6 + 2}{(6 - 8)(3 + 5) + 10} - \dfrac{18 - (9 + 8) + 6^0}{8(-1 - 3) + -2(-7 - 8)}$
10. $\left(\dfrac{1}{2} + \dfrac{-3}{4} - \dfrac{3}{-16}\right) \div \left(\dfrac{3}{4} + \dfrac{-5}{8}\right)$

TEST YOUR SKILLS

Do *not* write in this book. Use a separate sheet of paper to complete the following problems. Show your work and your final answer.

THE NUMBER LINE

Show each of the following figures as signed numbers. It is not necessary to show units.
1. A $50 deposit in a checkbook.
2. A $68 electric bill.
3. A room temperature of 20 degrees Celsius.
4. A 10 yard penalty in football.
5. A weight gain of 20 pounds.

Give the absolute value of the following:
6. $|-8|$
7. $|14|$
8. $|5 - 6|$
9. $|2 + 8|$
10. $|9 - 4|$

Insert the current inequality sign in the blank.
11. -15_____6
12. 10_____-2
13. 0_____9
14. $5 - 8$_____$7 + 1$
15. $-7 + 6$_____$0 - 3$

Positive and Negative Fractions

Determine if these fractions are positive (+) or negative (−).

16. $-\dfrac{1}{2}$

17. $-\dfrac{-3}{5}$

18. $-\dfrac{5}{-9}$

19. $-\dfrac{-4}{-7}$

20. $\dfrac{8}{5}$

21. $\dfrac{-6}{7}$

22. $\dfrac{1}{-4}$

23. $\dfrac{-2}{-5}$

24. $+\dfrac{+8}{-9}$

25. $+\dfrac{-7}{-2}$

ADDITION AND SUBTRACTION OF SIGNED NUMBERS

Perform the indicated operations.

26. $3 - 6$

27. $8 + 4$

28. $-5 + 9$

29. $-4 - 2$

30. $-7 - -5$

31. $12 - -6$

32. $-1 + -4$

33. $22 - 36$

34. $\dfrac{1}{2} + \dfrac{2}{3}$

35. $\dfrac{3}{8} - \dfrac{5}{8}$

36. $-\dfrac{1}{2} + \dfrac{5}{8}$

37. $\dfrac{6}{7} - \dfrac{5}{14}$

38. $-\dfrac{1}{6} - \dfrac{7}{12}$

39. $-\dfrac{3}{4} - \dfrac{2}{3}$

40. $0.7 + 1.3 - 4.05 - 5.6 + (-21.9)$

41. $2.01 - 4.06 + 6.15 + (-0.09)$

42. $-3 + 0 - 9 + 6 - (-2)$

43. $(-5) + (-8) - 13 + 6$

44. $0 - 3 - 7 + -6 - -5$

45. $9 - 0 + 3 - (-5)$

MULTIPLICATION AND DIVISION OF SIGNED NUMBERS

Perform the indicated multiplication.

46. 3×5

47. -4×6

48. 2.1×-1.5

49. -7.0×-9.3

50. $1.4(-4.02)$

51. $(-2.3)6.90$

52. $(0)(8)$

53. $(-10)(-5)$

54. $-\dfrac{2}{3} \times \dfrac{3}{4}$

55. $-\dfrac{1}{3} \times -\dfrac{5}{6}$

Perform the indicated division.

56. $\dfrac{24}{12}$

57. $-\dfrac{36}{6}$

58. $-\dfrac{-10}{2}$

59. $\dfrac{14}{-7}$

60. $-\dfrac{-15}{-5}$

61. $\dfrac{-21}{-7}$

62. $-4\,\overline{)12}$

63. $-2\,\overline{)-15}$

64. $-48 \div 8$

65. $30 \div -4$

APPLICATIONS OF SIGNED NUMBERS

Solve the following word problems.

66. A person leaves from home and drives 5 miles east to pick up a friend. They drive west on the same highway 8 miles to pick up a second friend. Then, they drive 2 miles east to the store. How far is the store from the driver's home?

67. What is the ending balance of a checkbook with the following transactions?

starting balance:	$250.50
check:	23.09
deposit:	125.00
check:	109.02
check:	1.45
check:	75.65
check:	21.75
deposit:	225.60

68. The temperature recorded at noon each day for one week was: -15, -10, 3, 8, -2, $+1$, and 0 degrees. What was the average temperature at noon for the week?

69. Five, 15-inch pieces are cut from an 8-foot-long board. The remaining piece is 27 inches too short for the next application. What is the correct length for the application?

70. Three boxes are each found to be short 5 books. Afterwards, an additional 6 books per box are removed. If there are a total of 9 books remaining, how many were in each box when they were full? (When full, each box holds the same amount.)

71. In a 1-week period, the highest recorded temperature in Chicago was $18°$, and the range was $30°$. What was the lowest temperature in that period?

72. In a 1-week period, a small business took in $1500. The expenses for that week came to $1750. What was the net income for that week?

73. A scuba diver descends to a depth of 52 feet, then, ascends 23 feet. At what depth is the scuba diver after ascending?

74. A mechanic buys a car to fix up and pays $320 for it. After sinking twice that amount into it for new parts, it is sold for $900. What is the net profit (or loss)?

75. An amount of $9000 is paid to take over a business. Debts of $6000 are then paid off. At the end of a year, the store records operating expenses of twice the amount that was used to pay off the debt. How much money must the store gross to break even?

POWERS AND ROOTS

Find the solution.

76. 6^2

77. $(-8)^2$

78. 3^3

79. $\sqrt{25}$

80. $\sqrt{49}$

81. $\sqrt[3]{-8}$

Approximate the square root to 2 significant figures.

82. $\sqrt{20}$

83. $\sqrt{40}$

84. $\sqrt{0.2}$

ORDER OF OPERATIONS

Find the solution.

85. $[4 + (3 - 5)]$

86. $2 + [-3 - (0 + -3) + 1]$

87. $-(-6 - -2)$

88. $7(-5) - [32 + 9 - 4(8)]$

89. $(-4)(-2) - (-2 + 3) - 5[2 - (3 - 7)]$

90. $6 + 5(2 - 4)$

91. $-5 - 2[3 - 4(-4 + 1)]$

13. $(4)(5)(-3)x + 1x^2 - x^3$
14. $5a^4 - (2)(-1)(-6)ab$
15. $0 + xyz + -1x - -2y$
16. $3y - -4x + -1xyz + 8(9)$
17. $4^2 + x^2 - x^0 + 18(-2)$
18. $a^2b^2 + 4ab - 5^2 + 12(0)$
19. $x^0 + y^1 - xy^2 - 4xyz^3$
20. $6^1 + 5x^0 - (5)(3)xy^4 - 0$

ADDITION AND SUBTRACTION WITH MONOMIALS

Monomials may actually be added or subtracted if they contain like terms. (Otherwise, addition or subtraction may only be *indicated*. Values must then be substituted for literal numbers before arithmetic can be performed.) In this way, an algebraic expression with like terms can be simplified. Follow the rules of addition/subtraction of monomials.

RULES OF ADDITION/SUBTRACTION OF MONOMIALS

- Add/subtract the coefficients of like terms following the rules of signed numbers. Keep in mind, if a coefficient is not written, it is 1.
- Addition/subtraction cannot be performed with unlike terms—it may only be indicated.

Examples:
$$6b + 4b = 10b$$
$$12x - 5x = 7x$$
$$-8xy + 4xy = -4xy$$
$$3x^2 - x^2 = 2x^2$$
$$4a^2 - a^3 + 5a^2 = 9a^2 - a^3$$

Practice Problems

On a separate sheet of paper, combine the like terms in the following monomials.

1. $5a + 3a$
2. $-3x + 8x$
3. $9y - 4y + 2y$
4. $12z - 15z + 5z$
5. $-7b - -6b + -3b + b$
6. $-9c + -c - 1c + 15c$
7. $12xy - 6xy + 3xy$
8. $-9ab - 6ab + -ab$
9. $2a + 3b + 4a - 3b$
10. $5x - 12y - 2x + 6y$
11. $7 - 3xy + 4x - 3y + 2xy - 9 - 4x - 3y$
12. $5x^2 - 2x + 3 - 4xy - 5 + 3x^2 - 4x + 13$
13. $-9a + -12ab - -5a - 6b + 4ab - a^2 + 1$
14. $6 - 0a + 13a - 1a + 6b - 13a + -3b - 6$
15. $-x^2 + 1 - 3x + 5x^2 - +4x + -2 - -3x^2 - 0$

THE LAWS OF EXPONENTS

Now that you are acquainted with the rules of adding and subtracting monomials, you are ready to learn the rules concerned with multiplying and dividing monomials. Before doing so, however, you should have an understanding of the laws of exponents. Their significance should soon become clear to you.

The laws of exponents are the rules used to simplify expressions containing numbers (including literal numbers) with exponents. Some of these rules may look familiar to you because they were presented in Chapter 10 in the rules of powers. After all, exponents are powers. The rules of exponents expand upon the rules of powers by explaining how to multiply and divide numbers with exponents.

RULES (AND LAWS) OF EXPONENTS

- Any number raised to the 1 power is the number itself.
  Examples:

  $$6^1 = 6$$
  $$a^1 = a$$

- Any number, except zero, raised to the zero power is 1. Zero cannot be raised to the zero power.
  Examples:

  $$100^0 = 1$$
  $$x^0 = 1$$

- A number written without an exponent has an exponent of 1.
  Example:

  $$5 = 5^1$$
  $$b = b^1$$

- When multiplying, if the bases are the same, add the exponents. The base stays the same. Remember, the base is the number being raised to a power. *Bases must be the same to apply this rule.*
  Examples:

  $$b^x \cdot b^y = b^{x+y}$$
  $$4^2 \cdot 4^4 = 4^{2+4} = 4^6 \text{ (proof: } 4^2 \cdot 4^4 = 4 \cdot 4 \cdot 4 \cdot 4 \cdot 4 \cdot 4 = 4^6)$$

- When dividing, if the bases are the same, subtract the bottom exponent from the top one. The base stays the same. *Bases must be the same to apply this rule.*
  Examples:

  $$\frac{a^x}{a^y} = a^{x-y}$$

  $$\frac{7^5}{7^2} = 7^{5-2} = 7^3 \text{ (Proof: } \frac{7^5}{7^2} = \frac{7 \cdot 7 \cdot 7 \cdot 7 \cdot 7}{7 \cdot 7} = 7 \cdot 7 \cdot 7 = 7^3)$$

  $$\frac{y^2}{y^5} = y^{2-5} = y^{-3}$$

- A number with a negative exponent may be rewritten as a fraction. The numerator will be 1. The denominator will be the base with a positive exponent.
  Example:

  $$y^{-3} = \frac{1}{y^3}$$

  Note: Proof of this is shown by the example:

  1) From before:

  $$\frac{y^2}{y^5} = y^{2-5} = y^{-3}$$

  2) By expanding and then canceling common factors:

  $$\frac{y^2}{y^5} = \frac{1 \cdot y \cdot y}{y \cdot y \cdot y \cdot y \cdot y} = \frac{1}{y^3} = y^{-3}$$

- When a negative exponent results in an answer, give the *final* answer in fractional form.

- When multiplying, if the exponents are the same and the bases are different, multiply the bases. The exponent stays the same. (The exponent is outside of grouping symbols.) *Exponents must be the same to apply this rule.*
  Examples:

  $$a^x \cdot b^x = (ab)^x$$
  $$2^3 \cdot 3^3 = (2 \cdot 3)^3 = 6^3$$

- When dividing, if the exponents are the same and the bases are different, divide the bases. The exponent stays the same. (The exponent is placed outside of grouping symbols.) *Exponents must be the same to apply this rule.*
  Examples:

  $$\frac{a^n}{b^n} = \left(\frac{a}{b}\right)^n$$

  $$\frac{4^3}{2^3} = \left(\frac{4}{2}\right)^3 = 2^3 = 8$$

- When factors inside of parentheses are raised to a power, remove the parentheses by *multiplying* the power of each factor inside the parentheses by the power outside.
  Examples:

  $$(3^2)^3 = 3^{2 \cdot 3} = 3^6$$
  $$(3x^2)^3 = 3^{1 \cdot 3}x^{2 \cdot 3} = 3^3x^6$$
  $$(x^2 yz^3)^4 = x^{2 \cdot 4}y^{1 \cdot 4}z^{3 \cdot 4} = x^8y^4z^{12}$$

- Fractional exponents may be written in root form.
  The numerator of the fraction is the power of the expression under the radical sign. The denominator is the amount given by the index.
  Examples:

  $$x^{1/2} = \sqrt{x}$$

  $$(x + 1)^{3/2} = \sqrt{(x + 1)^3}$$

- To find the root of an algebraic expression raised to a power, you can change it to a fractional exponent and then reduce the fraction.
  Examples:

  $$\sqrt{x^2} = x^{2/2} = x^1 = x$$

  $$\sqrt{x^4} = x^{4/2} = x^2$$

  $$\sqrt[8]{x^4} = x^{4/8} = x^{1/2}$$

  $$\sqrt{x + 1} = (x + 1)^{1/2}$$

  $$\sqrt{4x^2} = (4x^2)^{1/2} = 4^{1/2}x^{2/2} = 2x$$

  Note: One relationship of radicals is given by:

  $$\sqrt[n]{ab} = \sqrt[n]{a} \cdot \sqrt[n]{b} \quad \text{(General form)}$$

  Examples:

  $$\sqrt{4x^2} = \sqrt{4} \cdot \sqrt{x^2} = 2 \cdot \sqrt{x^2} = 2x^{2/2} = 2x$$
  $$\sqrt{32} = \sqrt{16 \cdot 2} = \sqrt{16} \cdot \sqrt{2} = 4\sqrt{2}$$

Practice Problems

On a separate sheet of paper, simplify the following using the laws of exponents. Exponents in the final answer must be positive.

1. 2^1
2. x^1
3. $-x^1 y^1$

4. a 3b^1
5. 8^0
6. $-$x^0
7. x^0 y^0
8. a^0 4b^1
9. 2^2 • $-$2^4 • 2^6 • $-$2^1
10. $-$x^1 • 3x^3 • $-$2x^0 • x^4
11. x^3 y^2 • x y^2 • $-$x^4 y
12. a^1 b^0 • $-$b • $-$a^4 b^4 • a^0

13. $\dfrac{5^6}{5^3}$ 14. $\dfrac{-x^7}{x^3}$ 15. $\dfrac{3x^4\ y^8}{\text{-x }4y^0}$ 16. $\dfrac{-a^3\ 2b^5\ c}{-a\ b\ 4c}$

17. $\dfrac{x}{x^2}$ 18. $\dfrac{-a^3\ b^6}{a^5\ b^2}$ 19. $\dfrac{-x^7\ y^{-3}}{-x^7\ y^3}$ 20. $\dfrac{2x^{-2}\ 3y^4}{-x^3\ 4y^{-5}}$

21. 2^3 • 4^3 22. 5^4 • 2^4 23. 4x^2 • y^2 24. 3a^{-1} • a^{-1}

25. $\dfrac{4^2}{2^2}$ 26. $\dfrac{x^3}{y^3}$ 27. $\dfrac{-a^6}{b^6}$ 28. $\dfrac{x^1}{-y^1}$

29. (3^2)^4 30. (x^3)^2 31. (4a^5)^3 32. (3x^2 y^3)^4

33. 16$^{1/2}$ 34. x$^{1/3}$ 35. ab$^{-1/2}$ 36. 4x$^{-1/2}$

37. 8$^{2/3}$ 38. a$^{2/2}$ 39. x$^{4/2}$ 40. 5y$^{-3/2}$

MULTIPLICATION AND DIVISION OF MONOMIALS

Monomials do not have to be like terms to be multiplied or divided. The arithmetic process is performed separately on coefficients and literal numbers. Follow the rules of multiplication/division of monomials.

RULES OF MULTIPLICATION/DIVISION OF MONOMIALS

- Multiply/divide all coefficients. Multiply/divide all literal numbers.
- Multiplication and division is performed on coefficients as would be with any numbers. Follow the rules of signed numbers. Follow the rules of order of operations and of exponents, where required.
- Apply the rules of exponents to literal numbers that are alike.
- Literal numbers that are unlike keep the same exponent.
- Combine coefficients and literal numbers for final answer, putting coefficients first. Put literal numbers in alphabetical order.

Examples:

$$(3x)(4x^2) = (3 \bullet 4)(x \bullet x^2) = 12x^{1+2} = 12x^3$$

$$(3x^2)(-2xy) = 3(-2)\ (x^2 \bullet x)y = -6x^{2+1}y = -6x^3y$$

$$\frac{(6x^7)}{(2x)} = \left(\frac{6}{2}\right)\left(\frac{x^7}{x}\right) = 3x^{7-1} = 3x^6$$

$$\frac{(-6y^2z)}{(2xyz)} = \left(\frac{-6}{2}\right)\left(\frac{1}{x}\right)\left(\frac{y^2}{y}\right)\left(\frac{z}{z}\right) = -3\left(\frac{1}{x}\right)\ (y^{2-1})\ (z^{1-1}) =$$

$$-3\ \frac{1}{x}\ (y^1)z^0 = -3\ \frac{1}{x}\ (y)\ (1) = \frac{-3y}{x}$$

$$(3x^2)^3(2z) = (3^{1 \bullet 3}x^{2 \bullet 3})(2z) = (3^3x^6)(2z) = (27x^6)(2z) = (27 \bullet 2)x^6z = 54x^6z$$

Practice Problems

On a separate sheet of paper, simplify the following monomials by performing the multiplication and division. Exponents in the answer should be positive.

1. $(-5x^2)\,(2x^3)$
2. $(6a^4)\,(-a^1)$
3. $(-4a^3)\,(-3a^5)\,(2a^2)$
4. $(10x^4)\,(-6x^6)\,(2x^2)$
5. $(11y^4)\,(4y^{-2})$
6. $(-12c^3)\,(5c^{-5})$
7. $(-7x^3\,y^2)\,(2x\,y)$
8. $(3x^2\,y)\,(-5x^5\,y^3)$
9. $(9a^4\,b^{-2})\,(-a^{-4}\,b^4)$
10. $(a^{-6})\,(3a^3\,b^{-3})$

11. $\dfrac{3x^3}{6x^2}$

12. $\dfrac{-9x^4}{3x^6}$

13. $\dfrac{-4a^{-4}\,b^5\,c^{-2}}{a^6\,8b^2\,5c}$

14. $\dfrac{5a^{-3}\,3b^2\,c^{-1}}{-15a\,2b^7\,4c^4}$

15. $\dfrac{-3x^{-2}\,4y^3}{-6x^{-4}\,2y^{-5}}$

16. $\dfrac{-8x^4\,2y}{4x^{-2}\,y^{-3}}$

17. $\dfrac{(4a^3)(-3b^{-2})}{(2a^{-1}\,4b^{-6})} \;\bullet\; \dfrac{(-4a^{-2}\,b^{-3})}{(3a^{-4})(2b^8)}$

18. $\dfrac{-x^3\,y^{-2}}{3y^5} \;\bullet\; \dfrac{-6y^3}{-4x^{-4}\,y^{-2}}$

19. $\dfrac{(6a^4\,b^0)(a^{-1}\,b^{-2}\,c^{-3})}{(a^0)(b^4)(c^{-1})(4a^{-3}\,b^4\,c^{-3})} \;\div\; \dfrac{-4a^2\,b^2\,c^2}{10a^{-3}\,b^2\,c^2}$

20. $\dfrac{-x^4\,y^{-3}}{2x^{-5}\,y^2} \;\div\; \dfrac{-3x^{-6}\,y^4}{6x^7\,y^{-1}}$

ADDITION AND SUBTRACTION OF POLYNOMIALS

When polynomials are to be added or subtracted, each will often be contained within parentheses. When removing parentheses, always work from the innermost first. The rules of removing grouping symbols, as described in Chapter 10, will apply. When adding or subtracting polynomials, follow the applicable rules.

RULES OF ADDITION OF POLYNOMIALS

- Addition of polynomials is indicated with a + sign outside of the parentheses.
- The parentheses can be removed without changing any of the signs.
- Combine like terms following the rules of signed numbers.
  Example:

$$(3x^2 + 2x - 5y) + (-4x^2 - 6x - 9y) =$$
$$3x^2 + 2x - 5y + -4x^2 - 6x - 9y =$$
$$-x^2 - 4x - 14y$$

RULES OF SUBTRACTION OF POLYNOMIALS

- Subtraction of polynomials is indicated with a − sign outside of the parentheses.
- Remove the parentheses by changing each of the signs inside. Note: If a number has both an arithmetic sign and a + or a − sign, change only the arithmetic sign.
- Combine like terms following the rules of signed numbers.
  Example:

$$(-2a + 4b) - (-5a + -3b) = -2a + 4b + 5a - (-3b) =$$
$$-2a + 4b + 5a + 3b = 3a + 7b$$

Practice Problems

On a separate sheet of paper, simplify the following polynomials by combining like terms.

1. $(-2x^2 + 5x - 3) + (4x^2 - 3x + 5)$
2. $(6a^3 + a - 4) + (-2a^3 - 3a) + (-4a^3 + a^2 - 3)$
3. $(3x^2 - 2x + 1) - (2x^2 + 4x - 3)$
4. $(-8xy^2 - 3xy - 4x) - (-5xy^2 - 3xy + 2y)$
5. $-(4a^2b + 5ab + 3a - b) - (a^2b + 5ab + 2a + -3b)$
6. $-(a^2 - 3ab^2 + 4a) + (3a^2 + ab^2 - 2a) - (3a^2 - 2a)$
7. $-(3x^2 + 4 - 5x) - -(4x^2 - 3x + 8) + -(-3 + x^2)$
8. $(3x)(4x) + (2x^2 - 12x + 3) - (2x)(5x)$
9. $(8a + 2a) - (4a^2 + 3a) + -(a^2 - 5a + 0) - +(4a^2 - 2a^2)$
10. $-(x^2 + 1) - -(5 - 3x^2 + 2x) - (2x + 3 - 2x^2) - (10 + x)$

MULTIPLICATION OF POLYNOMIALS

The arithmetic process involved in multiplying polynomials together is just like that for multiplying monomials. Although, when polynomials are involved, each term of one must be multiplied by each term of the other. The rules of multiplication of polynomials explain this.

RULES OF MULTIPLICATION OF POLYNOMIALS

- To multiply a polynomial by a monomial, multiply each term of the polynomial by the monomial. Follow the rules that apply to multiplication of monomials. Combine like terms.
  Example:

$$-3x(8 + 2x - 4y) = -3x \cdot 8 + -3x \cdot 2x - (-3x \cdot 4y) =$$
$$-24x - 6x^2 + 12xy$$

- To multiply a polynomial by a polynomial, multiply each term of the first polynomial by each term of the second polynomial. Follow the rules that apply to multiplication of monomials. Combine like terms.
  Example:

$$(2x + 3)(4x - 1) = 2x \cdot 4x - 2x \cdot 1 + 3 \cdot 4x - 3 \cdot 1$$
$$= 8x^2 - 2x + 12x - 3 = 8x^2 + 10x - 3$$

Practice Problems

Perform the indicated multiplication on a separate sheet of paper. Combine like terms.

1. $2x(3x + 4)$
2. $-6a(-2a^2 + 3a - 1)$
3. $-5xy(3x + 4x - 2xy - 3)$

4. $a^2b(-2a + 3b + 5a^2b - ab^2 - 2)$
5. $3x^2y^3(x - y + 3x^{-1} - 4y^{-3} - 2)$
6. $(2x + 1)(3x + 2)$
7. $(-4a - 2)(5a + 3)$
8. $(3x^2 - 2)(-3 - 5x)$
9. $(4a + 2b)(3a + 4b - 5ab + 1)$
10. $-2x[(3x - 4y)(5 - 2x^2 - y^2 + 2xy - 3x + 4y)]$

DIVISION OF POLYNOMIALS

Division of a polynomial by a monomial is accomplished in much the same way it is for monomials. Division of a polynomial by a polynomial is performed differently. This process is similar to the process of long division. For purposes here, we will be more concerned with division of polynomials by monomials. However, division of a polynomial by a polynomial is presented for your information. When the need arises, follow the rules of division of polynomials.

RULES OF DIVISION OF POLYNOMIALS

• To divide a polynomial by a monomial, divide each term of the polynomial by the monomial. Follow the rules that apply to division of monomials. Combine like terms.

Examples:

$$\frac{(4x^3 + 2x)}{2x} = \frac{4x^3}{2x} + \frac{2x}{2x} = 2x^2 + 1$$

$$\frac{(12x^2y^2 - 6x^3)}{3xy} = \frac{12x^2y^2}{3xy} - \frac{6x^3}{3xy} = 4xy - \frac{2x^2}{y}$$

• To divide a polynomial by a polynomial, divide by long division. First try to arrange both polynomials in descending order of some common letter. Divide first terms (answer is first term of quotient); multiply this term of quotient by all terms in divisor; subtract; bring down next term; repeat (as in long division).

Example:

$$
\begin{array}{r}
x - 2 \quad \text{(quotient)} \\
x + 4 \,\overline{)\, x^2 + 2x - 8} \\
\text{(divisor)} \quad \underline{x^2 + 4x} \\
-2x - 8 \\
\underline{-2x - 8} \\
0
\end{array}
$$

Practice Problems

Perform the indicated division on a separate sheet of paper. Combine like terms.

1. $\dfrac{(8x^2 + 4x)}{4x}$

2. $\dfrac{(12a^4 + 6a^2)}{-3a^2}$

3. $\dfrac{(-15x^3 - 5x^2 + 20x)}{-5x}$

4. $\dfrac{(-30x^2 y2 + 6xy)}{3xy}$

5. $\dfrac{(4a^2b - 8ab^2 + 16 a^2b^2)}{-8a^2b^2}$

6. $\dfrac{(-9a^4b^{-3} c^2 + 12a^{-2} bc^4 - a^3b^4c^{-2})}{3a^3 b^{-2} c^2}$

7. $\dfrac{(8x^3 y^2 + 4x^2y^{-3} + 6)}{-xy} \cdot \dfrac{(3x^2 + 6)}{2xy}$

8. $\dfrac{(2x^2 + 4y)}{4xy^2} \cdot \dfrac{(4x^4\ y^{-3})}{8x^{-2}\ y}$

9. $\dfrac{(2x^2 + 8x + 8)}{(x + 2)}$

10. $\dfrac{(3a^2 + 12a + 9)}{(a + 3)}$

FACTORING

Factoring is very useful in applications of math and so it is important to have a basic understanding of it. **Factoring** is reversing the process of finding a product. In factoring, a product is known. The object is to find factors of it. Two reasons for factoring are to simplify algebraic expressions, especially of fractional form, and to find solutions to equations.

Certain products in forms of algebraic expressions are encountered routinely. These *special products* have certain factors associated with them. Some of them are given in the rules for special products. Try to become familiar with them.

RULES FOR SPECIAL PRODUCTS

- *Common monomial factoring.* Common factors of a polynomial may be placed outside of parentheses. (The terms inside parentheses have the common factors removed. This step, which is not shown in the examples below, is done by dividing each original term by the common factors.)
  Examples:

  $ax + ay + az = a(x + y + z)$ (General form)
  $3x + 3y + 3z = 3(x + y + z)$
  $12x^2 + 24xy = 12x(x + 2y)$

- *Difference of two squares.* The difference of the squares of two numbers is equal to the sum of the two numbers times the difference of the two numbers.
  Examples:

  $x^2 - y^2 = (x + y)(x - y)$ (General form)
  $3^2 - 2^2 = (3 + 2)(3 - 2)$
  $64 - 49 = (8 + 7)(8 - 7)$
  $25x^2 - 16y^2 = (5x + 4y)(5x - 4y)$

- *Perfect square trinomials.* These are **trinomials** (polynomials of 3 terms) having two perfect squares, each preceded by a + sign. The third term is two times the product of the square roots of the other two. To factor, give the square roots of the perfect squares connected by the sign of the remaining term. Enclose in parentheses and indicate that the quantity is squared. Examples will help to explain.
  Examples:

  $x^2 + 2xy + y^2 = (x + y)^2$ (General form)
  $x^2 - 2xy + y^2 = (x - y)^2$ (General form)
  $16x^2 + 40x + 25 = (4x + 5)^2$

- Other products in trinomial form may be able to be factored according to the following equation. The key is to determine what numbers the letters a and b represent. Sometimes it is a trial-and-error process.
  Examples:

  $x^2 + (a + b)x + ab = (x + a)(x + b)$ (General form)
  $x^2 + 10x + 24 = (x + 6)(x + 4)$
  $x^2 + 2x - 15 = (x + 5)(x - 3)$

Now that you are somewhat familiar with factoring, let us see how factoring can be used to simplify algebraic expressions. (The importance of factoring in solving equations is again noted. Equation solving, however, will not be presented until the next chapter.)

Sample Problem 11-1.

Simplify the expression:

$$\frac{x^3 - 4x^2}{x^2 - 16}$$

Step 1. Inspect the numerator. Factor if possible.
Notice that x^2 is a common factor.

$$x^3 - 4x^2 = x^2(x - 4)$$

Step 2. Inspect the denominator. Factor if possible.
Notice that the denominator is the difference of two squares.

$$x^2 - 16 = (x + 4)(x - 4)$$

Step 3. Rewrite the expression with the factored forms. Cancel like factors in numerator and denominator.

$$\frac{x^2(x - 4)}{(x + 4)(x - 4)}$$

Step 4. Rewrite in simplified form.

$$\frac{x^2}{x + 4}$$

Sample Problem 11-2.

Simplify the expression:

$$\frac{4x + x^2 - 12 - 3x}{9 - x^2}$$

Step 1. Inspect the numerator. Factor if possible.
Notice that the numerator may be simplified by combining like terms.

$$x^2 + x - 12$$

This expression may be factored into:

$$(x - 3)(x + 4)$$

Step 2. Inspect the denominator. Factor if possible.
Notice the denominator is the difference of two squares.

$$9 - x^2 = (3 + x)(3 - x)$$

Step 3. Rewrite the expression with the factored forms.

$$\frac{(x - 3)(x + 4)}{(3 + x)(3 - x)}$$

Step 4. Notice the factors $x - 3$ and $3 - x$ differ only in sign. From the rules of grouping symbols, move the $-$ sign outside of the parentheses in one of the factors. Rewrite expression and cancel like factors.

$$\frac{-(-x + 3)(x + 4)}{(3 + x)(3 - x)} = \frac{-(3 - x)(x + 4)}{(3 + x)(3 - x)}$$

Step 5. Rewrite in simplified form.

$$\frac{-(x + 4)}{3 + x}$$

Practice Problems

On a separate sheet of paper, simplify the following algebraic expressions by factoring.

1. $6a^2 + 9a$
2. $12a - 6a^4$
3. $3x^3 + 6x - 9x^2$
4. $4x + 8x^2 + 12x - 16x^2$
5. $a^2 - b^2$
6. $4x^2 - y^2$
7. $36x^4 - 9y^2$
8. $49a^2 - 81b^4$
9. $a^2 + 2ab + b^2$
10. $x^2 + 6x + 9$
11. $x^2 + 8x + 16$
12. $4a^2 + 16a + 16$
13. $5x^3 + 10x^2 + 5x$
14. $2a^5 + 20a^4 + 50a^3$
15. $a^2 - 4a + 4$
16. $16x^2 - 24x + 9$
17. $a^2 + 8a + 15$
18. $a^2 - 8a + 12$
19. $\dfrac{x^2 - 9}{x^2 + 5x + 6}$
20. $\dfrac{a^2 + 9a + 8}{4a^2 + 32a}$

TEST YOUR SKILLS

Do *not* write in this book. Use a separate sheet of paper to complete the following problems. Show your work and your final answer.

BASIC TERMINOLOGY

With the algebraic expressions given, write each term. Also, identify the coefficients, variables, and constant terms. Include + or − signs. Give answers in tabular form as shown in the example below.
Example:

$$4x^2 + 5xy - 12$$

term	coefficient	variables	constant term
$+4x^2$	$+4$	x	
$+5xy$	$+5$	x,y	
-12			-12

1. $-5x + 6y^2 - 7$
2. $-3x - 4xy^2 + 15$
3. $9 + 8u - v$
4. $x^3 - 3x^2 + 8 - 12y + z$
5. $-3x^2y^3 - 10$

Write each term, then separate the factors. Only factor if multiplication is indicated. Include + or − sign.

Example: $2x^2 - 6xyz + 9$

term	factors
$+2x^2$	$+2 \cdot x \cdot x$
$-6xyz$	$-6 \cdot x \cdot y \cdot z$
$+9$	$-$

6. $-8a^3 + 3ab - 4$

7. $5 - (3)(4x) + 2x^2 - 10$

8. $6(x + 1) + 4 - 3xy$

9. $10 - 2(x + y) + (3x - 2)$

10. $x^2y^3z^2 + 3(a + 2b)$

OPERATIONS WITH MONMIALS AND POLYNOMIALS

Perform the indicated arithmetic operations. Simplify all answers.

11. $2a + 4y - 4a - y$

12. $x + 3x^2 - 3x + 2 - 5x$

13. $a + 3a - 5a + a$

14. $6b + 2b - b - 6b$

15. $-8 + 3d - 5d + 6d - 5d + 1$

16. $3y + (-5y) - (-2y)$

17. $(-3) + c + (-4) + (-5c)$

18. $2xy + 4x^2y - 3xy - 5x^2y$

19. $(3x)(2y)$

20. $2a + 4abc + 3abc - 5a$

21. $(5a)(a) - 8a^2 + (6a)(-2a) + 1$

22. $x^2 \cdot 4x^3 \cdot 3x^4$

23. $\dfrac{6x^2y^4z^3}{3xy^3z^2}$

24. $\dfrac{(-4x)(2xy^3)(-5x^2y^5)}{(8x^3)(-y^6)(10xy^4)}$

25. $\dfrac{(x + 2)(4x^2)(-3xy)}{(6x^3)(x + 2)(-y^3)}$

26. $a^0 \ (a \neq 0)$

27. b^1

28. $(x^2y^3)^4$

29. $(xyz)^2$

30. $\dfrac{y^3}{y^7}$

31. $(4a - 2b) + (2a + 3b)$

32. $(-3x + 3y) + (-4x - 6y)$

33. $(2a + 3b + 4c) + (3a - 4a - 3b)$

34. $-(3x - 4y) + (2z - 5z + 3x)$

35. $(-x + y - z) - (-3y + 2z - 5x)$

36. $-(-1 + 3xy - 4y) - (6xy + 5xy - 4x)$

37. $[4x + (3x - 4y) + 6] - [7 - (5y + x)]$

38. $(2x^3 - 3x^2) - (2x^0 - 2x^2) + (x^2 + 1)$

39. $(c - d) - (c + d) + (-c - d)$

40. $(x - y) - (3x + 4y) - (5x - 3y) + 2x$

41. $(2b^3)(4a^2bc)$

42. $(-6x^2y^3)(xy^2)$

43. $(-3ab)(4b^2c)(-2a^{-1})$

44. $2(a + 4b - c - 3d)$

45. $-4a(-a + 2a^2 - 3ab)$

46. $-2ab(-a + 3b)$

47. $2(a + b) - 3(a - b)$

48. $5[a + 3(a + 2b)] - 2a$

49. $(x + 2y)(3x - y)$

50. $2[(2x - 4x)(y + 3x)]$

51. $\dfrac{(12x + 8)}{2}$

52. $\dfrac{(16b - 4)}{4}$

53. $\dfrac{(x^4 - x^2)}{x}$

54. $\dfrac{12x^2y^2 - 8x^2y + 4xy^2)}{4xy}$

55. $\dfrac{(x^2 - x - 2)}{(x + 1)}$

FACTORING

Factor each of the following expressions completely.

56. $3x + 3xy$
57. $3x + 6y$
58. $64x^4 - 49y^4$
59. $4x^2 + 4x + 1$
60. $x^2 - 8x + 15$

Simplify each of the following fractions.

61. $\dfrac{x + y}{7x^2 + 7xy}$

62. $\dfrac{8u - 4v}{4u - 2v}$

63. $\dfrac{x^2 - 3x - 10}{x^2 + 2x}$

64. $\dfrac{x^4 - 81}{x + 3}$

65. $\dfrac{x^2 - 4x + 4}{x^2 - 4}$

Structural engineers used algebraic equations to calculate loads and material strengths when designing this steel bridge. (Jack Klasey)

Chapter 12

Algebraic Equations and Inequalities

OBJECTIVES

After studying this chapter, you will be able to:
- Solve algebraic equations.
- Solve algebraic inequalities.
- Solve simultaneous equations by the addition/subtraction method.
- Solve simultaneous equations by the substitution method.

An **equation** is a complete mathematical sentence stating that two algebraic expressions are equal. The two expressions are joined by an equal (=) sign. The expression to the left equals the expression to the right. When an algebraic equation is solved, the actual numerical value of a literal number, or unknown, is determined. The equation is solved by performing the arithmetic until the unknown equals a particular number.

ISOLATING THE UNKNOWN

When a particular unknown is **isolated,** all of the arithmetic in the equation that can be, has been performed. The isolated variable is written by itself, usually on the left side. It is equal to some expression on the right side. Moreover, if it is a simple, **linear equation** (equation with unknowns raised to the first power only—equations of the *first degree*) with one unknown, it is equal to a specific number. This numerical solution can be written in the original equation, in place of the unknown. If the arithmetic is performed using the solution, both sides of the equation should be equal. (Equations whose two sides are equal only when certain values are substituted for the unknowns are called **conditionial equations.** Some equations are equal for any value substituted for the unknowns. Such equations are called **identities.** Trying to isolate an unknown of an identity yields $0 = 0$.)

In the process of isolating an unknown, terms are transposed. **Transposing** involves reversing arithmetic operations in order to get all terms of the unknown being isolated, on one side. In transposing, the same operation is performed on both sides of an equation. If an operation involves addition of a term, the reverse operation is performed—the term is subtracted from both sides of the original equation. Likewise, a factor being multiplied is divided from both sides. This way, a factor or term is canceled on one side and shifted to the other side of the equation. The two equal sides remain equal.

In transposing, ask yourself what operations are being performed on the unknown, then reverse them. Priority of reversing operations is: grouping symbols first, addition/subtraction next, division, then multiplication. You may have to combine like terms. Remember: *Whatever is done to one side of an equation must be done to the other side.* Finally, check the solution by placing it in the original equation in place of the unknown.

Addition/subtraction

If an unknown is shown with addition, isolate it by subtracting from both sides. Refer to the following sample problems. Note: The reversing operation is underlined in these sample problems.

Sample Problem 12-1.

Solve this equation:

$$x + 5 = 18$$

Step 1. Transpose the equation to isolate the unknown.

$$x + 5 \underline{- 5} = 18 \underline{- 5} \quad \text{(subtract 5)}$$

Step 2. Simplify for the solution.

$$x = 13$$

Step 3. Check by substituting the value for the unknown.

$$(13) + 5 = 18$$
$$18 = 18$$

Sample Problem 12-2.

Solve his equation:
$$a + 4 = -15$$

Step 1. Transpose the equation to isolate the unknown.

$$a + 4 \underline{- 4} = -15 \underline{- 4} \quad \text{(subtract 4)}$$

Step 2. Simplify for the solution.

$$a = -19$$

Step 3. Check by substituting the value for the unknown.

$$(-19) + 4 = -15$$
$$-15 = -15$$

Sample Problem 12-3.

Solve this equation:
$$6 + c + 3 = 14$$

Step 1. Transpose the equation to isolate the unknown.

$$6 + c + 3 \underline{- 9} = 14 \underline{- 9} \quad \text{(subtract 9)}$$

Step 2. Simplify for the solution.

$$c = 5$$

Step 3. Check by substituting the value for the unknown.

$$6 + (5) + 3 = 14$$
$$14 = 14$$

If an unknown is shown with subtraction, isolate it by adding to both sides. Refer to the following sample problems.

Sample Problem 12-4.

Solve this equation:

$$y - 7 = 10$$

Step 1. Transpose the equation to isolate the unknown.

$$y - 7 \underline{+ 7} = 10 \underline{+ 7} \quad \text{(add 7)}$$

Step 2. Simplify for the solution.

$$y = 17$$

Step 3. Check by substituting the value for the unknown.

$$(17) - 7 = 10$$

$$10 = 10$$

Sample Problem 12-5.

Solve this equation:

$$b - 1 = -16$$

Step 1. Transpose the equation to isolate the unknown.

$$b - 1 \underline{+ 1} = -16 \underline{+ 1} \quad \text{(add 1)}$$

Step 2. Simplify for the solution.

$$b = -15$$

Step 3. Check by substituting the value for the unknown.

$$(-15) - 1 = -16$$

$$-16 = -16$$

Sample Problem 12-6.

Solve this equation:

$$9 - x = 12$$

Step 1. Transpose the equation to isolate the unknown.

$$9 - x \underline{- 9} = 12 \underline{- 9} \quad \text{(subtract 9)}$$

Step 2. Simplify for the solution.

$$-x = 3$$

$$x = -3 \text{ (change signs)}$$

Note: If an isolated unknown is preceded by a $-$ sign, remove it by changing the sign on both sides. (The negative sign in front of the letter is actually a coefficient of -1. It is removed by dividing both sides by -1.)

Step 3. Check by substituting the value for the unknown.

$$9 - (-3) = 12$$

$$12 = 12$$

Practice Problems

Solve the following equations on a separate sheet of paper. Show the steps necessary to transpose each equation, find the numerical value of the unknown, and check the solution by substituting the numerical value for the unknown.

1. $x + 8 = 12$
2. $6 + x = 9$
3. $5 + 1 = x$
4. $10 + 3 = y$
5. $a - 7 = 15$
6. $10 - x = 15$
7. $7 - 4 = b - 5$
8. $7 - 9 = a + 1$
9. $5 - 4 - 2 = -x + 6$
10. $8 + 12 - 10 = -a - 5$
11. $y + -9 = -15$

12. $-20 + b = -12$
13. $x + 5 = 3$
14. $-a - 3 = 1$
15. $-y + 7 = -10$
16. $-x - 5 = 10 - 3$
17. $15 - 6 - -9 = -5 + x$
18. $6 - 1 + 5 = 15 + a$
19. $-21 - 7 - 4 = -y + 1$
20. $5 - 6 = -a + 1$

Multiplication/division

If an unknown is shown with multiplication, isolate it by dividing both sides. Refer to the following sample problems.

Sample Problem 12-7.

Solve this equation:

$$2x = 18$$

Step 1. Transpose the equation to isolate the unknown.

$$\frac{2x}{2} = \frac{18}{2} \quad \text{(divide by 2)}$$

Step 2. Simplify for the solution.

$$x = 9$$

Step 3. Check by substituting the value for the unknown.

$$2(9) = 18$$
$$18 = 18$$

Sample Problem 12-8.

Solve this equation:

$$6a = -5$$

Step 1. Transpose the equation to isolate the unknown.

$$\frac{6a}{6} = \frac{-5}{6} \quad \text{(divide by 6)}$$

Step 2. Simplify for the solution.

$$a = -\frac{5}{6}$$

Step 3. Check by substituting the value for the unknown.

$$6\left(-\frac{5}{6}\right) = -5$$
$$-5 = -5$$

If an unknown is shown with division, isolate it by multiplying both sides. Refer to the following sample problem.

Sample Problem 12-9.

Solve this equation:

$$\frac{x}{2} = 12$$

Step 1. Transpose the equation to isolate the unknown.

$$\frac{x}{2} \cdot 2 = 12 \cdot 2 \quad \text{(multiply by 2)}$$

Step 2. Simplify for the solution.

$$x = 24$$

Step 3. Check by substituting the value for the unknown.

$$\frac{(24)}{2} = 12$$

$$12 = 12$$

If an unknown is in the denominator, first multiply both sides by the unknown to remove it from the denominator. Refer to the following sample problem.

Sample Problem 12-10.

Solve this equation:

$$\frac{3}{a} = 6$$

Step 1: Remove the unknown from the denominator.

$$\frac{3}{a} \cdot \underline{a} = 6 \cdot \underline{a} \quad \text{(multiply by a)}$$

$$3 = 6a$$

Step 2. Transpose the equation to isolate the unknown.

$$\frac{3}{6} = \frac{6a}{6} = \quad \text{(divide by 6)}$$

Step 3. Simplify for the solution.

$$a = \frac{3}{6} = \frac{1}{2} \quad \text{(reduce fraction)}$$

Step 4. Check by substituting the value for the unknown.

$$\frac{3}{\left(\frac{1}{2}\right)} = 6$$

$$3 \div \frac{1}{2} = 6$$

$$3 \cdot \frac{2}{1} = 6 \quad \text{(invert divisor)}$$

$$6 = 6$$

Practice Problems

Solve the following equations on a separate sheet of paper. Show the steps necessary to transpose each equation, find the numerical value of the unknown, and check the solution by substituting the numerical value for the unknown.

1. $3a = 6$
2. $5x = 10$
3. $4b = 8$
4. $2y = 12$
5. $-7x = 14$
6. $-8y = 32$
7. $-6y = -36$
8. $-9a = -81$
9. $2a = -8$
10. $3x = -18$
11. $15 = 5b$
12. $20 = 4y$

13. $-30 = 6c$
14. $-22 = 2a$
15. $5 = -5y$
16. $4 = -28x$
17. $-10 = -40a$
18. $-5 = -50b$
19. $4 = 8a$
20. $-3 = 9x$

21. $\dfrac{6}{x} = 3$ 22. $\dfrac{8}{y} = 2$

23. $\dfrac{a}{4} = 3$ 24. $\dfrac{b}{6} = 4$

25. $\dfrac{-16x}{2} = 4$ 26. $\dfrac{-45y}{5} = 3$

27. $\dfrac{14}{-2a} = 7$ 28. $\dfrac{16}{-4b} = 2$

29. $-6 = \dfrac{3y}{-6}$ 30. $8 = \dfrac{-5x}{-10}$

Combination equations

You have seen examples of transposing equations comprised strictly of addition/subtraction or strictly of multiplication/division. When combinations of operations, grouping symbols, or the unknown in more than one term exists, a few more steps are involved. The concept, however, is the same. The order of transposing such equations is given in the rules of transposing mixed equations.

RULES FOR TRANSPOSING MIXED EQUATIONS

1. Remove grouping symbols.
2. Combine like terms on each side of the equation.
3. If unknown is found on both sides of equation, transpose it so it is on only one side.
4. Transpose terms on the same side as the unknown that are being added or subtracted. Rewrite the equation in simplified terms.
5. Remove numbers on the same side as the unknown that are being divided by multiplying both sides, including everything on the other side, by these. Rewrite the equation in simplified terms.
6. Remove numbers on the same side as the unknown that are being multiplied by dividing both sides, including everything on the other side, by these. Rewrite the equation in simplified terms.
7. Write the solution so the unknown is in the left member, or on the left side of the equation.
8. Check the solution by substituting the value for the unknown.

Sample Problem 12-11.

Solve this equation:

$$5x + 1 = 3x + 7$$

Step 1. Transpose the equation to isolate the unknown. The unknown in this equation is on both side. From rule 3, transpose one so they are both on the same side.

$$5x + 1 \underline{- 3x} = 3x + 7 \underline{- 3x} \quad \text{(subtract } 3x)$$

$$5x - 3x + 1 = 7$$

Step 2. Transpose other terms on the unknown side. (Rule 4)

$$5x - 3x + 1 \underline{-1} = 7 \underline{-1} \quad \text{(subtract 1)}$$
$$5x - 3x = 7 - 1$$

Step 3. Rewrite the equation in simplified terms.

$$2x = 6$$

Step 4. Divide both sides. (Rule 6).

$$\frac{2x}{2} = \frac{6}{2} \quad \text{(divide by 2)}$$

Step 5. Rewrite the equation in simplified terms.

$$x = 3$$

Step 6. Check by substituting the value for the unknown. (Rule 8)

$$5(3) + 1 = 3(3) + 7$$
$$15 + 1 = 9 + 7$$
$$16 = 16$$

Sample Problem 12-12.

Solve this equation:

$$4(x + 3) = 8$$

Step 1. Transpose the equation to isolate the unknown. Remove grouping symbols. (Remove parentheses using multiplication.) (Rule 1)

$$\underline{4 \cdot}x + \underline{4 \cdot}3 = 8 \quad \text{(multiplication by 4)}$$
$$4x + 12 = 8$$

Step 2. Transpose terms from unknown side.

$$4x + 12 \underline{-12} = 8 \underline{-12} \quad \text{(subtract 12)}$$

Step 3. Rewrite the equation in simplified terms.

$$4x = -4$$

Step 4. Divide both sides.

$$\frac{4x}{4} = \frac{-4}{4} \quad \text{(divide by 4)}$$

Step 5. Rewrite the equation in simplified terms.

$$x = -1$$

Step 6. Check by substituting the value for the unknown.

$$4[(-1) + 3] = 8$$
$$4(2) = 8$$
$$8 = 8$$

Sample Problem 12-13.

Solve this equation:

$$4(x + 3) = 8$$

Step 1. Transpose the equation to isolate the unknown. Divide both sides. (Remove parentheses using division.) (Rule 6)

$$\frac{4(x + 3)}{4} = \frac{8}{4} \quad \text{(divide by 4)}$$

Step 2. Rewrite the equation in simplified terms.

$$x + 3 = 2$$

Step 3. Transpose the equation.

$$x + 3 \underline{\;-\; 3} = 2 \underline{\;-\; 3} \quad \text{(subtract 3)}$$

Step 4. Rewrite the equation in simplified terms.

$$x = 1$$

Step 5. Check by substituting the value for the unknown.

$$4[(-1) + 3] = 8$$

$$4(2) = 8$$

$$8 = 8$$

Note: Notice that this problem presented an alternate way of solving Sample Problem 12-12.

Practice Problems

Solve the following equations on a separate sheet of paper. Show the steps necessary to transpose each equation, find the numerical value of the unknown, and check the solution by substituting the numerical value for the unknown.

1. $4x + 3 = 2x + 1$
2. $5x - 6 = -3x + 10$
3. $-6y - 7 = 4 - 3y + 1$
4. $9 - 2a = 6 - 4a - 7$
5. $-6(x + 1) = 30$
6. $5(2a - 3) = 15$
7. $7a - 2(3 + a) = 9a + 2$
8. $10 - 4(x - 3) = -8 + 2x$

9. $\dfrac{3(y - 5)}{3} = \dfrac{-2(3y - 9)}{6}$

10. $\dfrac{32}{2(x + 2)} = \dfrac{16}{4(x - 4)}$

Fractional equations

Fractions in an equation are changed to whole numbers by multiplying each term, on both sides of the equation, by the least common denominator (LCD) of all fractions. The LCD is a number that all denominators will divide into exactly. When multiplying each term by the LCD, it becomes part of the numerator. The denominator will divide exactly into the LCD, which cancels the denominator.

Sample Problem 12-14.

Solve this equation:

$$\frac{2}{3} - \frac{x}{4} = \frac{1}{6}$$

Step 1. Multiply each term by the LCD. (LCD = 12)

$$\frac{12 \cdot 2}{3} - \frac{12 \cdot x}{4} = \frac{12 \cdot 1}{6} \quad \text{(multiply by 12)}$$

Step 2. Simplify.

$$\frac{24}{3} - \frac{12x}{4} = \frac{12}{6}$$

$$8 - 3x = 2 \quad \text{(reduce fractions)}$$

Step 3. Transpose the equation.

$$8 - 3x - 8 = 2 - 8 \quad \text{(subtract 8)}$$

$$-3x = -6$$

$$\frac{-3x}{-3} = \frac{-6}{-3} \quad \text{(divide by } -3)$$

Step 4. Simplify for the solution.

$$x = 2$$

Step 5. Check by substituting the value for the unknown.

$$\frac{2}{3} - \frac{(2)}{4} = \frac{1}{6}$$

(To add or subtract fractions, a common denominator is required.)

$$\frac{4 \cdot 2}{4 \cdot 3} - \frac{3 \cdot 2}{3 \cdot 4} = \frac{2 \cdot 1}{2 \cdot 6} \quad \text{(LCD} - 12)$$

$$\frac{8}{12} - \frac{6}{12} = \frac{2}{12} \quad \text{(perform arithmetic)}$$

$$\frac{2}{12} = \frac{2}{12}$$

Sample Problem 12-15.

Solve this equation:

$$\frac{1}{a} + \frac{3}{4} = 1$$

Step 1. Multiply each term by the LCD. (LCD = 4a)

$$\frac{4a \cdot 1}{a} + \frac{4a \cdot 3}{4} = \underline{4a} \cdot 1$$

Step 2. Simplify.

$$\frac{4a}{a} + \frac{12a}{4} = 4a$$

$$4 + 3a = 4a \quad \text{(reduce fractions)}$$

Step 3. Transpose the equation.

$$4 + 3a - 3a = 4a - 3a \quad \text{(subtract 3a)}$$

Step 4. Simplify. Write unknown on left side.

$$4 = a$$
$$a = 4$$

Step 5. Check by substituting the value for the unknown.

$$\frac{1}{(4)} + \frac{3}{4} = 1$$

$$\frac{1}{4} + \frac{3}{4} = \frac{4}{4}$$

$$\frac{4}{4} = \frac{4}{4}$$

Practice Problems

Solve the following equations on a separate sheet of paper. Show the steps necessary to transpose each equation, find the numerical value of the unknown, and check the solution by substituting the numerical value for the unknown.

1. $\dfrac{1}{2} + \dfrac{x}{3} = \dfrac{5}{6}$
2. $\dfrac{a}{3} + \dfrac{2}{5} = \dfrac{1}{3}$

3. $\dfrac{1}{2} - \dfrac{3y}{4} = 5$

4. $\dfrac{5}{6} - \dfrac{2b}{3} = -\dfrac{11}{6}$

5. $\dfrac{3}{a} + \dfrac{1}{2} = \dfrac{3}{4}$

6. $\dfrac{1}{3} - \dfrac{2}{y} = \dfrac{2}{5}$

7. $\dfrac{3}{4} + \dfrac{21}{2b} = 6$

8. $\dfrac{25}{5b} - \dfrac{2}{3} = -1$

9. $\dfrac{5}{x} + \dfrac{1}{4} = \dfrac{2}{3x}$

10. $3 + \dfrac{42}{3x} = \dfrac{-1}{2}$

Decimal equations

If an equation contains decimals, it is possible to solve it by working with the numbers as they are. However, without the aid of a calculator, it can be easier to remove the decimals. This is done by multiplying each term by the equivalent denominator given by the smallest place value. As a result, decimal numbers are changed to whole numbers.

Decimal numbers can be represented by fractions. The numerator gives the digits right of the decimal point. The denominator is a multiple of 10. It contains a zero for each decimal place. (For a detailed explanation of decimal numbers, refer to Chapter 3).

Examples of decimals and equivalent fractions include:

$$.2 = \frac{2}{10} \qquad\qquad .05 = \frac{5}{100}$$

$$.015 = \frac{15}{1000} \qquad\qquad .3016 = \frac{3016}{10,000}$$

Sample Problem 12-16.

Solve this equation:

$$0.04x + 3.1 = 0.39x + 3.2$$

Step 1. Smallest place value is 2 decimal places. Equivalent denominator is 100. Therefore, multiply each term by 100.

$$100 \cdot 0.04x + 100 \cdot 3.1 = 100 \cdot 0.39x + 100 \cdot 3.2$$

$$4x + 310 = 39x + 320$$

Step 2. Transpose the equation.

$$4x + 310 - 4x = 39x + 320 - 4x \quad \text{(subtract 4x)}$$

$$310 = 35x + 320$$

$$310 - 320 = 35x + 320 - 320 \quad \text{(subtract 320)}$$

$$-10 = 35x$$

$$\frac{-10}{35} = \frac{35x}{35} \quad \text{(divide by 35)}$$

Step 3. Simplify for the solution in fractional form.

$$x = -\frac{2}{7} \quad \text{(change order and reduce fraction)}$$

Step 4. Change fraction to a decimal.

$$x = -0.286 \quad \text{(rounded)}$$

Step 5. Check by substituting the value for the unknown.

$$0.04(-0.286) + 3.1 = 0.39(-0.286) + 3.2$$

$$-0.011 + 3.1 = -0.111 + 3.2$$

$$3.089 = 3.089$$

Practice Problems

Solve the following equations on a separate sheet of paper. Show the steps necessary to transpose each equation, find the numerical value of the unknown, and check the solution by substituting the numerical value for the unknown.

1. $.008 + .032x - .005 = .015x - .014$

2. $4.5a - 9.40 = -1.200a - 4.0 + 3.00a + 2.7$

3. $-.05y + .35 + .80y - .15 = .02 - .10y - .05 - 2.32$

4. $.03(4b + 5) - .50 = 1.00 - .05(-3 + 3b) + .20 - .08$

5. $-2.1c - .2(15c + 150) = .3c(3 - 60) - 1.2(c + 20) + 10.1$

Equations with powers and roots

The unknown in an equation is isolated by performing arithmetic opposite to that indicated in the equation. If the unknown is raised to a power, remove the power by taking the root of both sides of the equation. If the unknown is shown with a root, remove it by raising both sides of the equation to the power given by the index. (Remember, if no index is given, the power is 2.)

Sample Problem 12-17.

Solve this equation:

$$x^2 = 16$$

Step 1. Transpose the equation.

$$\sqrt{x^2} = \sqrt{16} \quad \text{(take square root)}$$

Step 2. Simplify for the solution.

$$x = 4$$

Step 3. Check by substituting the value for the unknown.

$$(4)^2 = 16$$
$$16 = 16$$

Sample Problem 12-18.

Solve this equation:

$$\sqrt{3x} = 6$$

Step 1. Transpose the equation.

$$(\sqrt{3x})^2 = 6^2 \quad \text{(square both sides)}$$
$$3x = 6^2 \quad \text{(square root removed)}$$
$$3x = 36 \quad \text{(square 6)}$$
$$\frac{3x}{3} = \frac{36}{3} \quad \text{(divide by 3)}$$

Step 2. Simplify for the solution.

$$x = 12$$

Step 3. Check by substituting the value for the unknown.

$$\sqrt{3(12)} = 6$$
$$\sqrt{36} = 6$$
$$6 = 6$$

Practice Problems

Solve the following equations on a separate sheet of paper. Show the steps necessary to transpose each equation, find the numerical value of the unknown, and check the solution by substituting the numerical value for the unknown.

1. $a^2 = 9$

2. $y^2 = 64$

3. $3x^2 = 75$

4. $5a^2 = 180$

5. $x^3 = 8$

6. $b^3 = 27$

7. $\sqrt{y} = 7$

8. $\sqrt{a} = 9$

9. $\sqrt{4x} = 10$

10. $\sqrt{6y} = 12$

INEQUALITIES

An **inequality** is an algebraic statement in which one quantity is *less than, less than or equal to, greater than, greater than or equal to,* or *not equal to* another quantity. In Chapter 10, the concept of an inequality was shown on the number line. In this chapter, this concept is expanded to include solutions to inequalities that have unknowns. Follow the rules of inequalities when trying to find their solutions.

RULES OF INEQUALITIES

- The signs of inequalities include:
  Greater than $>$
  Less than $<$
  Greater than or equal to $\geq$
  Less than or equal to $\leq$
  Not equal to $\neq$
- To solve, transpose the inequality the same as an equation to isolate the unknown.
- If transposing involves multiplying or dividing both sides of the inequality by a negative number, change the direction **(sense)** of the inequality sign. Otherwise, the direction stays the same.
- The solution is all numbers that satisfy the inequality. This may include whole numbers, fractions, and decimals.

Sample Problem 12-19.

Solve this inequality:

$$3x > 12$$

Step 1. Transpose the equation.

$$\frac{3x}{3} > \frac{12}{3} \quad \text{(divide by 3)}$$

Step 2. Simplify for the solution.

$$x > 4 \text{ (all numbers greater than 4)}$$

Step 3. Check by substituting a value for the unknown. Since the unknown, x, is any values greater than 4, try 5.

$$3(5) > 12$$

$$15 > 12$$

Sample Problem 12-20.

Solve this inequality:

$$y - 6 < 10$$

Step 1. Transpose the equation.

$$y - 6 + 6 < 10 + 6 \quad \text{(add 6)}$$

Step 2. Simplify for the solution.

$$y < 16 \quad \text{(all numbers less than 16)}$$

Step 3. Check by substituting a value for the unknown.

$$(0) - 6 < 10 \quad \text{(try 0)}$$

$$-6 < 10$$

Sample Problem 12-21.

Solve this inequality:

$$8 - a \geq 25$$

Step 1. Transpose the equation.

$$8 - a \underline{- 8} \geq 25 \underline{- 8} \quad \text{(subtract 8)}$$

Step 2. Simplify for the solution. To get the unknown by itself, you must divide out the -1 coefficient. Since dividing both sides by the same negative number, change direction of the inequality sign.

$$-a \geq 17$$

$$a \leq -17 \quad \text{(change signs and inequality)}$$

Step 3. Check by substituting the value equal and a value less than for the unknown.

$$8 - (-17) \geq 25 \quad \text{(try equal value: 17)}$$

$$25 \geq 25$$

$$8 - (-20) \geq 25 \quad \text{(try value less than: } -20\text{)}$$

$$28 \geq 25$$

Practice Problems

Solve the following equations on a separate sheet of paper. Show the steps necessary to transpose each equation, find the numerical value of the unknown, and check the solution by substituting the numerical value for the unknown.

1. $5x > 15$
2. $4a < 24$
3. $-6y < 72$
4. $-3b > 51$
5. $-18 \geq 2a$
6. $-30 \leq 5x$
7. $-12 \leq -4b$
8. $-48 \geq -6y$
9. $3x > -42$
10. $7a < -49$
11. $2y^2 < 50$
12. $5b^2 > 245$
13. $a - 5 > 20$
14. $x + 3 < 15$
15. $7 + b \leq 2b + 4$
16. $5 + 5y \geq -3y + 37$
17. $8x - 20 \geq 4x + 20$
18. $6x + 6 \leq 5x + 12$
19. $(3x - 4)^2 > 4$
20. $(2a - 10)^2 < 16$

EQUATIONS WITH TWO UNKNOWN/SYSTEMS OF EQUATIONS

While procedures discussed to this point focused on solving equations with one unknown, equations and inequalities may contain two (or more) unknowns. Equations of two unknowns have numerous solutions. For example, in the equation

$y = 2x + 2$, solutions include $x = 0$, $y = 2$; $x = 1$, $y = 4$; $x = 2$, $y = 6$; $x = -3$, $y = -4$, and so on. This type of equation is known as *an equation in two unknowns.* The **solutions,** the pairs of numbers that satisfy this equation, are infinite in number. Choose any value of x, and a value for y may be found that makes the equation true.

When two equations, each in two unknowns, are given, they make up a **system of simultaneous equations.** In such a system—two equations in two unknowns—the solution is limited. Values must satisfy both equations. The unknowns have the same value in both equations. If the equation $y = 2x + 2$ was a member of a system whose other member was the equation $y = x - 1$, the solution would be limited to $x = -3$, $y = -4$. Only these numbers satisfy both equations within the system.

When solving two equations with two unknowns, the goal is to eliminate one of them from one of the equations. The value of that unknown, then, is substituted into the other equation. The second unknown remains and its value is readily found. Two methods of solving simultaneous equations are explored.

Solution by the addition/subtraction method

Using the addition/subtraction method, the two equations are added or subtracted together. This will result in a new equation with only one unknown. Follow the rules for addition/subtraction method.

RULES FOR ADDITION/SUBTRACTION METHOD

1. Add or subtract equations.
   a. If the two equations have an unknown whose coefficients are of equal value value (disregarding the signs):
      - Coefficients having opposite signs: Add the equations, eliminating the variable. Refer to Sample Problem 12-22.
      - Coefficients having the same sign: Subtract the equations, eliminating the variable. Refer to Sample Problem 12-23.
   b. If the two equations do *not* have an unknown whose coefficients are of equal value (disregarding the signs):
      - Multiply all terms of either equation by a factor to make one unknown have the same coefficient in both equations. Refer to Sample Problem 12-24.
      - It may be necessary to use one factor for one equation and a different factor for the other equation. This is a process similar to changing fractions to a common denominator. Refer to Sample Problem 12-25.
      - After adjusting the coefficients, follow step a.
2. Solve the equation resulting from the addition/subtraction for the unknown remaining.
3. Substitute the value of this unknown into either original equation and solve for the other unknown.
4. Check the solutions to both variables by substituting them into both original equations.

Sample Problem 12-22.

Solve these equations:

$$x + y = 8$$

$$x - y = 2$$

Step 1. Add the equations.

$$\begin{array}{r} x + y = 8 \\ x - y = 2 \\ \hline 2x \quad\;\; = 10 \end{array}$$

Step 2. Solve the resulting equation.

$$2x = 10$$

$$2x = 5$$

Step 3. Substitute into either original equation and solve for the other unknown.

$$x + y = 8$$
$$(5) + y = 8$$
$$y = 3$$

Step 4. Check the solutions in both original equations.

Solutions: $x = 5$ and $y = 3$

$$x + y = 8$$
$$(5) + (3) = 8$$
$$8 = 8$$

$$x - y = 2$$
$$(5) - (3) = 2$$
$$2 = 2$$

Sample Problem 12-23.

Solve these equations:

$$5a - 3b = 12$$
$$2a - 3b = 3$$

Step 1. Subtract the equations.

$$\begin{array}{rcl} 5a - 3b &=& 12 \\ 2a - 3b &=& 3 \\ \hline 3a &=& 9 \end{array}$$

Step 2. Solve the resulting equation.

$$3a = 9$$
$$a = 3$$

Step 3. Substitute into either original equation and solve for the other unknown.

$$5a - 3b = 12$$
$$5(3) - 3b = 12$$
$$15 - 3b = 12$$
$$-3b = -3$$
$$b = 1$$

Step 4. Check the solutions in both original equations.

Solutions: $a = 3$ and $b = 1$

$$5a - 3b = 12$$
$$5(3) - 3(1) = 12$$
$$15 - 3 = 12$$
$$12 = 12$$

$$2a - 3b = 3$$
$$2(3) - 3(1) = 3$$
$$6 - 3 = 3$$
$$3 = 3$$

Sample Problem 12-24.

Solve these equations:

$$7x - 2y = 2$$

$$3x + 4y = 30$$

Step 1. To eliminate y, multiply first equation by 2:

$$7x - 2y = 2$$

$$2 \cdot 7x - 2 \cdot 2y = 2 \cdot 2$$

$$14x - 4y = 4$$

Add the equations:

$$14x - 4y = 4$$
$$\underline{3x + 4y = 30}$$
$$17x \qquad = 34$$

Step 2. Solve the resulting equation.

$$17x = 34$$

$$x = 2$$

Step 3. Substitute into either original equation and solve for the other unknown.

$$3x + 4y = 30$$

$$3(2) + 4y = 30$$

$$6 + 4y = 30$$

$$4y = 24$$

$$y = 6$$

Step 4. Check the solutions in both original equations.

Solutions: $x = 2$ and $y = 6$

$$7x - 2y = 2$$

$$7(2) - 2(6) = 2$$

$$14 - 12 = 2$$

$$2 = 2$$

$$3x + 4y = 30$$

$$3(2) + 4(6) = 30$$

$$6 + 24 = 30$$

$$30 = 30$$

Sample Problem 12-25.

Solve these equations:

$$2x + 3y = 6$$

$$3x + 4y = 7$$

Step 1. "Common denominator" to eliminate y is 12. Multiply first equation by 4 and second equation by 3.

$$2x + 3y = 6$$

$$4 \cdot 2x + 4 \cdot 3y = 4 \cdot 6$$

$$8x + 12y = 24$$

$$3x + 4y = 7$$

$$3 \cdot 3x + 3 \cdot 4y = 3 \cdot 7$$

$$9x + 12y = 21$$

Subtract the adjusted equations:

$$\begin{array}{rcl} 8x + 12y &=& 24 \\ 9x + 12y &=& 21 \\ \hline -1x &=& 3 \end{array}$$

Step 2. Solve the resulting equation.

$$-1x = 3$$

$$x = -3$$

Step 3. Substitute into either original equation and solve for the other unknown.

$$3x + 4y = 7$$

$$3(-3) + 4y = 7$$

$$-9 + 4y = 7$$

$$4y = 16$$

$$y = 4$$

Step 4. Check the solutions in both original equations.

Solutions: x = 3 and y = 4

$$2x + 3y = 6$$

$$2(-3) + 3(4) = 6$$

$$-6 + 12 = 6$$

$$6 = 6$$

$$3x + 4y = 7$$

$$3(-3) + 4(4) = 7$$

$$-9 + 16 = 7$$

$$7 = 7$$

Practice Problems

On a separate sheet of paper, solve the systems of equations using the addition/subtraction method. Show the steps involved and check the solutions in both original equations.

1. $x + y = 9$
   $2x - 3y = -2$

2. $4x + 5y = -7$
   $6x - 5y = 27$

3. $x + 2y = -10$
   $4x - 5y = -14$

4. $5x = 5y$
   $-3x - 7y = -20$

5. $7x + 5y = -19$
   $-7x + 3y = 11$

6. $-9x + 5y = 52$
   $3x - 2y = -19$

$$7. \quad 10a - 2b = -18$$
$$-5a + 3b = 27$$

$$8. \quad 2a + 4b = -14$$
$$5a - 8b = -35$$

$$9. \quad -3x + 5y = -32$$
$$6x - 5y = 44$$

$$10. \quad 10x + y = -9$$
$$x + 8y = 7$$

Solution by substitution method

An equation containing two unknowns can be arranged so that one of them is isolated and equal to an expression containing the second unknown. When the expression is substituted into the other equation, the equation will contain only one unknown. The value of the unknown can be found, then substituted into the first equation. The value of second variable can then be found.

RULES FOR SUBSTITUTION METHOD

1. Select one equation. Solve this equation for one unknown. This yields an expression that is in terms of the second unknown.
2. In the other equation, replace the unknown solved for in step 1 with the expression that resulted from step 1.
3. Solve this combined equation, finding the number value of the remaining unknown.
4. Substitute the number value into the first equation and solve for the number value of the first unknown.
5. Check by substituting both values into both equations.

Sample Problem 12-26.

Solve these equations.

$$x + 2y = 10$$
$$2x - y = 5$$

Step 1. Select an equation and solve for one unknown.

$$x + 2y = 10$$
$$x = 10 - 2y$$

Step 2. Substitute for x into the other equation.

$$2x - y = 5$$
$$2(10 - 2y) - y = 5$$

Step 3. Solve this equation for y.

$$2(10 - 2y) - y = 5$$
$$20 - 4y - y = 5$$
$$20 - 5y = 5$$
$$-5y = 5 - 20$$
$$-5y = -15$$
$$y = 3$$

Step 4. Substitute into the first equation.

$$x + 2y = 10$$
$$x + 2(3) = 10$$
$$x + 6 = 10$$
$$x = 4$$

Step 5. Check both unknowns in both equations.

Solutions: x = 4 and y = 3

x + 2y = 10

(4) + 2(3) = 10

4 + 6 = 10

10 = 10

2x − y = 5

2(4) − (3) = 5

8 − 3 = 5

5 = 5

Sample Problem 12-27.

Solve these equations.

x − y = −4

4y − 3x = 14

Step 1. Select an equation and solve for one unknown.

x − y = −4

x = y − 4

Step 2. Substitute for x into the other equation.

4y − 3x = 14

4y − 3(y − 4) = 14

Step 3. Solve this equation for y.

4y − 3(y − 4) = 14

4y − 3y + 12 = 14

y + 12 = 14

y = 2

Step 4. Substitute into the first equation.

x − y = −4

x − (2) = −4

x − 2 = −4

x = −2

Step 5. Check both unknowns in both equations.

Solutions: x = −2 and y = 2

x − y = −4

(−2) − (2) = −4

−2 − 2 = −4

−4 = −4

4y − 3x = 14

4(2) − 3(−2) = 14

8 − 6 = 14

14 = 14

Practice Problems

On a separate sheet of paper, solve the systems of equations using the substitution method. Show the steps involved and check the solutions in both original equations.

1. $3x + y = 0$
    $-5x - 2y = 1$

2. $6a + 2b = 40$
    $-a - 3b = -20$

3. $4x - y = 16$
    $x + 5y = -38$

4. $x + 10y = 14$
    $-2x - 3y = 6$

5. $-10x - 5y = -25$
    $5x + 2y = 4$

6. $-x - y = 0$
    $-5x + 10y = 5$

7. $-3x + 8y = -20$
    $4x - 2y = -8$

8. $7x + 3y = 4$
    $14x - 10y = -88$

9. $4x - 5y = 43$
    $6x + 5y = 27$

10. $-9x - 2y = -10$
    $18x + 3y = 15$

TEST YOUR SKILLS

Do *not* write in this book. Use a separate sheet of paper to complete the following problems. Show your work and your final answer.

ISOLATING THE UNKNOWN

Solve the following equations. Show the steps necessary to transpose, solve for the unknown, and check the solution with substitution.

Example: $x + 12 = 6$

Transpose: $x + 12 - 12 = 6 - 12$ (subtract 12)

Solution: $x = -6$

Check: $(-6) + 12 = 6$

$6 = 6$

1. $x + 4 = 9$

2. $3 + a = 7$

3. $y - 4 + 10 = -4$

4. $b - 5 = 0$

5. $-8 + z = 4$

6. $10 - c = 5$

7. $2a + 3a = 10$

8. $2 + 3x = 6 - x$

9. $7d = 3d + 2$

10. $\dfrac{y}{4} = 20$

11. $\dfrac{3x}{2} = -15$

12. $\dfrac{12}{2a} = 6$

13. $12x + 3 = 6x - 21$

14. $7(5x - 2) = 6(6x - 1)$

15. $3 - (y + 2) = 5 + 3(y + 2)$

16. $8(2a + 1) = 4(7a + 8)$

17. $5 + y(2 + 4) = 2(y + 3) - 4$

18. $\dfrac{x}{2} + \dfrac{2x}{3} - \dfrac{5}{6} = \dfrac{4x}{12}$

19. $\dfrac{3}{4a} - \dfrac{1}{2a} + \dfrac{2}{8} = \dfrac{2}{a}$

20. $\dfrac{3}{8} = \dfrac{5}{8x} + \dfrac{3}{2x} - \dfrac{1}{x}$

21. $0.25x - 1.5 + 0.05x = 0.15x$

22. $x - 0.25x = 1.5$

23. $x^2 = 25$

24. $3a^2 = 12$

25. $\sqrt{y^2} = 100$

INEQUALITIES

Solve the following inequalities. Show the steps necessary to transpose, solve for the unknown, and check the solution with substitution.

26. $x + 2 > 7$

27. $y - 3 < 1$

28. $3a - 5 \geq 10$

29. $2 + 3b \leq 14$

30. $6 - 4a > 4 - 2a$

EQUATIONS WITH TWO UNKNOWNS/SYSTEMS OF EQUATONS

Solve the systems of equations using the addition/subtraction method. Show the steps involved and check the solutions in both equations.

31. $3x + y = 10$
    $4x - y = 4$

32. $2x + y = 10$
    $x + y = 6$

33. $3x - y = 5$
    $x - y = 1$

34. $2a - b = 8$
    $a + 2b = 9$

35. $2x + 3y = 5$
    $3x - 2y = 1$

Solve the systems of equations using the substitution method. Show the steps involved and check the solutions in both equations.

36. $x + y = 5$
    $2x + 5y = 16$

37. $2x + y = 7$
    $4x - 2y = -6$

38. $x + 3y = 16$
    $7x + 4y = 10$

39. $x - y = 5$
    $3x - 2y = 14$

40. $4x - y = 0$
    $6x - 3y = 12$

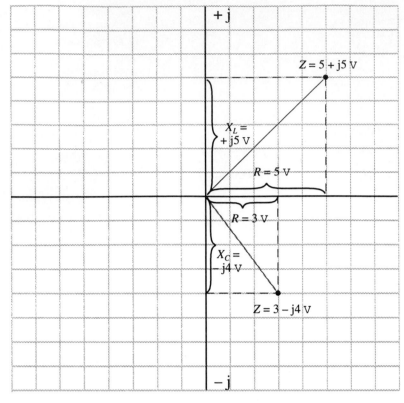

A

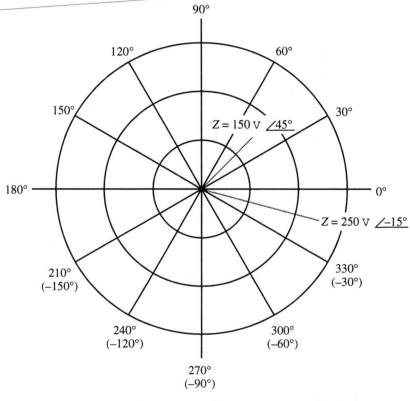

B

In this illustration, electrical impedance is expressed in two different coordinate systems.
A—Rectangular coordinate system. B—Polar coordinate system.

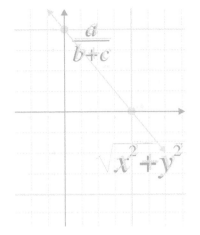

Chapter 13

Graphs of Linear Equations and Inequalities

OBJECTIVES

After studying this chapter, you will be able to:
- Explain basic terminology of the rectangular coordinate system.
- Draw graphs of one-variable equations.
- Draw graphs of two-variable equations.
- Give the equation of a straight line from its graph.
- Solve inequalities graphically.
- Solve simultaneous equations graphically.

A graph gives a "picture" of an equation or inequality with two variables. It shows the relationship between the two. (A **functional relationship** exists if for every value of the independent variable, there is only one corresponding value of the dependent variable. Remember from Chapter 6, a dependent variable is a variable whose value depends on the value of another variable—the independent variable. Generally, in algebraic equations, y is the dependent and x is the independent variable.)

All of the graphs presented here are *linear,* which means they are straight lines. The equations depicted by these graphs are **linear equations**. Linear equations are equations of the first degree. Their general form is $Ax + By + C = O$. Graphs are quite useful when finding the one solution that is common to two equations with two variables. They also provide a method of finding variable values without actual computation.

RECTANGULAR COORDINATE SYSTEM

The **rectangular coordinate system** is a standardized system for drawing a graph. Refer to Fig. 13-1. The **x-axis** is the main horizontal line. It is the same as the number line discussed in Chapter 10. The x-axis is labeled with $+x$ on the right and $-x$ on the left. The **y-axis** is drawn vertically, with $+y$ on the top and $-y$ on the bottom. Just like the x-axis, it is also a number line. The **origin** is where the x-axis and the y-axis cross. The origin has both x and y values of zero.

The x- and y-axes divide a rectangular coordinate system into four quarters or **quadrants**. The quadrants are defined as follows:
- *First quadrant:* x is positive and y is positive.
- *Second quadrant:* x is negative and y is positive.
- *Third quadrant:* x is negative and y is negative.
- *Fourth quadrant:* x is positive and y is negative.

An **ordered pair** locates a point in the rectangular coordinate system. An ordered pair is written inside parentheses, with x always first, followed by $y,$ and separated by a comma: (x,y). In this ordered pair, x and y are **coordinates**.

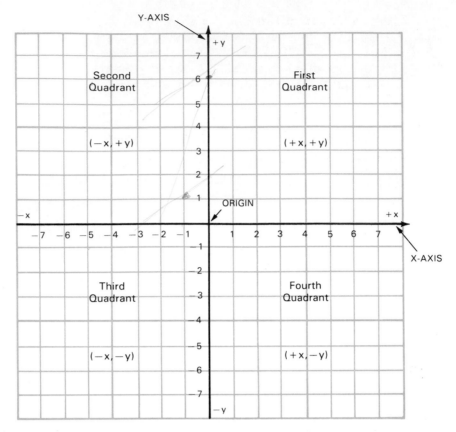

Fig. 13-1. The rectangular coordinate system.

Refer to Fig. 13-2. In this figure, there are five points: one in the center and one in each quadrant. To locate a point in a quadrant, count right or left on the x-axis as directed by the first number in the ordered pair. Then, count up or down on the y-axis as directed by the second number of the ordered pair. In this figure, a dotted line is used to help visualize each point and its relationship to the axis.

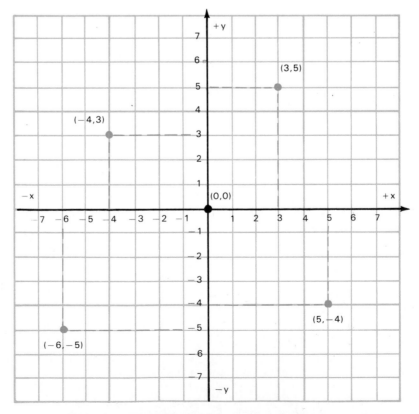

Fig. 13-2. Locating points in rectangular coordinates, (x,y).

Points located in Fig. 13-2 are given by:
- The center point, or the origin: (0,0) x = 0 and y = 0.
- First quadrant: (3,5) x = +3 and y = +5.
- Second quadrant: (−4,3) x = −4 and y = +3.
- Third quadrant: (−6,−5) x = −6 and y = −5.
- Fourth quadrant: (5,−4) x = 5 and y = −4.

Practice Problems

Using a piece of graph paper, draw lines for the x- and y-axes. Place each of the following ordered pairs on the graph and label each point.

(2,2)	(−3,5)	(6,−7)	(−4,−2)	(5,8)
(−5,8)	(2,−2)	(−1,−8)	(0,5)	(0,−3)
(1,0)	(−2,0)	(3,5)	(−8,3)	(4,−5)
(−3,−6)	(2,8)	(−1,2)	(1,−8)	(−5,−1)
(8,1)	(−2,7)	(3,−4)	(−6,−6)	(0,0)

GRAPHS OF ONE-VARIABLE EQUATIONS

When an equation has only one variable, it is plotted on the rectangular coordinate system as a straight line, parallel to one of the axes. The graph of a single-variable equation intersects (crosses through) the axis of the variable at a point equal to its value. The second variable, not used in the equation, can be any value, and the equation is still valid.

Now, you might be thinking, that if such an equation has only one variable, what is this *second* variable just mentioned? As previously stated, a graph gives a picture of two variables. Here is a special case. The second variable, in this instance, can be considered to have a coefficient of zero. Thereby, it is eliminated. For example, Fig. 1-33 shows the graphs of x = −3 and x = 5. Consider the original equations to be x + 0y = −3 and x + 0y = 5. Here, the y terms drop out of the equations, leaving the initial equations. The graphs of such equations still have coordinates for two variables.

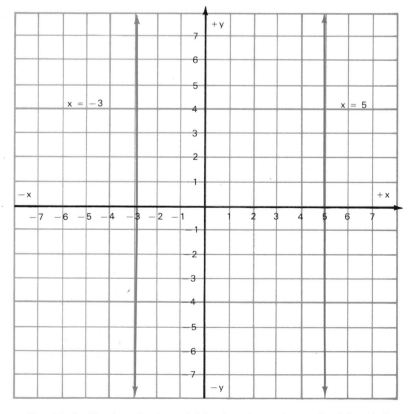

Fig. 13-3. Plotting single variable equations: x = −3 and x = 5.

Look again at the graphs of x = −3 and x = 5 in Fig. 13-3. These lines are drawn parallel to the y-axis. The graph of x = −3 intersects the x-axis at −3. The graph of x = 5 intersects the x-axis at +5. In either of these, y can be any value and have no effect on the equation.

Fig. 13-4 shows the graphs of y = 6 and y = −4. These lines are drawn parallel to the x-axis. The graph of y = 6 intersects the y-axis at 6. The graph of y = −4 intersects the y-axis at −4. These equations are valid with any value of x.

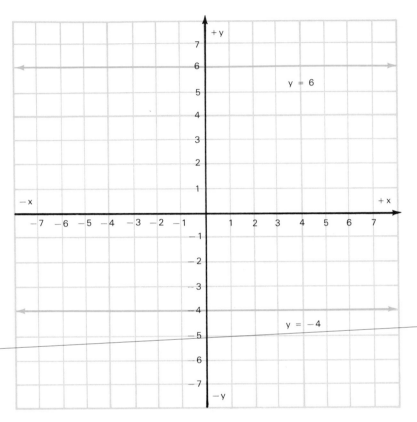

Fig. 13-4. Plotting single variable equations: y = 6 and y = −4.

GRAPHS OF TWO-VARIABLE EQUATIONS

An equation containing two variables has an infinite combination of (x,y) values that will solve the equation. The linear graph demonstrates this by being a straight line, with no ending points. For every value of x, there is a value of y to solve the equation.

Fig. 13-5 is the graph of the equation y = 2x − 1. To plot this graph, five different values are arbitrarily assigned to x. These values are substituted into the equation, which is solved to find the values of y. These (x,y) combinations are the ordered pairs used to draw the graph. The process of finding the y-values of the equation y = 2x − 1 is as follows:

$$\text{let } x = -1: y = 2(-1) - 1$$
$$y = -2 - 1$$
$$y = -3$$
$$\text{Ordered pair: } (-1, -3)$$

$$\text{let } x = 0: \quad y = 2(1) - 1$$
$$y = -1$$
$$\text{Ordered pair: } (0, -1)$$

$$\text{let } x = 1: \quad y = 2(1) - 1$$
$$y = 2 - 1$$
$$y = 1$$
$$\text{Ordered pair: } (1, 1)$$

let x = 2: y = 2(2) − 1
y = 4 − 1
y = 3
Ordered pair: (2,3)

let x = 3: y = 2(3) − 1
y = 6 − 1
y = 5
Ordered pair: (3,5)

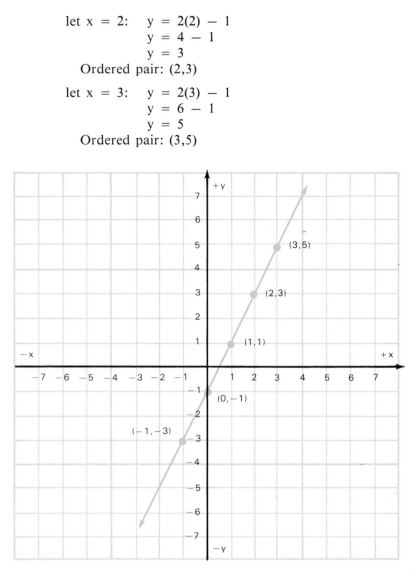

Fig. 13-5. Graph of the equation: y = 2x − 1.

Intercepts and slope

Only two points are needed to draw a straight line. Therefore, to plot a straight line, it is only necessary to find two ordered pairs. By using the value of zero for one variable, the location where the graph will cross the axis of the other variable can be found. Use zero for x to find where it crosses the y-axis. Use zero for y to find where it crosses the x-axis. (The ordered pairs that fall where the graph of a line crosses the axes are called the **x-** and **y-intercepts.**) Finding the intercepts provides another way of determining the graph of a line. The advantage of this method is that it will often greatly simplify the arithmetic. This method cannot be used, however, if the variable is in the denominator.

Fig. 13-6 is a graph of the equation x + y = 4. To find the ordered pairs for the graph of this equation, zero was selected—first for x, then for y. A third point was selected along the line, just to verify the line is correct. The process of using the intercept values to plot the line of x + y = 4 is as follows:

let x = 0: (0) + y = 4
y = 4
Ordered pair: (0,4)

let y = 0: x + (1) = 4
x = 4
Ordered pair: (4,0)

$$\text{let } x = 2: (2) + y = 4$$
$$y = 2$$
$$\text{Ordered pair: } (2,2)$$

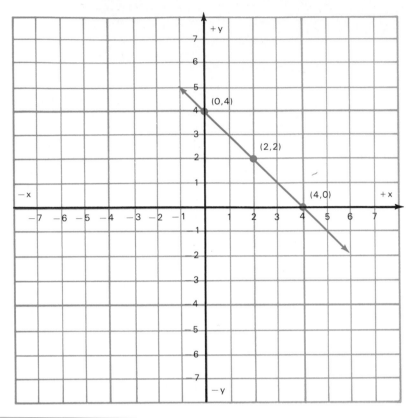

Fig. 13-6. Graph of the equation: x + y = 4.

The general form of a linear equation was mentioned. Two other forms are worthy of mention. They are:
• Point-slope form.
• Slope-intercept form.

The **slope (m)** of a line is a measure of the upward or downward slant of a line. It is equal to the *rise* over the *run*. In other words, it is defined as the vertical distance from one point to another, divided by the horizontal distance from the first point to the second. The equation for slope is:

$$M = \frac{\text{rise}}{\text{run}} = \frac{y - y_1}{x - x_1}$$

On a graph, a line rising to the right has a positive slope. A line falling to the right has a negative slope. A horizontal line has a slope of zero, and the slope of a vertical line is undefined. See Fig. 13-7.

Slope is also used in practice. In carpentry, slope indicates the incline of a roof. It is expressed as "x [inches] in 12." For example, a roof that rises at the rate of 4 inches for each foot of run is said to have a 4 in 12 slope. See Fig. 13-8.

The **point-slope form** of a straight line is useful for determining the equation of a line when the slope of the line and some point through which the line passes is known. The point-slope form of a linear equation is:

$$y - y_1 = m(x - x_1)$$

Direct substitution of slope and ordered pair (x_1, y_1) will give the equation for the line.

The **slope-intercept form** of a linear equation is useful because, written in this form, the slope of a line and its y-intercept are readily found. The slope-intercept form of a linear equation is:

$$y = mx + b$$

where m is the slope and b is the y-intercept. Fig. 13-7 is a graph of the equation $y = -\frac{1}{2}x + \frac{1}{2}$.

From this equation, the slope is readily determined as $-1/2$; the y-intercept as $1/2$.

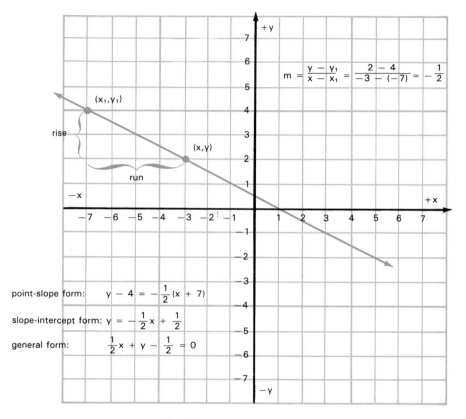

$$m = \frac{y - y_1}{x - x_1} = \frac{2 - 4}{-3 - (-7)} = -\frac{1}{2}$$

point-slope form: $y - 4 = -\frac{1}{2}(x + 7)$

slope-intercept form: $y = -\frac{1}{2}x + \frac{1}{2}$

general form: $\frac{1}{2}x + y - \frac{1}{2} = 0$

Fig. 13-7. The meaning of slope.

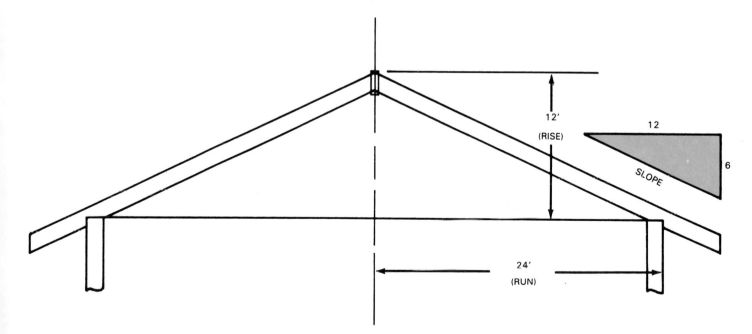

Fig. 13-8. A roof with a 6 in 12 slope. (Slope equals 1/2.)

Sample Problem 13-1.

Velocity of an object under constant acceleration is plotted as a function of time, resulting in a straight line. After 10 seconds, the object has a velocity of 20 feet per second. The slope of the line is 1/2. Determine the equation relating velocity and time and graph the equation. From the graph, determine the initial velocity and velocity after 20 seconds.

Step 1. Use the point-slope form to state the relationship.

$$y - y_1 = m(x - x_1)$$
$$x_1 = 10 \text{ sec.}$$
$$y_1 = 20 \text{ ft./sec.}$$
$$y - 20 = \frac{1}{2}(x - 10)$$

Note: Velocity is represented by the variable *y* because it is the dependent variable. The independent variable is time, so it is represented by *x*.

Step 2. Simplify and put in slope-intercept form.

$$y - 20 = \frac{1}{2}(x - 10)$$

$$y = \frac{1}{2}(x - 10) + 20$$

$$y = \frac{1}{2}x - 5 + 20$$

$$y = \frac{1}{2}x + 15$$

Step 3. Determine ordered pairs to plot the graph. One ordered pair is known: (10,20). The other can be found by inspecting the slope-intercept equation. The y-intercept is at 15 ft./sec. The value of x at the y-intercept is zero. Therefore, another ordered pair is found at (0.15).

Step 4. Plot the graph. See Fig. 13-9.

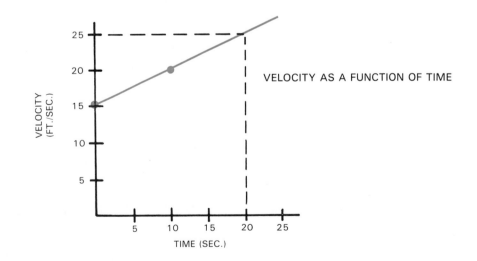

Fig. 13-9. *This graph gives the relationship of velocity to time of a certain object.*

Step 5. The initial velocity is the velocity at 0 seconds. It is the same as the y-intercept.
initial velocity = 15 ft./sec.
From the graph, the velocity at 20 seconds is found.
velocity at 20 seconds = 25 ft./sec.

Graphs of inequalities

Inequalities can be *greater than, greater than or equal to, less than,* or *less than or equal to.* They can also be *not equal to,* but this type will be excluded from this discussion. The solutions to an inequality are not just along a line, but rather every point above or below and, sometimes, including the line. The line is used as a boundary. If the boundary line is solid, the line is included in the solutions. This is the case only when an inequality includes *equal to.* If the boundary is a dashed line, the points on the line are excluded from the solutions.

Fig. 13-10 shows two, independent inequalities $y \geq x$ and $y \leq x$ on one graph. The boundary line is solid because the solutions of the inequalities include $y = x$. To the left of the line are the solutions to $y > x$. To the right of the line are the solutions to $y < x$. Values can be tested by selecting any ordered pair and testing to see if the inequality is true.

Just as simultaneous equations may be solved graphically, so too may simultaneous inequalities. This is done by drawing the graph of each inequality on one graph, and finding the region in which the graphs overlap. The ordered pairs lying in the overlapping region satisfy both inequalities. This region, therefore, is the solution to the system of inequalities. Consider if the two inequalities graphed in Fig. 13-10 were simultaneous inequalities, and not independent, as stated. Their solution would be the line of $y = x$. All ordered pairs along this line satisfy both equations.

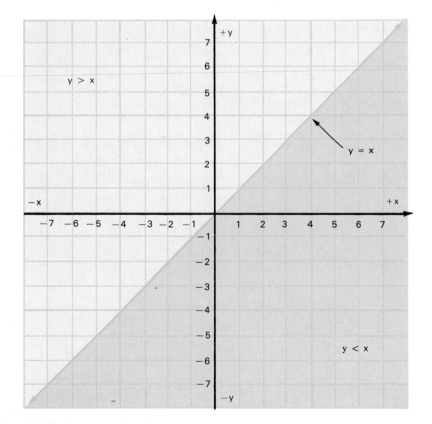

Fig. 13-10. A comparison between $y \geq x$ and $y \leq x$ with $y = x$ as a boundary.

Sample Problem 13-2.

Draw the graph of this inequality:

$$4x - 2y < 8$$

Step 1. Determine the boundary line by replacing the inequality sign with an equal sign and finding ordered pairs for the graph of the equation.

$$4x - 2y < 8$$
$$4x - 2y = 8$$

Using the intercept method:

$$\text{let } x = 0: 4(x) - 2y = 8$$
$$-2y = 8$$
$$y = -4$$

Ordered pair: $(0, -4)$

$$\text{let } y = 0: 4x - 2(0) = 8$$
$$4x = 8$$
$$x = 2$$

Ordered pair: $(2, 0)$

Step 2. Draw a line to represent the inequality. Since the inequality does not contain *equal to,* it is a dashed line. See Fig. 13-11.

Step 3. Select an ordered pair one side of the line and another pair the other side of the line. Substitute the values in the inequality. One pair will be a correct solution. The other pair will be wrong.

Pair to the left—select: $(0,0)$
$$4x - 2y < 8$$
$$4(0) - 2(0) < 8$$
$$0 < 8 \text{ (correct)}$$

Pair to the right—select: $(4, -4)$
$$4x - 2y < 8$$
$$4(4) - 2(-4) < 8$$
$$16 - (-8) < 8$$
$$24 < 8 \text{ (wrong)}$$

Step 4. Shade the graph on the side of the line containing the correct solutions.

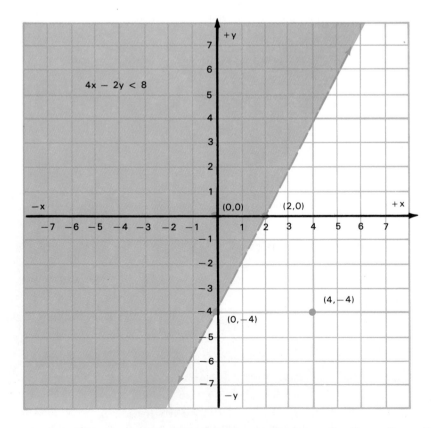

Fig. 13-11. The shaded area is the solution to the inequality $4x - 2y < 8$.

Practice Problems

Graph the following equations and inequalities on graph paper. Include the points where the line crosses the x- and y-axes. Determine the slope of the line.

1. $x = 2$
2. $x = -6$
3. $y = -2$
4. $y = 4$
5. $x > 3$
6. $x \leq 2$
7. $y < -4$
8. $y \geq -1$
9. $x > 0$
10. $y < 0$
11. $x + y = 0$
12. $-x + y = -3$
13. $2x - 3y = -13$
14. $-4x + 2y = -12$
15. $3x - 6y + 30 = 0$
16. $-2x + 4y - 6 = 0$
17. $5x + y < -8$
18. $-7x + 3y < 2$
19. $-4x - 2y > -10$
20. $x - 5y > 9$

GRAPHIC SOLUTION OF SIMULTANEOUS LINEAR EQUATIONS

One equation with two variables has an infinite number of solutions. These can be found along a graph of the equation. Simultaneous linear equations have only one solution common to both equations. Three methods of finding their solution include the addition/subtraction method, the substitution method, and the graphics method. The first two are presented in Chapter 12. To solve by the graphics method, the graphs of the two equations are placed on the same rectangular coordinate system. The single point where the lines intersect is the only solution common to both equations. Equations of lines that are parallel have no common solution.

Sample Problem 13-3.

Solve this system of equations graphically:

$$2x - y = 5$$
$$x + y = 4$$

Step 1. Find ordered pairs for the graph of $2x - y = 5$. Using the intercept method:

$$\text{let } x = 0: 2(0) - y = 5$$
$$-y = 5$$
$$y = -5$$

Ordered pair: $(0, -5)$

$$\text{let } y = 0: 2x - (0) = 5$$
$$2x = 5$$
$$x = 2\frac{1}{2}$$

Ordered pair: $(2\frac{1}{2}, 0)$

Step 2. Find ordered pairs for the graph of $x + y = 4$. Using the intercept method:

$$\text{let } x = 0: (0) + y = 4$$
$$y = 4$$

Ordered pair: $(0, 4)$

$$\text{let } y = 0: x + (0) = 4$$
$$x = 4$$

Ordered pair: $(4, 0)$

Step 3. Plot the graphs. Refer to Fig. 13-12.

Step 4. Examine the graphs of these equations. The point where the two lines intersect is the common solution to both equations. The solution is $(3, 1)$.

Step 5. Check the solution by substituting into both equations.

Solutions: x = 3, y = 1

2x − y = 5	x + y = 4
2(3) − (1) = 5	(3) + (1) = 4
6 − 1 = 5	4 = 4
5 = 5	

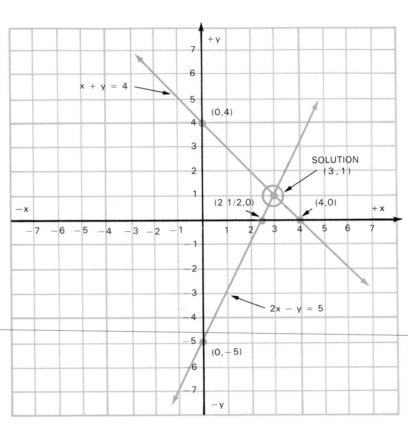

Fig. 13-12. Graph of the equations: 2x − y = 5
x + y = 4

Sample Problem 13-4.

Solve this system of equations graphically:

$$x - y = 2$$
$$2x + 3y = 9$$

Step 1. Find ordered pairs for the graph of x − y = 2. Using the intercept method:

let x = 0: (0) − y = 2
−y = 2
y = −2
Ordered pair: (0,−2)

let y = 0: x − (0) = 2
x = 2
Ordered pair: (2,0)

Step 2. Find ordered pairs for the graph of 2x + 3y = 9. Using the intercept method:

let x = 0: 2(0) + 3y = 9
3y = 9
y = 3
Ordered pair: (0,3)

$$\text{lct } y - 0:\ 2x + 3(0) = 9$$
$$2x = 9$$
$$x = 4\frac{1}{2}$$

Ordered pair: $(4\frac{1}{2}, 0)$

Step 3. Plot the graphs. Refer to Fig. 13-13.

Step 4. Examine the graphs of these equations. The point where the two lines intersect is the common solution to both equations. The solution is (3,1).

Step 5. Check the solution by substituting into both equations.

$$\begin{array}{ll}
\text{Solutions: } x = 3,\ y = 1 & 2x + 3y = 9 \\
\qquad\quad x - y = 2 & 2(3) + 3(1) = 9 \\
\qquad\quad (3) - (1) = 2 & 6 + 3 = 9 \\
\qquad\qquad\quad 2 = 2 & 9 = 9
\end{array}$$

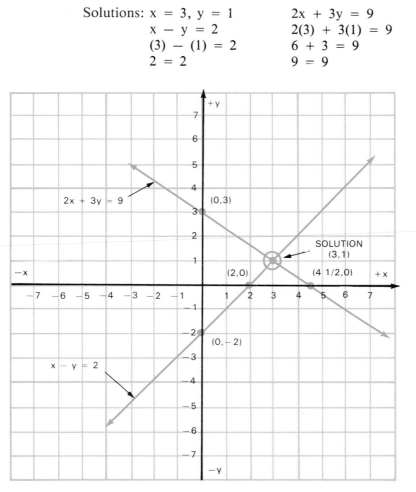

Fig. 13-13. Graph of the equations: x − y = 2
2x + 3y = 9

Sample Problem 13-5.

Solve this system of equations graphically:

$$x + y = 0$$
$$3x - 2y = 10$$

Step 1. Find ordered pairs for the graph of $x + y = 0$. Using the intercept method:

$$\text{let } x = 0:\ (0) + y = 0$$
$$y = 0$$

Ordered pair: (0,0)

Note: If a straight line passes through the origin, its x- and y-intercepts are the same ordered pairs, (0,0). Since two different points are needed to define a line, the intercept method cannot be used to find the second ordered pair.

Therefore:

$$\text{let } y = -4: \quad x + (-4) = 0$$
$$x = 4$$

Ordered pair: $(4, -4)$

Step 2. Find ordered pairs for the graph of $3x - 2y = 10$. Using the intercept method:

$$\text{let } x = 0: \quad 3(0) - 2y = 10$$
$$-2y = 10$$
$$y = -5$$

Ordered pair: $(0, -5)$

$$\text{let } y = 0: \quad 3x - 2(0) = 10$$
$$3x = 10$$
$$x = 3\frac{1}{3}$$

Ordered pair: $(3\frac{1}{3}, 0)$

Step 3. Plot the graphs. Refer to Fig. 13-14.

Step 4. Examine the graphs of these equations. The point where the two lines intersect is the common solution to both equations. The solution is $(2, -2)$.

Step 5. Check the solution by substituting into both equations.

Solutions: $x = 2$, $y = -2$

$x + y = 0$	$3x - 2y = 10$
$(2) + (-2) = 0$	$3(2) - 2(-2) = 10$
$0 = 0$	$6 - (-4) = 10$
	$10 = 10$

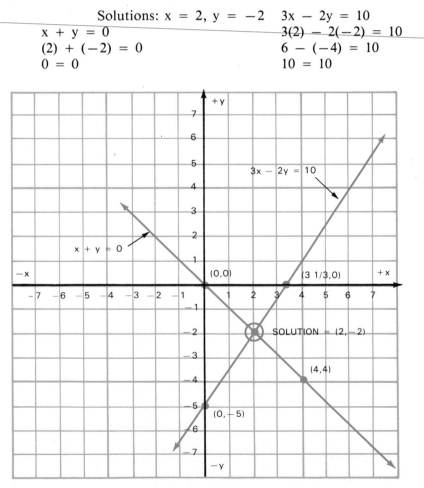

Fig. 13-14. Graph of the equations: $x + y = 0$
$3x - 2y = 10$

Sample Problem 13-6.

Solve this system of equations graphically:

$$y = -x + 5$$
$$y - 2x = -1$$

Step 1. Find ordered pairs for the graph of $y = -x + 5$. Using the intercept method:

$$\text{let } x = 0: y = -(0) + 5$$
$$y = 5$$

Ordered pair: (0,5)

$$\text{let } y = 0: (0) = -x + 5$$
$$x = 5$$

Ordered pair: (5,0)

Step 2. Find ordered pairs for the graph of $y - 2x = 1$. Using the intercept method:

$$\text{let } x = 0: y - 2(0) = -1$$
$$y = -1$$

Ordered pair: $(0, -1)$

$$\text{let } y = 0: (0) - 2x = -1$$
$$-2x = -1$$
$$x = \frac{1}{2}$$

Ordered pair: $(\frac{1}{2}, 0)$

Step 3. Plot the graphs. Refer to Fig. 13-15.

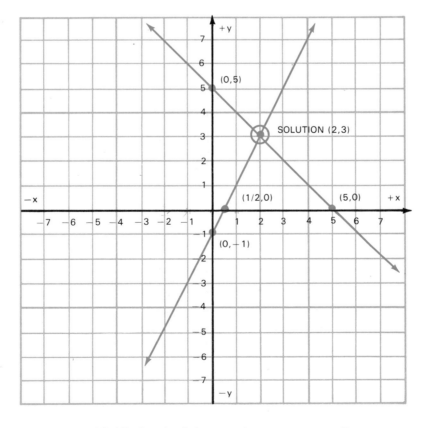

Fig. 13-15. Graph of the equations: $y = -x + 5$
$$y - 2x = -1$$

Step 4. Examine the graphs of these equations. The point where the two lines intersect is the common solution to both equations. The solution is (2,3).

Step 5. Check the solution by substituting into both equations.

Solutions: x = 2, y = 3   y − 2x = −1
         y = −x + 5   (3) − 2(2) = −1
         (3) = −(2) + 5   3 − 4 = −1
         3 = 3   −1 = −1

Practice Problems

Graph the following pairs of equations on graph paper. Include the points where each line crosses the x- and y-axes. Determine the point that is the common solution to both equations.

1. 2x + 3y = 7
   x − 4y = −13

2. −2x − 5y = 1
   −4x + 2y = −10

3. 3x − 4y = −3
   5x + y = 18

4. 6x + y = −26
   −3x + 3y = 6

5. −5x + 2y = −16
   5x + 7y = 34

6. 4x + 3y = −5
   x + y = 0

7. −2x + 3y = −6
   x − y = 1

8. −3x − 2y = 3
   7x + 2y = 1

9. x + 10y = 14
   −2x − 3y = 6

10. 4x − y = 2
    −5x + y = −2

TEST YOUR SKILLS

Do *not* write in this book. Use a separate sheet of paper to complete the following problems. Show your work and your final answer.

1. Give the ordered pair for points a through l on the graph in Fig. 13-16.

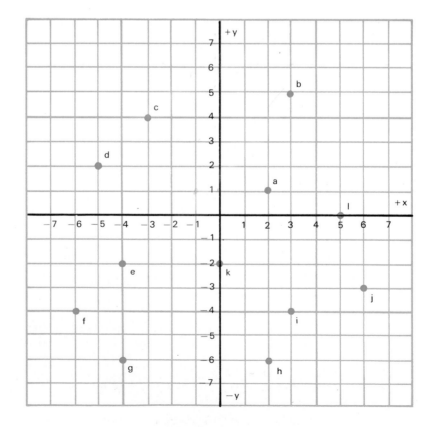

Fig. 13-16. Give the ordered pairs.

Use graph paper similar to that shown in Fig. 13-17 for Problems 2 through 5.

2. With the equation $4x - 2y = 14$:
   a. Show your work to find three ordered pairs that will solve the equation.
   b. On graph paper, label the three ordered pairs and draw the graph of the equation.
   c. Give the equation in slope-intercept form.
   d. Give the slope and y-intercept.

3. Solve these equations using a graph. Show your work to find the ordered pairs. Check the solution in both equations.

$$y = x + 3$$
$$x + y = 6$$

4. Solve these equations using a graph. Show your work to find the ordered pairs. Check the solution in both equations.

$$x = y + 3$$
$$2x + y = 6$$

5. Shade the area giving the solutions to the following inequality. Show your work to find the ordered pairs. Select an ordered pair to prove the inequality.

$$4x + 2y < 10$$

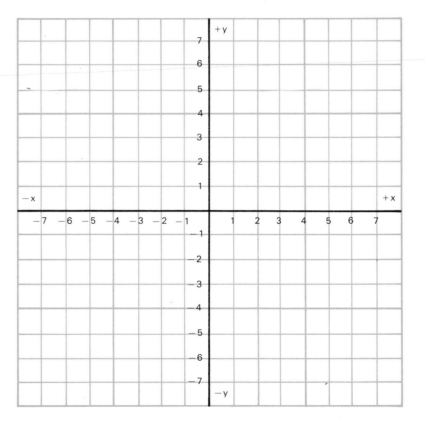

Fig. 13-17. Graph paper for a rectangular coordinate system.

PROBLEM-SOLVING ACTIVITIES
Activity 13-1.
Classroom-size rectangular coordinate system

Objective: To create a classroom-size rectangular coordinate system.
Instructions:
1. Use masking tape to mark the x- and y-axes on the floor of a classroom. The origin of the axes should be near the middle of the room.

2. Place tape marks at 1' intervals along each axis. (If the floor has square tiles, use the size of the tile as the interval measurement.)
3. Place pieces of tape at the points where the markings along the x-axis intersect the markings along the y-axis. When finished, the tape on the floor will form a grid (set of points) that resembles a piece of graph paper.

Activity 13-2.
Graphing a linear equation

    Objective: To use a string to show the graph of a linear equation using the classroom-size rectangular coordinate system created in Activity 13-1.

    Instructions:

1. Select an equation from the practice exercises in this chapter.
2. Solve the equation for the x- and y-intercept values. Mark these values on the floor.
3. Pull a string tightly across the grid so that it crosses the calculated intercept values. Every point along the string is a possible solution.
4. If two equations are graphed, the point where the two strings cross represents the one common solution.

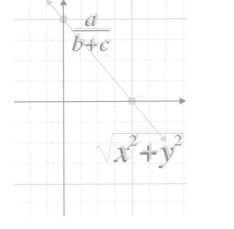

Chapter 14

Solving Problems with Algebra

OBJECTIVES

After studying this chapter, you will be able to:
- Change word expressions into algebraic expressions.
- Solve word problems using algebra.
- Solve problems using ratio and proportion.

The task of solving word problems can be eased by the use of algebra. In these problems, many of the numbers needed to perform arithmetic are not given. Instead, a relationship between numbers is given. From this relationship, mathematical expressions and equations can be formed using known values and literal numbers to express unknown values.

CHANGING WORD EXPRESSIONS INTO ALGEBRAIC EXPRESSIONS

Correctly translating from word phrases, or expressions, to algebraic expressions is the key to solving word problems. Key to this is identifying the proper arithmetic operations to be performed. Also key is representing the unknown numbers.

Identifying the operations

In Chapter 1, the technique of solving word problems was presented. Although algebra was not used, some "key" words were introduced. From these, the operations to be performed could be identified. Look for these same key words when translating word phrases to algebraic expressions. The following are examples:

- *Addition:*

    Phrase: x *plus* 10
    Translate: x + 10

    Phrase: The *sum* of 3 and y
    Translate: 3 + y

    Phrase: 6 *is increased by* 2x
    Translate: 6 + 2x

- *Subtraction:*

    Phrase: y *minus* 4
    Translate: y − 4

    Phrase: 7a *less* 9
    Translate: 7a − 9

    Phrase: The *difference* between 6 and 5b
    Translate: 6 − 5b

- *Multiplication:*

  Phrase: 5 *times* a
  Translate: 5a or 5•a

  Phrase: The *product* of 9 and y
  Translate: 9y or 9•y

  Phrase: *Twice* x
  Translate: 2x or 2•x

- *Division:*

  Phrase: The *quotient* of x and 5

  Translate: $x \div 5$ or $\dfrac{x}{5}$

  Phrase: 6 *divided by* a

  Translate: $6 \div a$ or $\dfrac{6}{a}$

  Phrase: *Divide* x *by* y

  Translate: $x \div y$ or $\dfrac{x}{y}$

Representing unknown numbers

A word problem expresses a relationship between numbers. When translated, the unknown numbers of a problem are written using literal numbers. Oftentimes, one unknown number in the problem is represented by a literal number, and any other unknown numbers in the problem are expressed in terms of it. In this case, it is best to allow the literal number to represent the easiest unknown number in the relationship. See the examples that follow:

- A number is 10 times a second number:
    let second number = x
    then first number = 10x
- A number is 5 smaller than a second number:
    let second number = x
    then first number = x − 5
- The sum of two numbers is 12:
    let first number = x
    then second number = 12 − x

Sample Problem 14-1.

Use the relationship given to determine the unknown number in each pair.

Relationship: There are 4 times as many 2 x 4 boards in stack #1 as stack #2.

	Stack #1 (boards)	Stack #2 (boards)
(a)	24	?
(b)	?	8
(c)	x	?
(d)	?	2x
(e)	12x	?

Step 1. Translate the relationship into equation form.
        number in stack #2 = #2
        number in stack #1 = #1 = 4(#2)

Step 2. Calculate the missing values.

(a) $24 = 4(\#2)$

$\dfrac{24}{4} = \#2 = 6$

(b) $\#1 = 4(8) = 32$

(c) $x = 4(\#2)$

$\dfrac{x}{4} = \#2 = \dfrac{x}{4}$

(d) $\#1 = 4(2x) = 8x$

(e) $12x = 4(\#2)$

$\#2 = \dfrac{12x}{4} = 3x$

Step 3. Fill in the missing values.

	Stack #1 (boards)	Stack #2 (boards)
(a)	24	6
(b)	32	8
(c)	x	$\dfrac{x}{4}$
(d)	8x	2x
(e)	12x	3x

Sample Problem 14-2.

Use the relationship given to determine the unknown number in each pair.
Relationship: There are 3 gallons more yellow paint than brown paint.

	Yellow (gallons)	Brown (gallons)
(a)	6	?
(b)	?	10
(c)	x	?
(d)	?	5y
(e)	12z	?

Step 1. Translate the relationship into equation form.

gallons in brown $= B$

gallons in yellow $= Y = B + 3$

Step 2. Calculate the missing values.

(a) $6 = B + 3$
$B = 3$

(b) $Y = 10 + 3 = 13$

(c) $x = B + 3$
$B = x - 3$

(d) $Y = 5y + 3$

(e) $12z = B + 3$
$B = 12z - 3$

Step 3. Fill in the missing values.

	Yellow (gallons)	Brown (gallons)
(a)	6	3
(b)	13	10
(c)	x	x − 3
(d)	5y + 3	5y
(e)	12z	12z − 3

Sample Problem 14-3.

Use the relationship given to determine the unknown number in each pair.
Relationship: A 15 foot long board is cut into 2 pieces.

	1st Piece (feet)	2nd Piece (feet)
(a)	8	?
(b)	?	3
(c)	x	?
(d)	?	2y
(e)	9z	?

Step 1. Translate the relationship into equation form.

total feet = 15
feet of 1st piece = 1st = 15 − 2nd
feet of 2nd piece = 2nd = 15 − 1st

Step 2. Calculate the missing values.

(a) 2nd = 15 − 8 = 7

(b) 1st = 15 − 3 = 12

(c) 2nd = 15 − x

(d) 1st = 15 − 2y

(e) 2nd = 15 − 9z

Step 3. Fill in the missing values.

	1st Piece (feet)	2nd Piece (feet)
(a)	8	7
(b)	12	3
(c)	x	15 − x
(d)	15 − 2y	2y
(e)	9z	15 − 9z

Practice Problems

Complete the following problems on a separate sheet of paper. Use the relationships given to determine the unknown number in each pair.

1. Relationship: A husband is 5 years older than his wife.

	Husband's Age (years)	Wife's Age (years)
(a)	28	?
(b)	?	31
(c)	43	?
(d)	?	42
(e)	62	?

2. Relationship: The tip given is 10% of the cost of the meal.

	Meal Cost (dollars and cents)	Tip Amount (dollars and cents)
(a)	$15.00	?
(b)	?	$1.20
(c)	$25.00	?
(d)	?	$2.50
(e)	$9.00	?

3. Relationship: A board is cut into 6 equal lengths.

	Board Length (inches)	Each Piece (inches)
(a)	96"	?
(b)	?	36"
(c)	168"	?
(d)	?	24"
(e)	120"	?

4. Relationship: The Pacific time zone is 3 hours behind the Eastern time zone.

	Pacific Time Zone (time)	Eastern Time Zone (time)
(a)	9:30 a.m.	?
(b)	?	4:00 a.m.
(c)	11:00 p.m.	?
(d)	?	2:15 p.m.
(e)	5:25 a.m.	?

5. Relationship: There are 4 quarts in 1 gallon.

	Gallons (quantity)	Quarts (quantity)
(a)	3	?
(b)	?	12
(c)	0.5	?
(d)	?	18
(e)	6	?

SOLVING WORD PROBLEMS

To solve a word problem algebraically, an equation is set up. To write an equation, enough information must be given to:
- Write expressions using the given relationships. (It is best if the unknown can represent the question asked.)
- Equate the expressions to a result.

Once the equation is set up, solve the problem.

After the problem is solved, it is necessary to check the solution. The solution should not be checked in the equation because this does not prove that it answers the question asked. Check it by placing it into the written statement to see if it answers the question. The stated relationship should be clear when replaced with numbers. The sample problems that follow will demonstrate this. They follow the rules of solving word problems.

RULES OF SOLVING WORD PROBLEMS

- Write expressions using the given relationships.
- If two unknown quantities, represent one with some letter (generally x). Then, try to express the other in terms of that letter.
- Equate the expressions to a result.
- Solve the equation. Give the solution. Remember, many problems ask for more than one answer.
- Check the solution with the word problem.

Sample Problem 14-4.

Geometry problem. The length of a room is 4 feet longer than the width. The perimeter is 40 feet. Find the length and width of this room.

Step 1. Write expressions using the given relationships.

width $= x$
length $= x + 4$
perimeter $= 40$ ft.

Step 2. Set up the equation. Use the formula for perimeter. Then, substitute the relationships.

$$p = 2l + 2w$$
$$40 = 2(x + 4) + 2x$$

Step 3. Solve the equation.

$$40 = 2(x + 4) + 2x$$
$$40 = 2x + 8 + 2x$$
$$40 = 4x + 8$$
$$32 = 4x$$
$$x = 8$$

Solutions to the problem:

width = x   *width = 8 ft.*
length = x + 4   *length = 12 ft.*

Step 4. Check the solutions in the written problem.

length = 4 ft. + width
12 ft. = 4 ft. + 8 ft.
12 ft. = 12 ft.
perimeter = twice length + twice width
40 ft. = 2(12 ft.) + 2(8 ft.)
40 ft. = 24 ft. + 16 ft.
40 ft. = 40 ft.

Sample Problem 14-5.

Age Problem. Mr. Jones is 4 times as old as his son. In 16 years, he will be only twice as old. What are their present ages?

Step 1. Write expressions using the given relationships.

At present:
son's present age = x
father's present age = 4x
In 16 years:
son's age = x + 16
father's age = 4x + 16

Step 2. Set up the equation. After 16 years:

father's age = 2 times son's age
4x + 16 = 2(x + 16)

Step 3. Solve the equation.

$$4x + 16 = 2(x + 16)$$
$$4x + 16 = 2x + 32$$
$$2x = 16$$
$$x = 8$$

Solutions to the problem:
son's present age = x   *son's age = 8 yr.*
father's present age = 4x   *father's age = 32 yr.*

Step 4. Check the solutions in the written problem.

At present: Father is 4 times older than son.
father's age = 4(son's age)
32 yr. = 4(8 yr.)
32 yr. = 32 yr.
After 16 years: Father is twice as old as son.
father's age = 2 times son's age
32 yr. + 16 yr. = 2(8 yr. + 16 yr.)
48 yr. = 2(24 yr.)
48 yr. = 48 yr.

Sample Problem 14-6.

Coin problem. Joyce has $1.55 in nickels and dimes. She has 7 more nickels than dimes. Find the number of each.

Step 1. Write expressions using the given relationships.

Quantity:

dimes = x

nickels = x + 7

Dollar value:

dimes = 10¢

nickels = 5¢

$1.55 = 155¢

Step 2. Set up the equation.

total value = value of dimes + values of nickels

155 = 10x + 5(x + 7)

Step 3. Solve the equation.

155 = 10x + 5(x + 7)

155 = 10x + 5x + 35

155 = 15x + 35

120 = 15x

x = 8

Solutions to the problem:

dimes = x   *dimes = 8*

nickels = x + 7   *nickels = 15*

Step 4. Check the solutions in the written problem.

number of nickels = number of dimes + 7

15 = 8 + 7

15 = 15

8 dimes = 80¢

15 nickels = 75¢

total value = value of nickels + value of dimes

155¢ = 75¢ + 80¢

155¢ = 155¢

Sample Problem 14-7.

Motion Problem. Two cars, car #1 and car #2, leave from the same point at the same time. Car #1 travels north at an average speed of 60 miles per hour. Car #2 travels south at an average speed of 54 miles per hour. How many hours will it take for the two cars to be 342 miles apart?

Step 1. Write expressions using the given relationships.

number of hours = x

miles traveled by car #1 = 60x

miles traveled by car #2 = 54x

Step 2. Set up the equation.

total miles = miles of car #1 + miles of car #2

Step 3. Solve the equation.

342 = 60x + 54x

Solution to the problem:

number of hours = x   *travel time = 3 hr.*

Step 4. Check the solutions in the written problem.

distance for car #1 = (60 mph) (3 hr.)
car #1 = 180 mi.

distance for car #2 = (54 mph) (3 hr.)
car #2 = 162 mi.

total = car #1 + car #2
342 mi. = 180 mi. + 162 mi.
342 mi. = 342 mi.

Sample Problem 14-8.

Number problem. One number is five times a second number. The sum of the two numbers is 36. Find the two numbers.

Step 1. Write expressions using the given relationships.

second number = x
first number = 5x
sum = 36

Step 2. Set up the equation.

sum = first number + second number
36 = x + 5x

Step 3. Solve the equation.

36 = x + 5x
36 = 6x
x = 6

Solutions to problem:

second number = x   *second number = 6*
first number = 5x   *first number = 30*

Step 4. Check the solutions in the written problem.

first number = 5 times second number
30 = 5(6)
30 = 30
sum = first number + second number
36 = 30 + 6
36 = 36

Consecutive integer problems

As mentioned before, integers include all whole numbers and their opposites. Consecutive integers follow right after the other. For example, 5, 6, 7 are consecutive integers. Relationships for positive, consecutive integers are given as follows:

- First integer = x
- Second integer = x + 1
- Third integer = x + 2

Consecutive odd and consecutive even integers fall every other number. For example, 1, 3, and 5 are consecutive odd numbers, and 2, 4, and 6 are consecutive even numbers. Relationships for positive odd or even integers are given by the same expressions, including:

- First integer = x
- Second integer = x + 2
- Third integer = x + 4

Sample Problem 14-9.

Consecutive integer problem. Find three, positive, consecutive odd integers such that twice the smallest added to the largest is 67.

Step 1. Write expressions using the given relationships.

first (smallest) integer = x
second integer = x + 2
third (largest) integer = x + 4

Step 2. Set up the equation.

twice the smallest + largest = 67
2x + (x + 4) = 67

Step 3. Solve the equation.

2x + (x + 4) = 67
2x + x + 4 = 67
3x = 63
x = 21

Solutions to the problem:

first integer = x   *first integer = 21*
second integer = x + 2   *second integer = 23*
third integer = x + 4   *third integer = 25*

Step 4. Check the solutions in the written problem.

twice the smallest + largest = 67
2(21) + 25 = 67
42 + 25 = 67
67 = 67

Mixture problems: cost basis

Mixture problems are common. Some mixture problems are based on cost. This particular problem has units of cost and weight. The equation must contain relationships using the same measurements. The same concept would apply if, instead of weight, units of volume were used.

Sample Problem 14-10.

Mixture problem. A brand of wild bird seed contains a mixture of 18¢ per pound and 30¢ per pound seed. If a 60 pound mixture sells for 20¢ per pound, how much of each kind of seed is mixed?

Step 1. Write expressions using the given relationships.

Amount of seed in final mixture:
    total weight = 60 lb.
pounds of 18¢ seed = x
pounds of 30¢ seed = 60 − x
Cost of seed in final mixture:
    total cost = 20(60) = 1200¢
        (20¢ per pound times 60 pounds)
cost of 18¢ seed = 18x
cost of 30¢ seed = 30(60 − x)

Step 2. Set up the equation.

total cost = cost of 18¢ seed + cost of 30¢ seed
1200 = 18x + 30(60 − x)

Step 3. Solve the equation.

$$1200 = 18x + 30(60 - x)$$
$$1200 = 18x + 1800 - 30x$$
$$-600 = -12x$$
$$x = 50$$

Solutions to the problem:

weight of 18¢ seed = x   *weight of 18¢ seed = 50 lb.*
weight of 30¢ seed = 60 − x   *weight of 30¢ seed = 10 lb.*

Step 4. Check the solutions in the written problem.

total weight = weight of 18¢ seed + weight of 30¢ seed
60 lb. = 50 lb. + 10 lb.
60 lb. = 60 lb.
total cost = cost of 18¢ seed + cost of 30¢ seed
1200¢ = (50 lb.)(18¢/lb.) + (10 lb.)(30¢/lb.)
1200¢ = 900¢ + 300¢
1200¢ = 1200¢

Mixture problems: percent basis

Mixtures can be made by adding one item to another, such as adding ink to water. The strength of the mixture or solution is given as a percentage. The percentage concentration can be increased or decreased. Certain relationships can be established and equations can be written to determine the new percentage of concentration.

Sample Problem 14-11.

Mixture problem. How many ounces of water must be added to 16 ounces of 25% sulfuric acid to make a 10% solution? (Solution diluted.)

Step 1. Write expressions using the given relationships.

Starting solution:
acid = 25%
ounces of solution = 16
ounces of acid (25% of 16) = 4
New solution:
ounces of water to be added = x
ounces of solution = 16 + x
ounces of acid (10% of solution) = 0.1(16 + x)
Note: Review chapter on percent if necessary.

Step 2. Set up the equation.

acid in new = acid in starting
$$0.1(16 + x) = 4$$

Step 3. Solve the equation.

$$0.1(16 + x) = 4$$
$$16 + x = 40 \text{ (multiply both sides by 10)}$$
$$x = 40 - 16$$
$$x = 24$$

Solution to the problem:

ounces water to be added = x   *water added = 24 oz.*

Step 4. Check the solution in the written problem.

starting solution + added water = new solution
16 oz. + 24 oz. = 40 oz.
10% of 40 oz. = 4 oz. acid
0.1 (40 oz.) = 4 oz.
4 oz. = 4 oz.

Sample Problem 14-12.

Mixture problem. How many quarts of alcohol must be added to 40 quarts of 20% solution to make a 33.3% solution? (Solution strengthened.)

Step 1. Write expressions using the given relationships.

Starting solution:
   alcohol = 20%
   quarts of solution = 40
   quarts of alcohol (20% of 40) = 8
New solution:
   quarts of alcohol to be added = x
   quarts of solution = 40 + x
   total quarts of alcohol = 8 + x
   quarts of alcohol (33.3% of solution) = $\frac{1}{3}$(40 + x)

Step 2. Set up the equation.

   quarts of alcohol = total quarts of alcohol
   $\frac{1}{3}$(40 + x) = 8 + x

Step 3. Solve the equation.

   $\frac{1}{3}$(40 + x) = 8 + x
   40 + x = 3(8 + x)
   40 + x = 24 + 3x
   40 − 24 = 3x − x
   16 = 2x
   x = 8

Solution to the problem:

   alcohol to be added = x   *alcohol added = 8 quarts*

Step 4. Check the solution in the written problem.

   starting solution + alcohol added = new solution
      40 qt. + 8 qt. = 48 qt.
   alcohol in start + alcohol added = total alcohol
      8 qt. + 8 qt. = 16 qt.
   33.3% of 48 qt. = 16 qt.
      $\frac{1}{3}$(48 qt.) = 16 qt.
   16 qt. = 16 qt.

"Work" problems

In "work" problems, the efforts of more than one person or machine are combined to complete a job in a specific length of time. Relationships are based on how much "work" is produced in the stated length of time.

Sample Problem 14-13.

Work problem. Pump #1 can remove the water from a tank in 20 minutes. Pump #2 can empty the tank in 30 minutes. If the two pumps are used together, how long will it take to empty the tank?

Step 1. To help visualize the problem, draw a sketch. See Fig. 14-1.

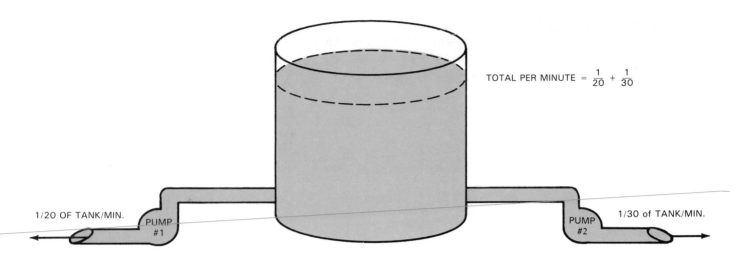

TOTAL PER MINUTE = $\frac{1}{20} + \frac{1}{30}$

1/20 OF TANK/MIN.

PUMP #1

1/30 of TANK/MIN.

PUMP #2

Fig. 14-1. This sketch shows the contents of a tank being pumped out by two pumps working at the same time. A sketch such as this can help in solving problems.

Step 2. Write expressions using the given relationships.

minutes to empty the tank = x
fraction of tank emptied per minute:

$$\text{pump \#1} = \frac{1}{20}$$

$$\text{pump \#2} = \frac{1}{30}$$

$$\text{combined} = \frac{1}{x}$$

Step 3. Set up the equation.

pump #1 + pump #2 = combined

$$\frac{1}{20} + \frac{1}{30} = \frac{1}{x}$$

Step 4. Solve the equation. (LCD = 60x)

$$\frac{1}{20} + \frac{1}{30} = \frac{1}{x}$$

$$\frac{1 \cdot 60x}{20} + \frac{1 \cdot 60x}{30} = \frac{1 \cdot 60x}{x}$$

$$3x + 2x = 60$$
$$5x = 60$$
$$x = 12$$

Solution to the problem:

minutes to empty tank = x   *time to empty = 12 min.*

Step 5. Check the solution in the written problem.

$$\text{pump \#1} = \frac{1}{20} \times 12 = \frac{3}{5} \text{ of total emptied or total work}$$

$$\text{pump \#2} = \frac{1}{30} \times 12 = \frac{2}{5} \text{ of tank emptied or total work}$$

$$\text{pump \#1 and pump \#2} = \frac{3}{5} + \frac{2}{5}$$

$$= \frac{5}{5} \text{ of tank emptied or total work}$$

Sample Problem 14-14.

"Work" problem. Paychecks of a certain company are printed by a new computer in 7 hours, where the old computer takes 14 hours. After the new computer prints paychecks for 3 hours, the old computer is put into operation and, together, they finish the job. What is the total time to print the paychecks?

Step 1. Write expressions using the given relationships.

combined operation: total hours to print = x

fraction of checks printed in 1 hour:

$$\text{new computer} = \frac{1}{7}$$

$$\text{old computer} = \frac{1}{14}$$

$$\text{combined operation} = \frac{1}{x}$$

Step 2. Set up the equation and solve for x.

combined operation = new computer + old computer

$$\frac{1}{x} = \frac{1}{7} + \frac{1}{14}$$

$$x = 4.667 \text{ hr.}$$

Step 3. Find the total time to print the checks.

From the value of x obtained in Step 2, we find that the total amount of time needed to print the paychecks with both computers running at the same time is 4.667 hours. However, the problem statement said that the new computer ran for 3 hours before the old one began operation. The fraction of checks printed prior to operation of the old computer, then, was 3/7ths, for 3 hours out of 7 hours. This means that 4/7ths of the paychecks remained to be printed when the old computer came on.

The total time to print the lot of paychecks is the time in which the new computer ran by itself *plus* the time when the new and old ran together. The new computer printed, by itself, 3/7ths of the lot of paychecks at the rate of 7 hours per lot. Combined, they printed 4/7ths of the lot, at a rate of 4.667 hours per lot. Therefore:

$$\textit{total time to print paychecks} = \frac{3}{7} \text{ lot} \left(\frac{7 \text{ hr.}}{\text{lot}} \right) + \frac{4}{7} \text{ lot} \left(\frac{4.667 \text{ hr.}}{\text{lot}} \right)$$

$$= 5.667 \textit{ hr., or 5 hr., 40 min.}$$

RATIO AND PROPORTION

A **ratio** expresses a relationship between two or more numbers. A **proportion** is a comparison of two equal ratios, with different numbers. A **variation** is an equation that relates one variable to one or more other variables.

Ratio

The relationship expressed by a ratio compares two numbers using division. There are three ways in which to write a ratio. In general form, these include:

A to B   A:B $\frac{A}{B}$

Each of these three forms are equal. In general, they are all read "A to B" or "A is to B." Fractions, perhaps, provide the easiest form in which to write ratios and to see the relationship.

The numbers used to express a ratio are not important. Provided their ratios (quotients) are equivalent, any two numbers can be used. The ratio of 1:2, for example, can be expressed by any fraction that can be reduced to 1/2 (or 0.5). Each of the following examples are equivalent fractions with a ratio of 1 to 2:

$$\frac{1}{2} = \frac{2}{4} = \frac{3}{6} = \frac{4}{8} = \frac{5}{10} = \frac{15}{30}$$

To find an equivalent ratio, a ratio should be in form of A/B or A:B. Both numbers of the ratio are then multiplied (or divided) by the same number. In this way, the numbers written in the ratio are larger or smaller but the ratio itself remains the same. For example, on a certain machine, pulleys between motor shaft and machine shaft have an RPM ratio of 3:5. (RPM is revolutions per minute.) This means that for every 3 revolutions of the motor shaft, the machine shaft makes 5.

Among equivalent ratios are:

$$\frac{3}{5} \times \frac{10}{10} = \frac{30}{50} \qquad \frac{3}{5} \times \frac{100}{100} = \frac{300}{500}$$

For every 30 turns of the motor shaft, the machine shaft turns 50 times. For every 300 turns, the machine shaft turns 500 times. The same thought would apply to any equivalent ratio.

A ratio may contain more than two numbers, for example, the ratio of 3:5:8. An equivalent ratio is found in the same manner as before—multiply (or divide) *all* numbers of the ratio by the same number.

Sample Problem 14-15.

A 72-foot beam is cut into 3 pieces in a ratio of 2:3:7. Find the length of each piece.

Step 1. Write expressions using the given relationship.

common multiplier $= x$
first length $= 2x$
second length $= 3x$
third length $= 7x$

Step 2. Set up the equation.

first length + second length + third length $= 72$
$2x + 3x + 7x = 72$

Step 3. Solve the equation.

$2x + 3x + 7x = 72$
$12x = 72$
$x = 6$

Solutions to the problem:

first length $= 2x = 2 \cdot 6 = 12\ ft.$
second length $= 3x = 3 \cdot 6 = 18\ ft.$
third length $= 7x = 7 \cdot 6 = 42\ ft.$

Step 4. Check the solution in the written problem.

$$\frac{12}{6} : \frac{18}{6} : \frac{42}{6} = 2{:}3{:}7 \text{ (equivalent ratios)}$$

12 ft. + 18 ft. + 42 ft. $= 72$ ft.

Proportion

A proportion is an equation stating that two ratios are equal. Proportions are written in either of the following general forms:

$$\frac{a}{b} = \frac{c}{d} \qquad\qquad a{:}b :: c{:}d$$

Both of these forms are equal and are read: "a is to b as c is to d."

Proportions often contain two different units of measure. The units in one ratio should always be written exactly the same as the units in the other ratio. For example:

$$\frac{dollars}{hour} = \frac{dollars}{hour} \qquad \frac{cents}{pound} = \frac{cents}{pound} \qquad \frac{miles}{gallon} = \frac{miles}{gallon}$$

Proportions may be solved as algebraic equations. Proportions indicate two equal operations of division. Therefore, a number on one side is transposed to the other side of an equation by multiplying both sides of the equation by that number. The task of solving proportions via this method can sometimes be cumbersome. A procedure called *cross multiplication* can often simplify the task.

Cross multiplication is the name given to the method used to transpose numbers in a proportion. Cross multiplication removes the division from an equation. To cross multiply, multiply each denominator by the opposite numerator; then equate the two products as shown:

$$\frac{a}{b} = \frac{c}{d} \text{ (proportion)}$$

$$\frac{a}{b} \diagdown \frac{c}{d}$$

a•d = b•c (cross multiplication)

Note: If both ratios were **inverted** (positions of numerator and denominator reversed), the cross multiplication would be the same. For example:

$$\frac{a}{b} = \frac{c}{d} \text{ (proportion)}$$

$$\frac{b}{a} = \frac{d}{c} \text{ (inverted)}$$

a•d = b•c (cross multiplication)

Sample Problem 14-16.

A car uses 18 gallons to travel 270 miles. At this rate, how many miles can be driven using 24 gallons? Also, what is the average miles per gallon?

Step 1. Set up a proportion.

let new distance = x mi.

$$\frac{270 \text{ mi.}}{18 \text{ gal.}} = \frac{x \text{ mi.}}{24 \text{ gal.}}$$

Step 2. Solve the equation using cross multiplication.

$$\frac{270 \text{ mi.}}{18 \text{ gal.}} \diagup\!\!\!\!\diagdown \frac{x \text{ mi.}}{24 \text{ gal.}}$$

(270) (24) = 18x
6480 = 18x
x = 360

Solution to the problem:

new distance = x mi.   *distance = 360 mi.*

Step 3. Check the solution in the written problem.

Check the solution by substituting the x value into the proportion. If solution is correct, both ratios will be equal.

$$\frac{270 \text{ mi.}}{18 \text{ gal.}} \diagup\!\!\!\!\diagdown \frac{360 \text{ mi.}}{24 \text{ gal.}}$$

15 mi./gal. = 15 mi./gal.

Notice that reducing the ratio also gives the mpg. Therefore, in answer to the other question:

car's mileage = 15 mpg

Note: The equation could also have been solved by transposing "24" to isolate x. The problem could have been solved without algebra using a conversion bar:

$$\left(\frac{24 \text{ gal.}}{}\right) \quad \left(\frac{270 \text{ mi.}}{18 \text{ gal.}}\right) = 360 \text{ mi.}$$

Sample Problem 14-17.

On a certain machine, pulleys between motor shaft and machine shaft have an RPM ratio of 3:5. If the motor turns at 300 RPM, how fast does the machine shaft turn?

Step 1. Set up a proportion.

$$\frac{3 \text{ RPM}}{5 \text{ RPM}} = \frac{300 \text{ RPM}}{x \text{ RPM}}$$

Step 2. Solve the equation using cross multiplication.

$$\frac{3 \text{ RPM}}{5 \text{ RPM}} = \frac{300 \text{ RPM}}{x \text{ RPM}}$$

$$3x = 5(300)$$
$$3x = 1500$$
$$x = 500$$

Solution to the problem:

machine shaft = x RPM   *machine shaft = 500 RPM*

Step 3. Check the solution in the written problem.

$$\frac{3 \text{ RPM}}{5 \text{ RPM}} = \frac{300 \text{ RPM}}{500 \text{ RPM}}$$

$$\frac{3}{5} = \frac{3}{5}$$

TEST YOUR SKILLS

Do *not* write in this book. Use a separate sheet of paper to complete the following problems. Show your work and your final answer.

CHANGING WORD EXPRESSIONS INTO ALGEBRAIC EXPRESSIONS

Use the relationship given to determine the unknown number in each pair.

1. Relationship: The length of a rectangle is 3 times its width.

	length (feet)	width (feet)
(a)	6	?
(b)	?	3
(c)	x	?
(d)	?	y + 4
(e)	4z	?

2. Relationship: Airplane #1 travels 40 miles per hour faster than plane #2.

	Plane #2 (mph)	Plane #1 (mph)
(a)	240	?
(b)	?	120
(c)	a	?
(d)	?	3b
(e)	4c + 1	?

3. Relationship: There are 38 boards, some 8-footers and others, 10-footers.

	8-footers (quantity)	10-footers (quantity)
(a)	7	?
(b)	?	25
(c)	x	?
(d)	?	2y
(e)	z + 2	?

SOLVING WORD PROBLEMS

With each of the following word problems:
   (a) State the relationships.
   (b) Write the equation.
   (c) Show the steps in solving the equation.
   (d) Identify the solution to the problem.
   (e) Check your work.

4. *Geometry problem.* The perimeter of a rectangle is 168 feet. Its length is 5 times its width. What are the measurements of the sides of this rectangle?

5. *Geometry problem.* Side *a* of a triangle is two inches longer than side *b*. Side *b* is three times as long as side *c*. Find the length of each side if the perimeter is 37 inches.

6. *Age problem.* Mrs. Smith is three times as old as Melissa. In 12 years, Mrs. Smith will be only twice as old. What are their ages at the present time?

7. *Weight problem.* Before starting work, a box of roofing nails is twice as heavy as a box of drywall screws. At the end of the day, 2 pounds had been used from each box and the roofing nails were 6 pounds heavier than the drywall screws. What is the weight of each box at the start of the day?

8. *Coin problem.* In a coin box is $1.25 in nickels and dimes. If there are 3 times as many nickels as dimes, how many are there of each?

9. *Coin problem.* The value of a handfull of coins is $2.34 in pennies, nickels, and dimes. There are three times as many dimes as nickels and 6 more pennies than dimes. Determine how many of each coin.

10. *Motion problem.* A truck leaves a warehouse, traveling north at an average speed of 35 mph. At the same time, a second truck leaves a store 150 miles away, traveling south at an average speed of 40 mph. How many miles from the warehouse will they meet, and how long will it take?

11. *Motion problem.* A bus departs from a station at 9:00 a.m., traveling at an average speed of 30 mph. At 12:00 noon, a car leaves from the same point. It catches up to the bus at 2:00 p.m. What was the car's average rate of speed?

12. *Number problem.* One number is 4 times a second number and their sum is 30. Find the numbers.

13. *Number problem.* When one-half of a certain number is added to one-third of the same number, the sum is 5. What is the number?

14. *Consecutive integer problem.* Find three consecutive even integers whose sum is 90.

15. *Consecutive integer problem.* Find three consecutive odd integers where the sum of the first two is 3 greater than the third.

16. *Mixture problem.* There were 3000 bricks used to build a fireplace. One style of brick cost $2 apiece. The other style cost $1 apiece. With the total cost of bricks costing $4,850, determine how many of each style brick was used.
17. *Mixture problem.* To paint an apartment building, a total of 70 gallons of paint was used. For each gallon of red, 4 times as many gallons of white were used. How many gallons of each were needed for this building?
18. *Mixture problem.* How many ounces of water must be added to 20 ounces of 30% sulphuric acid solution to make a 15% solution?
19. *Mixture problem.* There are 2 cups of salt in a 20-cup solution. How many cups of water must be added to make an 8% salt solution?
20. *Mixture problem.* How many quarts of antifreeze must be added to 25 quarts of a 20% solution to make a 60% solution?
21. *Mixture problem.* How many gallons of insecticide must be added to 20 gallons of 10% solution to make a 33.3% insecticide solution?
22. *"Work" problem.* Truck #1 can haul 480 tons of steel in 24 trips. Truck #2 can haul the same amount in 48 trips. If the two trucks are used together, how many trips would it take to haul the steel?
23. *"Work" problem.* Using a large tractor, a field can be plowed in 12 hours. Using a small tractor, the field can be plowed in 36 hours. If both tractors are used, how long will it take to plow the field?

RATIO AND PROPORTION

With each of the following problems:
- (a) Write the equation.
- (b) Show the steps to solve the equation.
- (c) Identify the solution.
- (d) Check the answer.

24. The ratio of photo developing fluid to water is given as 2:5. If the mixing container holds 49 ounces, what is the maximum amount of each that can be used?
25. A picture measures 2 1/4" high by 3 1/4" wide. If it is enlarged to have a width of 13 inches, what will be its height?
26. A fruit punch is mixed in a ratio of 3 parts orange juice, 1 part pineapple juice, and 2 parts ginger ale. If the punch bowl holds 3 quarts, how many ounces of each is needed to fill the bowl? (Note: 1 quart = 32 ounces.)
27. At a certain time of day, a flagpole casts a shadow 27 feet. A 5-foot-tall person casts a shadow 3 feet. What is the height of the flagpole?
28. A motorist travels 152 miles in 4 hours. With the same average rate of speed, how long will it take to travel 247 miles?
29. A sample test of 184 transistors contained 6 defective devices. With this failure rate, how many transistors can be expected to be unacceptable in a selection of 1288?
30. On a map, 3/4 inches represents 10 miles. How many miles does 6 inches represent?

Part IV

Advanced Applied Math

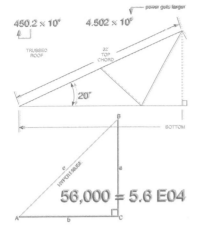

Chapter 15

Applied Trigonometry

OBJECTIVES

After studying this chapter, you will be able to:
- Explain basic terminology of trigonometry.
- Solve right triangles using the Pythagorean theorem and trigonometric functions.
- Solve oblique triangles using the law of sines and the law of cosines.

Trigonometry is the mathematics used to solve triangles. *Solving a triangle* means finding the length of the three sides and three angles. Triangular shapes are involved in many different applications. They come in all sizes and shapes.

TYPES OF TRIANGLES

Triangles were briefly discussed in Chapter 9 of this book. Therefore, some of the material in the paragraphs to follow might look familiar. Triangles may be classified according to their angles or their sides. There are three basic types of angles—*obtuse, right,* and *acute.* Triangles may be named after one of these three by the types of angles they have. Triangles named by their sides fall into three groups—*equilateral, isosceles,* and *scalene.* Triangles have other names and classifications to further describe their particular characteristics.

CLASSIFICATION BY ANGLES

Fig. 15-1 shows the three types of angles. An angle can range from 0°, the starting point, to 360°, a complete rotation. An angle of 180° produces a straight line. An **obtuse angle** is greater than 90°. An **acute angle** is less than 90°.

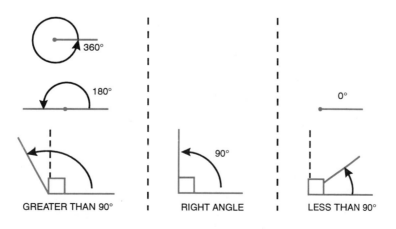

Fig. 15-1. Types of angles.

A **right angle** is exactly 90° — with the two lines perpendicular to each other. The small square drawn in the corner is the symbol of a right angle. When two lines form a right angle, they are said to be *square*.

A triangle with a right angle is a **right triangle**. See Fig. 15-2. Right triangles are especially useful in trigonometry. A triangle that does not have a right angle is an **oblique triangle**. There are two types of oblique triangles — obtuse and acute.

The primary characteristic of an **obtuse triangle** is that one angle is greater than 90°. Examine the triangles shown in Fig. 15-3. The sum of the three angles of any triangle is 180°. Therefore, a triangle will never have more than one 90° (or greater) angle. The other two angles of an obtuse triangle must always be acute.

A triangle having all three angles less than 90° is an **acute triangle**. Fig. 15-4 shows acute triangles of different shapes. In all cases, the three angles have a sum of 180°.

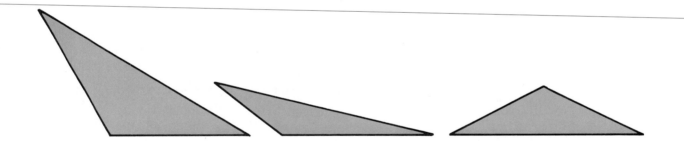

Fig. 15-2. Right triangles have one 90° angle.

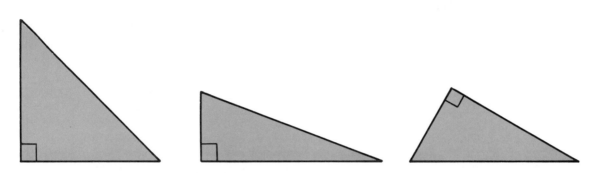

Fig. 15-3. Obtuse triangles have one angle greater than 90°.

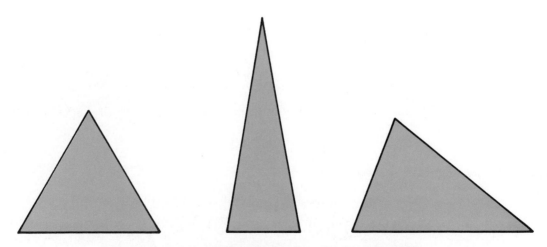

Fig. 15-4. Acute triangles have all angles less than 90°.

Classification by sides

Triangles may be classified in a different manner. They can be classified by the relationship of their sides. The three classifications of triangles, as shown in Fig. 15-5 are *equilateral, isosceles,* and *scalene.* It is possible for a triangle classified by angle to fit into more than one classification by side.

An **equilateral triangle** has all sides and, therefore, all angles, equal. The only type, by angle, that can fall into this classification is an acute triangle. The three angles of an equilateral triangle are 60°, since they must have a sum of 180°.

An **isosceles triangle** has two equal sides and, therefore, two equal angles. Any of the three types, by angle, may fit this description. If a right triangle is isosceles, the two equal angles are each 45° because a right triangle has only one angle equal to 90°, and the sum of the angles, as always, is 180°.

A **scalene triangle** has no two equal sides and, therefore, no two equal angles. All three types, by angle, may fit into this classification.

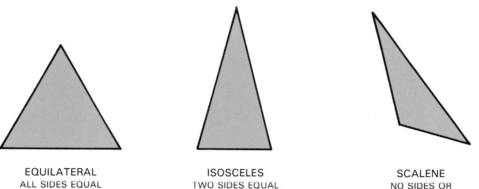

EQUILATERAL
ALL SIDES EQUAL
ALL ANGLES EQUAL

ISOSCELES
TWO SIDES EQUAL
TWO ANGLES EQUAL

SCALENE
NO SIDES OR
ANGLES EQUAL

Fig. 15-5. Classification of triangles by sides.

Similar triangles

Two triangles that have the exact same angles are said to be **similar triangles**. These triangles will have the same basic shape but may be of different scale, or size. The sides between similar triangles are proportional. One of the triangles may appear as if it grew or shrank. See Fig. 15-6.

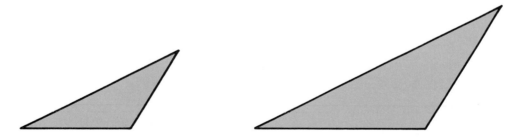

Fig. 15-6. Similar triangles have the same corresponding angles and their sides are in proportion.

SOLVING RIGHT TRIANGLES

In a right triangle, one angle is always 90°; the other two are always acute. The side opposite the right angle is the **hypotenuse**. See Fig. 15-7. *Trigonometric functions* are used to calculate the values of the side lengths and the acute angles. The *Pythagorean theorem* is also used. These will be explained in greater detail.

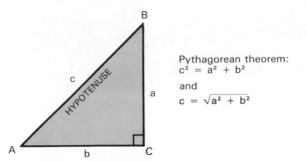

Pythagorean theorem:
$$c^2 = a^2 + b^2$$

and

$$c = \sqrt{a^2 + b^2}$$

Fig. 15-7. A right triangle. The side opposite the right angle is called the hypotenuse.

Trigonometric functions

There are six trigonometric functions. *Sine (sin), cosine (cos),* and *tangent (tan)* are the three that will be discussed. (The other three are their respective inverses — consecant, secant, and cotangent.) Each of these functions is a different ratio of the sides of a right triangle for a given angle, θ. (The Greek letter θ, or *theta,* is used here to represent a generalized angle. It is used like a variable. Angles are commonly symbolized by Greek letters.)

The terms *opposite side* and *adjacent side* are key in the discussion of solving right triangles. To help explain these terms, refer again to Fig. 15-7. The angles are labeled A, B, and C; the sides — a, b, and c. The opposite side is the side across from an angle θ. The adjacent side in a right triangle is the side between the right angle and an angle θ. It is *not* the hypotenuse. For example, the opposite side of angle A is side a. The adjacent side of angle A is side b. Likewise, the opposite side of angle B is side b. The adjacent side is side a.

As mentioned, the trigonometric functions are ratios of sides of a right triangle for a given angle. They can be written as formulas. The sine, cosine, and tangent of the angle is equal to the ratio. If the angle and a side are known, with the help of the formulas, an unknown side can be found. Also, if the ratio of an unknown angle can be found, the angle can be determined.

Keep in mind, the three angles have a sum of 180°. Since one angle is 90°, the other two have a sum of 90°. Therefore, after finding one angle, the other angle can be found using subtraction. Working with this knowledge, the trigometric functions, and the Pythagorean theorem (to follow), a right triangle can be solved. The trigonometric functions are given in the rules of solving triangles.

The Pythagorean theorem

The **Pythagorean theorem** has wide application in the solution of problems and is widely used in technology. Basically, it is the relationship of the three sides of a right triangle. The theorem states that the square of the hypotenuse is equal to the sum of the squares of the other two sides. It is given in general form by the equation $c^2 = a^2 + b^2$, where c is the hypotenuse and a and b are the other two sides. The hypotenuse, then, is given as $c = \sqrt{a^2 + b^2}$. Look again at Fig. 15-7. Now you have the "tools" needed to solve right triangles. They are summarized under the rules of solving triangles.

RULES OF SOLVING RIGHT TRIANGLES

- The trigonometric functions are listed below. They are used when two of three values are known.

$$\sin \theta = \frac{\text{length of opposite side}}{\text{length of hypotenuse}}$$

$$\cos \theta = \frac{\text{length of adjacent side}}{\text{length of hypotenuse}}$$

$$\tan \theta = \frac{\text{length of opposite side}}{\text{length of adjacent side}}$$

- A well-known mnemonic code to help remember these formulas is: SOH-CAH-TOA (pronounced so-cuh-toe-uh).

  SOH: **S**ine is **O**pposite over **H**ypotenuse.

  $$\sin \theta = \frac{\text{opposite}}{\text{hypotenuse}}$$

  CAH: **C**osine is **A**djacent over **H**ypotenuse.

  $$\cos \theta = \frac{\text{adjacent}}{\text{hypotenuse}}$$

  TOA: **T**angent is **O**pposite over **A**djacent.

  $$\tan \theta = \frac{\text{opposite}}{\text{adjacent}}$$

- When two sides of a triangle are known, the third side can be found by the Pythagoren theorem. The Pythagoren theorem is:

  $$c^2 = a^2 + b^2$$

  where c is the hypotenuse.
- The sum of the angles of any triangle is 180°.
- The sum of the two acute angles of a right triangle is 90°.
- A triangle without a right angle (an oblique triangle) can be solved by first resolving (splitting up) into two right triangles. Working from the known values, the two triangles can be solved and results combined to give the desired angles and sides of the oblique triangle. See Fig. 15-8.

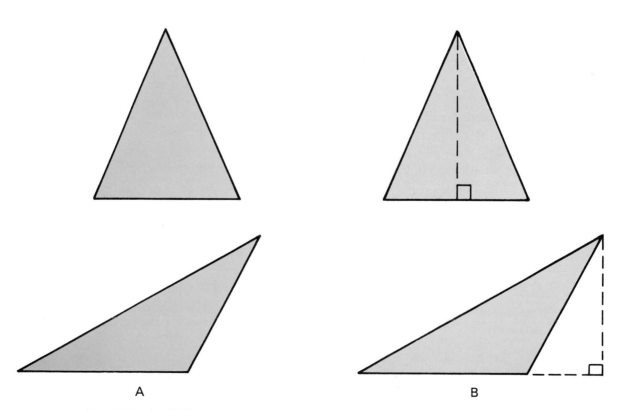

A   B

Fig. 15-8. A—Oblique triangles. B—Oblique triangles resolved into right angles.

Sample Problem 15-1.

Solve the triangle of Fig. 15-9 given one angle and one side.

Step 1. Find angle B by subtracting angle A from 90°.

Formula: B = 90 — A
Substitution: B = 90° — 53°
Solution: B = 37°

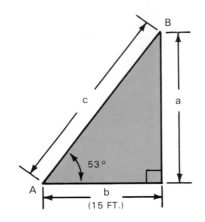

Fig. 15-9. Solve this triangle.

Step 2. The given side is adjacent to the given angle. Select a function containing the adjacent side: cosine or tangent. Tangent is selected for this example. The formula will need to be rearranged to solve for the unknown value (opposite).

Formula: $\tan A = \dfrac{\text{opposite}}{\text{adjacent}}$

Substitution: $\tan 53° = \dfrac{\text{opposite}}{15 \text{ ft.}}$

$1.3270 = \dfrac{\text{opposite}}{15 \text{ ft.}}$

Note: The values of the ratios given by the trigonometric functions are found in Fig. B-10 of the Technical Section. These ratios are for angles of 1° to 90° and for functions of sine, cosine, and tangent. To use the table, find the desired angle and read across to the desired function. For example: $\tan 53° = 1.3270$.

Another way to get the value is with a calculator. Instructions for using the calculator can be found in the Technical Section at the back of this book. An important note is that the calculator must be in degree mode. Consult your *calculator user's guide.*

Transposition: opposite = 1.3270 (15 ft.)
Solution: opposite = a = 20 feet (approximately)

Step 3. Solve for the hypotenuse using the Pythagorean theorem.

Formula: $c = \sqrt{a^2 + b^2}$

Substitution: $c = \sqrt{(15 \text{ ft.})^2 + (20 \text{ ft.})^2}$

Intermediate step: $c = \sqrt{225 \text{ ft.}^2 + 400 \text{ ft.}^2} = \sqrt{625 \text{ ft.}^2}$

Solution: c = 25 feet

Sample Problem 15-2.

Solve the triangle of Fig. 15-10 given two sides.

Step 1. Find the hypotenuse using the Pythagorean theorem.

Formula: $c = \sqrt{a^2 + b^2}$

Substitution: $c = \sqrt{(4 \text{ ft.})^2 + (3 \text{ ft.})^2}$

Intermediate step: $c = \sqrt{16 \text{ ft.}^2 + 9 \text{ ft.}^2} = \sqrt{25 \text{ ft.}^2}$

Solution: $c = 5$ feet

Note: This triangle has hypotenuse of 5 feet. The other sides are 3 feet and 4 feet. A right triangle of this proportion is referred to as a *3-4-5 triangle*. This concept is used in construction to check if two walls are square. In application, units of 3 and 4 are measured on the walls, and the diagonal (hypotenuse) will be 5 units if the walls are square.

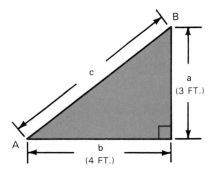

Fig. 15-10. Solve this triangle.

Step 2. Find either angle by selecting a trigonometric function. The cosine function will be used here, finding angle A first.

Formula: $\cos A = \dfrac{\text{adjacent}}{\text{hypotenuse}}$

Substitution: $\cos A = \dfrac{4 \text{ ft.}}{5 \text{ ft.}}$

Transposition: $A = \cos^{-1} 0.8$

Solution: $A = 36.9°$

Note: To isolate A, the **inverse cosine** ($\cos^{-1}$, or arccos) is taken of both sides. The inverse function, in general, gives the angle whose function (sin, cos, or tan) is the ratio given. The inverse functions include: arcsin, or $\sin^{-1}$; arccos, or $\cos^{-1}$; arctan, or $\tan^{-1}$.

The angle may be found using Fig. B-11 of the Technical Section. To find the angle, go to the column of the respective function and find the value of the ratio. Then, read back across to the angle of that ratio. For example: $\cos^{-1} 0.8 = 37°$ (approximately) — so, $\cos 37° = 0.8$. The inverse function of a calculator will also give the angle. Using a calculator, in general, will give more accurate results than Fig. B-11.

Step 3. The angle B is found by subtracting angle A from 90°.

Formula: $B = 90° - A$

Substitution: $B = 90° - 36.9°$

Solution: $B = 53.1°$

Sample Problem 15-3.

Solve the triangle of Fig. 15-11 given the hypotenuse and one angle.

Step 1. Find angle A by subtracting angle B from 90°.

Formula: $A = 90° - B$

Substitution: $A = 90° - 45°$

Solution: $A = 45°$

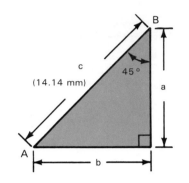

Fig. 15-11. Solve this triangle.

Step 2. Select a trignometric function to find one of the sides. The sine function is chosen with angle A.

Formula: $\sin A = \dfrac{\text{opposite}}{\text{hypotenuse}}$

Substitution: $\sin 45° = \dfrac{\text{opposite}}{14.14 \text{ mm}}$

$0.7071 = \dfrac{\text{opposite}}{14.14 \text{ mm}}$

Transportation: opposite $= 0.7071 \,(14.14 \text{ mm})$

Solution: opposite $= b = 10 \text{ mm}$

Step 3. Find the remaining side using the Pythagorean theorem.

Formula: $c^2 = a^2 + b^2$

Substitution: $(14.14 \text{ mm})^2 = a^2 + (10 \text{ mm})^2$

Transposition: $a^2 = (14.14 \text{ mm})^2 - (10 \text{ mm})^2$

Intermediate step: $a = \sqrt{199.94 \text{ mm}^2 - 100 \text{ mm}^2} = \sqrt{99.94 \text{ mm}^2}$

Solution: $a = 10 \text{ mm}$ (approximately)

Note: This is an isosceles triangle.

Sample Problem 15-4.

Solve the oblique triangle in Fig. 15-12 using right triangles.

Step 1. The oblique triangle is divided into two right triangles. The height is given as 10'. The unknown angles, B_1 and B_2, are found by subtracting angles A_1 and A_2 from 90°.

Formula: $B_1 = 90° - A_1$

Substitution: $B_1 = 90° - 70°$

Solution: $B_1 = 20°$

Solution: $B_2 = 40°$

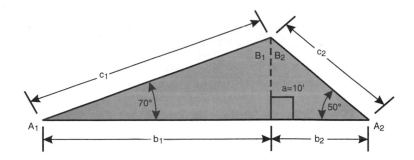

Fig. 15-12. The oblique triangle is divided into two right triangles.

Step 2. Find the lengths of b_1 and b_2 using the tangent of the A_1 and A_2 angles.

Formula: $\tan A_1 = \dfrac{\text{opposite}}{\text{adjacent}}$

Substitution: $\tan 70° = \dfrac{10'}{b_1}$

$$0.940 = \dfrac{10'}{b_1}$$

$$b_1 = \dfrac{10'}{0.940}$$

Solution: $b_1 = 10.6'$

Solution: $b_2 = 13.1'$

Step 3. Find the lengths of sides c_1 and c_2 using the Pythagorean theorem.

Formula: $c = \sqrt{a^2 + b^2}$

Substitution: $c_1 = \sqrt{10^2 + 10.6^2}$

Solution: $c_1 = 14.6'$

Solution: $c_2 = 16.5'$

Sample Problem 15-5.

Use right triangles to solve the oblique triangle in Fig. 15-13.

Step 1. Find angle B_1 of the large right triangle by subtracting angle A_1 from 90°.

Formula: $B_1 = 90° - A_1$

Substitution: $B_1 = 90° - 15°$

Solution: $B_1 = 75°$

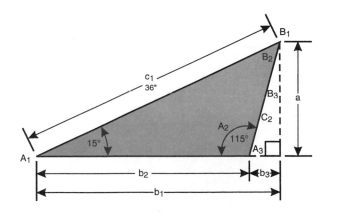

Fig. 15-13. Use right triangles to solve the oblique triangle.

Step 2. Find the length of side b_1 of the large right triangle using the given angle A_1 and the hypotenuse.

Formula: $\cos A_1 = \dfrac{\text{adjacent}}{\text{hypotenuse}}$

Substitution: $\cos 15° = \dfrac{b_1}{36"}$

$$0.966 = \dfrac{b_1}{36"}$$

$$b_1 = 0.966 \times 36"$$

Solution: $b_1 = 34.8"$

Step 3. Find side a of the right triangle using the Pythagorean theorem.

Formula: $c_1^2 = a^2 + b_1^2$

Rearranging: $a^2 = c_1^2 - b_1^2$

$$a = \sqrt{c_1^2 - b_1^2}$$

Substitution: $a = \sqrt{36^2 - 34.8^2}$

Solution: $a = 9.22"$

Step 4. Find angle A_3 of the small right triangle by subtracting angle A_2 from 180°. Note: This is possible because the two combined angles form a straight line, which has an angle of 180°.

Formula: $A_3 = 180° - A_2$

Substitution: $A_3 = 180° - 115°$

Solution: $A_3 = 65°$

Step 5. Angle B_3 of the small right triangle is found by subtracting angle A_3 from 90°.

Formula: $B_3 = 90° - A_3$

Substitution: $B_3 = 90° - 65°$

Solution: $B_3 = 25°$

Step 6. Use side a of the small right triangle to find side c_2. Use the sine function with angle A_3.

Formula: $\sin A_3 = \dfrac{\text{opposite}}{\text{hypotenuse}}$

Substitution: $\sin 65° = \dfrac{9.22"}{c_2}$

$$0.906 = \dfrac{9.22"}{c_2}$$

Solution: $c_2 = 10.2"$

Step 7. Find side b_3 of the small right triangle using the Pythagorean theorem.

Formula: $c_2^2 = a^2 + b_3^2$

Rearranging: $b_3^2 = c_2^2 - a^2$

$$b_3 = \sqrt{c_2^2 - a^2}$$

Substitution: $b_3 = \sqrt{10.2^2 - 9.22^2}$

Solution: $b_3 = 4.36"$

Step 8. Side b_2 of the oblique triangle is found by subtracting side b_3 from side b_1.

Formula: $b_2 = b_1 - b_3$

Substitution: $b_2 = 34.8" - 4.36"$

Solution: $b_2 = 30.4"$

Step 9. Find angle B_2 in the oblique triangle by adding the two given angles, A_1 and A_2, and then subtracting the sum of the angles from $180°$.

Formula: $B_2 = 180° - (A_1 - A_2)$

Substitution: $B_2 = 180° - (15° + 115°)$
$B_2 = 180° - 130°$

Solution: $B_2 = 50°$

Practice Problems

Complete the following problems on a separate sheet of paper.

Use the right triangle shown in Fig. 15-14 to find the unknown information in the following table.

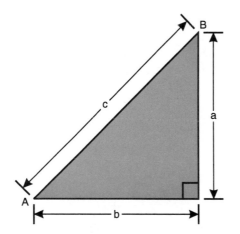

Fig. 15-14. Sample right triangle for practice problems.

	Angles		Sides		
	A	B	a	b	c
1.	35°		6'		
2.	40°			12"	
3.	60°				5 cm
4.		20°	10 m		
5.		45°		15 yd	
6.		75°			8"
7.			7'	9'	
8.			14"		20"
9.				35 cm	50 cm
10.	45°				60"

Part B. Using Right Triangles to Solve Oblique Triangles

Use the oblique triangles shown in Fig. 15-15 to find the unknown information in the following tables.

Note: These problems ask only for the information on the oblique triangle. However, it is necessary to solve for two right triangles to arrive at the answers.

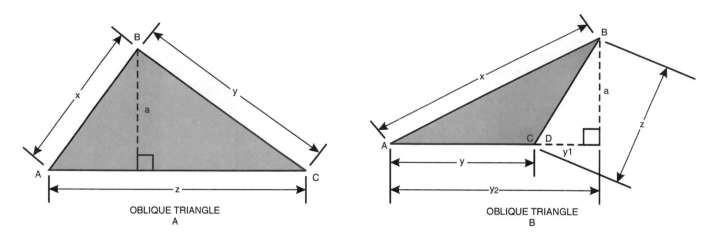

Fig. 15-15. Sample oblique triangles for practice problems. These triangles should be solved using right triangles.

Oblique Triangle A

	Angles			Sides		
	A	B	C	x	y	z
1.	25°	105°		4'		
2.	40°		60°		16"	
3.		135°	25°	15 cm		
4.		75°	45°		25'	
5.	45°		60°		15 yd	

Oblique Triangle B

	Angles			Sides		
	A	B	C	x	y	z
6.		35°	105°	21"		
7.	20°			8'	5'	
8.	30°		115°	36.3"		20"
9.	35°		100°			35 cm
10.		60°	110°			60"

SOLVING OBLIQUE TRIANGLES

As mentioned, oblique triangles are any triangles not containing a right angle. Unless the oblique triangle is resolved into two right triangles, the rules of solving right triangles cannot be used to solve oblique triangles. However, the rules of solving oblique triangles can be used with right and oblique triangles alike.

Law of Sines

The Law of Sines states that in any triangle, the sides are proportional to the sines of the opposite angles. The Law of Sines, then, provides the solutions of a triangle by means of a proportion. The Law of Sines is given in the rules for solving oblique triangles.

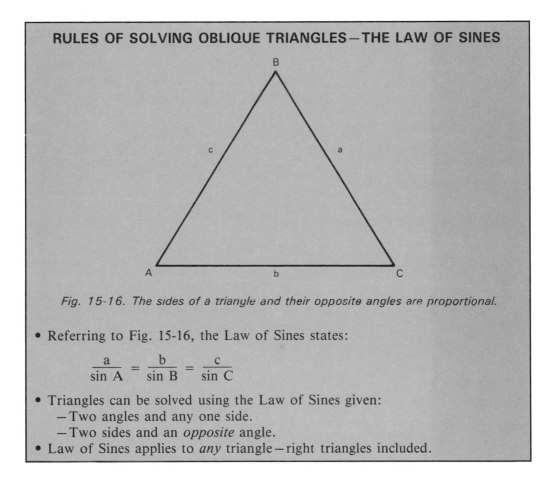

RULES OF SOLVING OBLIQUE TRIANGLES—THE LAW OF SINES

Fig. 15-16. The sides of a triangle and their opposite angles are proportional.

- Referring to Fig. 15-16, the Law of Sines states:

$$\frac{a}{\sin A} = \frac{b}{\sin B} = \frac{c}{\sin C}$$

- Triangles can be solved using the Law of Sines given:
  - Two angles and any one side.
  - Two sides and an *opposite* angle.
- Law of Sines applies to *any* triangle—right triangles included.

Sample Problem 15-6.

Solve the triangle of Fig. 15-17 given two angles and a side.

Step 1. Find angle A by subtracting the sum of the two known angles from 180°.

Formula: A = 180 − (C + B)

Substitution: A = 180° − (37.3° + 24.5°)

Solution: A = 118.2°

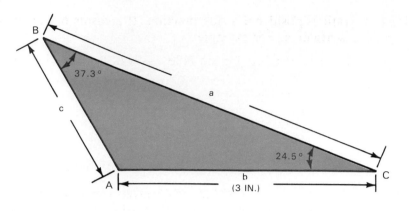

Fig. 15-17. Solve this triangle.

Step 2. The given side is opposite angle B. Therefore, the ratio with angle B must be used with either of the other two ratios. The ratio with angle C is selected here because it is a given angle. The proportion will contain one unknown.

Formula: $\dfrac{b}{\sin B} = \dfrac{c}{\sin C}$

Substitution: $\dfrac{3 \text{ in.}}{\sin 37.3°} = \dfrac{c}{\sin 24.5°}$

$\dfrac{3 \text{ in.}}{0.6060} = \dfrac{c}{0.4147}$

Note: The ratios of the trigonometric functions of angles to fractional degrees may be obtained from Fig. B-11 by interpolation. **Interpolation** is an estimation of a value between two known values. In interpolating trigonometric functions, it is assumed that the change in the ratio is proportional to the change in the angle.

Take for example, the sine of 37.3°. Ratios are known for 37° and 38°. To find the ratio of 37.3°, set up the proportion.

$$\dfrac{\text{change in ratio}}{\text{change in angle}} = \dfrac{0.6157 - 0.6018}{38° - 37°} = \dfrac{x - 0.6018}{37.3° - 37°}$$

Solving for x, the unknown ratio:

$x = 0.6060$

There is an easier way to find the functions of such angles — the calculator!

Transposition: $c = \dfrac{0.4147 \ (3 \text{ in.})}{0.6060}$

Solution: c = 2 inches (approximately)

Step 3. Solve for the remaining side using the third ratio with either of the other two ratios. It is usually best to use the ratio containing given values.

Formula: $\dfrac{b}{\sin B} = \dfrac{a}{\sin A}$

Substitution: $\dfrac{3 \text{ in.}}{\sin 37.3°} = \dfrac{a}{\sin 118.2°}$

$\dfrac{3 \text{ in.}}{0.6060} = \dfrac{a}{0.8813}$

Note: To use Fig. B-11 for angles between 90° and 180°, subtract the angle from 180°. Find the value of the function for *that* angle. The sign of the ratio is positive if a sine function. If a cosine or tangent function, the sign is negative. For example:

$$\sin 120° = \sin (180° - 120°) = \sin 60° = 0.8660$$

Transposition: $a = \dfrac{0.8813\ (3\ in.)}{0.6060}$

Solution: a = 4.4 inches

Sample Problem 15-7.

Solve the triangle of Fig. 15-18 given two sides and one opposite angle.

Step 1. Determine the unknown angle opposite side a. Use the proportion containing the two known sides.

Formula: $\dfrac{a}{\sin A} = \dfrac{c}{\sin C}$

Substitution: $\dfrac{2.5\ in.}{\sin A} = \dfrac{1.5\ in.}{\sin 32°}$

$$\dfrac{2.5\ in.}{\sin A} = \dfrac{1.5\ in.}{0.5299}$$

Transposition: $\sin A = \dfrac{0.5299\ (2.5\ in.)}{1.5\ in.}$

sin A = 0.8832
A = sin⁻¹ 0.8832

Solution: A = 62°

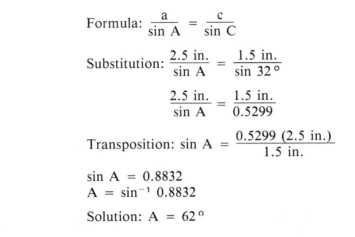
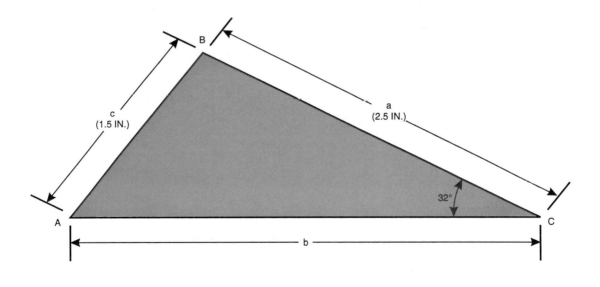

Fig. 15-18. Solve this triangle.

Step 2. Angle B is found by subtracting the sum of the known angles from 180°.

Formula: B = 180 − (A + C)

Substitution: B = 180° − (62° + 32°)

Solution: B = 86°

Step 3. Find the remaining side using the proportion containing the unknown side.

Formula: $\dfrac{b}{\sin B} = \dfrac{c}{\sin C}$

Substitution: $\dfrac{b}{\sin 86°} = \dfrac{1.5 \text{ in.}}{\sin 32°}$

$\dfrac{b}{0.9976} = \dfrac{1.5 \text{ in.}}{0.5299}$

Transposition: $b = \dfrac{0.9976 \, (1.5 \text{ in.})}{0.5299}$

Soluton: b = 2.8 inches

Law of Cosines

The Law of Cosines provides a means of solving an oblique triangle knowing two sides and their included angle or all three sides. The *included angle* is the angle that is formed by two sides. Stated in words, the Law of Cosines is rather lengthy. It is easier understood by looking at the equations that make up the Law of Cosines. These equations are given in the rules of solving oblique triangles.

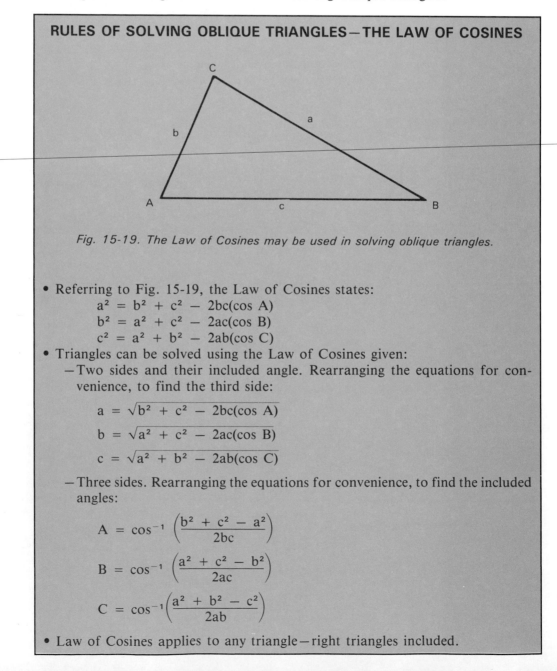

RULES OF SOLVING OBLIQUE TRIANGLES — THE LAW OF COSINES

Fig. 15-19. The Law of Cosines may be used in solving oblique triangles.

- Referring to Fig. 15-19, the Law of Cosines states:
  $a^2 = b^2 + c^2 - 2bc(\cos A)$
  $b^2 = a^2 + c^2 - 2ac(\cos B)$
  $c^2 = a^2 + b^2 - 2ab(\cos C)$
- Triangles can be solved using the Law of Cosines given:
  - Two sides and their included angle. Rearranging the equations for convenience, to find the third side:

    $a = \sqrt{b^2 + c^2 - 2bc(\cos A)}$

    $b = \sqrt{a^2 + c^2 - 2ac(\cos B)}$

    $c = \sqrt{a^2 + b^2 - 2ab(\cos C)}$

  - Three sides. Rearranging the equations for convenience, to find the included angles:

    $A = \cos^{-1}\left(\dfrac{b^2 + c^2 - a^2}{2bc}\right)$

    $B = \cos^{-1}\left(\dfrac{a^2 + c^2 - b^2}{2ac}\right)$

    $C = \cos^{-1}\left(\dfrac{a^2 + b^2 - c^2}{2ab}\right)$

- Law of Cosines applies to any triangle — right triangles included.

Sample Problem 15-8.

Solve the triangle of Fig. 15-20 given two sides and their included angle.

Step 1. Select the formula needed to solve for the side opposite angle B.

Formula: $b = \sqrt{a^2 + c^2 - 2ac(\cos B)}$

Substitution: $b = \sqrt{(2 \text{ in.})^2 + (4 \text{ in.})^2 - 2(2 \text{ in.})(4 \text{ in.})(\cos 120°)}$

$b = \sqrt{(2 \text{ in.})^2 + (4 \text{ in.})^2 - 2(2 \text{ in.})(4 \text{ in.})(-0.5)}$

Intermediate step: $b = \sqrt{4 \text{ in.}^2 + 16 \text{ in.}^2 + 8 \text{ in.}^2} = \sqrt{28 \text{ in.}^2}$

Solution: $b = 5.3$ inches

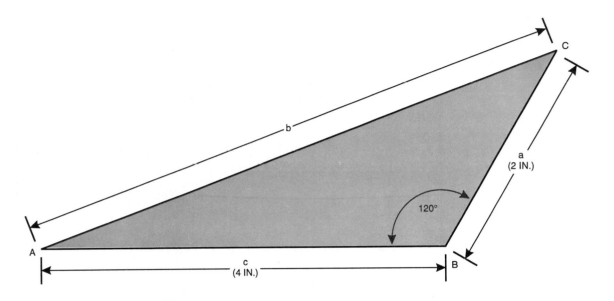

Fig. 15-20. Solve this triangle.

Step 2. Knowing two sides and an opposite angle, the Law of Sines may be used. (The Law of Sines is used because it generally provides a simpler method of solving.) Find angle A using the Law of Sines.

Formula: $\dfrac{a}{\sin A} = \dfrac{b}{\sin B}$

Substitution: $\dfrac{2 \text{ in.}}{\sin A} = \dfrac{5.3 \text{ in.}}{\sin 120°}$

$\dfrac{2 \text{ in.}}{\sin A} = \dfrac{5.3 \text{ in.}}{0.866}$

Transposition: $\sin A = \dfrac{0.866 \, (2 \text{ in.})}{5.3 \text{ in.}}$

$\sin A = 0.3268$
$A = \sin^{-1} 0.3268$

Solution: $A = 19°$

Step 3. Find the third angle by subtracting the sum of the other two angles from 180°.

Formula: $C = 180 - (A + B)$

Substitution: $C = 180° - (19° + 120°)$

Solution: $C = 41°$

Sample Problem 15-9.

Solve the triangle of Fig. 15-21 given three sides.

Step 1. Use the Law of Cosines to find any one of the three angles.

Formula: $A = \cos^{-1}\left(\dfrac{b^2 + c^2 - a^2}{2bc}\right)$

Substitution: $A = \cos^{-1}\left(\dfrac{(75 \text{ mm})^2 + (80 \text{ mm})^2 - (48 \text{ mm})^2}{2(75 \text{ mm})(80 \text{ mm})}\right)$

Intermediate step:

$A = \cos^{-1}\left(\dfrac{5625 \text{ mm}^2 + 6400 \text{ mm}^2 - 2304 \text{ mm}^2}{12{,}000 \text{ mm}^2}\right)$

$A = \cos^{-1}\left(\dfrac{9721 \text{ mm}^2}{12{,}000 \text{ mm}^2}\right)$

Solution: $A = 35.9°$

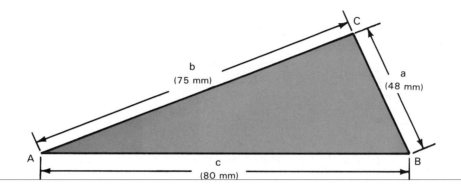

Fig. 15-21. Solve this triangle.

Step 2. Knowing two sides and an opposite angle, the Law of Sines may be used. Find angle B using the Law of Sines.

Formula: $\dfrac{a}{\sin A} = \dfrac{b}{\sin B}$

Substitution: $\dfrac{48 \text{ mm}}{\sin 35.9°} = \dfrac{75 \text{ mm}}{\sin B}$

$\dfrac{48 \text{ mm}}{0.5864} = \dfrac{75 \text{ mm}}{\sin B}$

Transposition: $\sin B = \dfrac{0.5864 (75 \text{ mm})}{48 \text{ mm}}$

$\sin B = 0.9162$

$B = \sin^{-1} 0.9162$

Solution: $B = 66.4°$

Step 3. Find the third angle by subtracting the two known angles from 180°.

Formula: $C = 180 - (A + B)$

Substitution: $C = 180° - (35.9° + 66.4°)$

Solution: $C = 77.7°$

Practice Problems

Complete the following table on a separate sheet of paper. Use the Law of Sines to solve for the missing numbers. Use Figs. 15-16 to 15-18 for reference.

	Angles			Sides		
	A	B	C	a	b	c
1.	30°	40°		6'		
2.	100°		40°	12"		
3.		105°	25°		10 cm	
4.		75°		25'	40'	
5.			45°		30 yd	45 yd

Complete the following table on a separate sheet of paper. Use the Law of Cosines to solve for the missing numbers. Use Figs. 15-19 to 15-21 for reference.

	Angles			Sides		
	A	B	C	a	b	c
6.				21"	28"	10"
7.			80°	8'	5'	
8.		30°		18"		24"
9.	40°				25 cm	60 cm
10.				40"	50"	60"

THE POLAR COORDINATE SYSTEM

The polar coordinate system uses the length of a line at an angle from the horizontal axis to locate a point. Zero degrees is to the right of the origin, as shown in Fig. 15-22. All angle measurements are made in reference to zero degrees. Positive angles are rotated counterclockwise from zero; negative angles are rotated clockwise from zero.

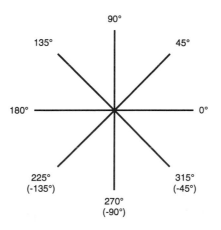

Fig. 15-22. Polar coordinate system.

When compared to the rectangular coordinate system, 0° is + x, 180° is − x, 90° is + y, and −90° is −y. See Fig. 15-23. Note that 180° is always referred to as positive. A complete circle is 360° but is shown only as 0°.

Referring again to Fig. 15-23, two points are shown plotted in both the rectangular coordinate system and the polar coordinate system. Point **a** has rectangular coordinates of (6,8). In the polar system, point **a** is expressed as 10/53.1°. The line used in the polar system is equal in length to the hypotenuse of the triangle formed by the rectangular coordinates.

Point **b** in Fig. 15-23 has rectangular coordinates of (−5, −5). In the polar system, it is 7.07/225°. Even though the right triangle forms an angle of 45°, the polar angle is always in reference to 0°.

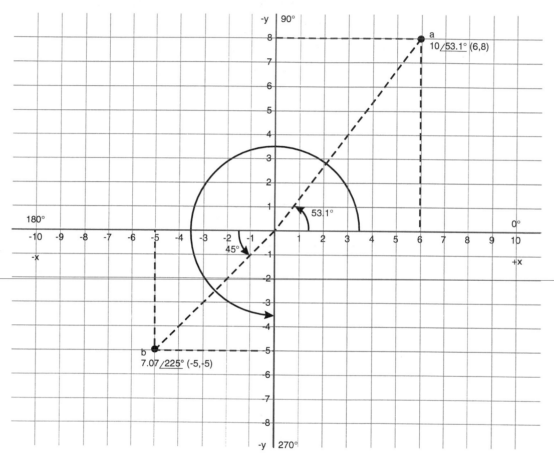

Fig. 15-23. Example of combined polar and rectangular coordinate systems.

Practice Problems

Use a sheet of graph paper to plot the points represented by the following polar coordinates. It would be helpful to use a protractor to locate the angles as accurately as possible.

1. 5/15°
2. 6/100°
3. 10/240°
4. 8/300°
5. 7/−30°
6. 4/−150°
7. 3/150°
8. 9/60°
9. 8/180°
10. 5/0°
11. 6/−75°
12. 10/−90°
13. 4/100°
14. 7/190°
15. 9/80°
16. 3/135°
17. 5/270°
18. 6/360°

TEST YOUR SKILLS

Do *not* write in this book. Use a separate sheet of paper to complete the following problems. Show your work and your final answer.

TYPES OF TRIANGLES

Identify each triangle as isosceles, scalene, or equilateral.

1. (Fig. 15-24) _____

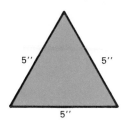

Fig. 15-24. Identify the triangle.

2. (Fig. 15-25) _____

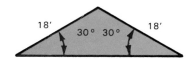

Fig. 15-25. Identify the triangle.

3. (Fig. 15-26) _____

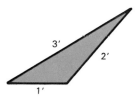

Fig. 15-26. Identify the triangle.

4. What is the relationship between the two triangles of Fig. 15-27?

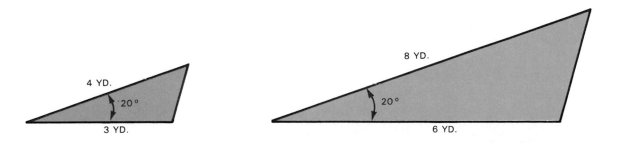

Fig. 15-27. Two obtuse and similar triangles.

SOLVING RIGHT TRIANGLES

Use trigonometric functions and the Pythagorean theorem to determine the requested measurements. Round all answers to one decimal place, where necessary.

5. Fig. 15-28: length of roof rafter

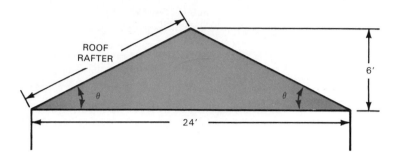

Fig. 15-28. Find the requested unknowns.

6. Fig. 15-29.
   - a. height of building
   - b. path of sunlight
   - c. angle B

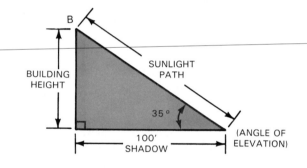

Fig. 15-29. Find the requested unknowns.

7. Fig. 15-30.
   - a. length of guy wire
   - b. height of antenna
   - c. angle B

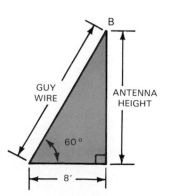

Fig. 15-30. Find the requested unknowns.

8. Fig. 15-31.
   a.   length of run
   b.   length of road surface
   c.   angle A

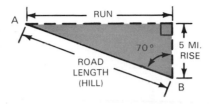

Fig. 15-31. Find the requested unknowns.

9. Fig. 15-32.
   a.   travel distance
   b.   width of river
   c.   angle B

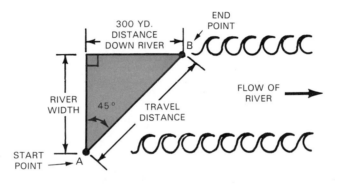

Fig. 15-32. Find the requested unknowns.

10. Fig. 15-33.
    a.   length of roof rafter
    b.   height of attic
    c.   angle A

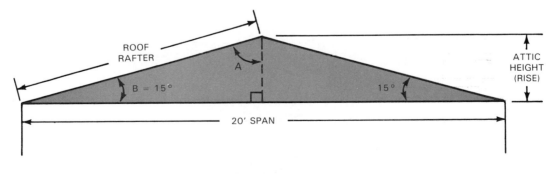

Fig. 15-33. Find the requested unknowns.

11. Fig. 15-34.
    a.   length of ladder
    b.   angle A
    c.   angle B

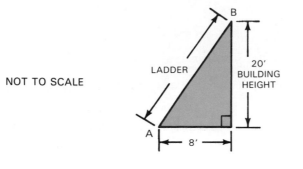

Fig. 15-34. Find the requested unknowns.

12. Fig. 15-35.
    a.   length of diagonal
    b.   angle A
    c.   angle B

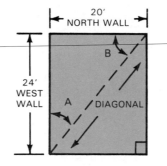

Fig. 15-35. Find the requested unknowns.

13. Fig. 15-36.
    a.   length of stairway stringer
    b.   angle A
    c.   angle B

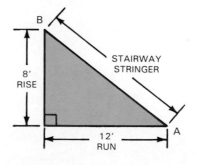

Fig. 15-36. Find the requested unknowns.

14. Fig. 15-37.
   a. length of rafter a
   b. length of rafter b
   c. angle A

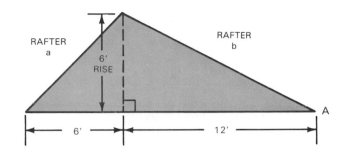

Fig. 15-37. Find the requested unknowns.

15. Fig. 15-38.
   a. length of cross bridging
   b. angle A
   c. angle B

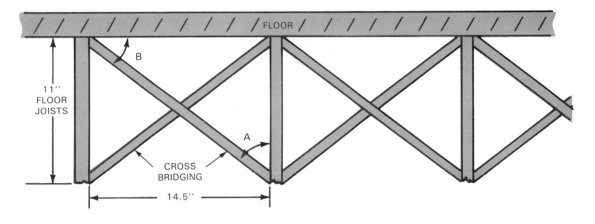

Fig. 15-38. Find the requested unknowns.

16. Fig. 15-39.
   a. height of vertical support
   b. length of base support
   c. angle B

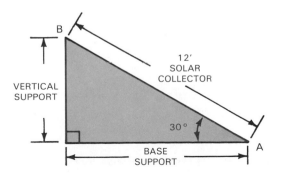

Fig. 15-39. Find the requested unknowns.

17. Fig. 15-40.
    a. height of wall
    b. length of base
    c. angle B

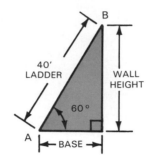

Fig. 15-40. Find the requested unknowns.

18. Fig. 15-41.
    a. length of run
    b. height of rise
    c. angle A

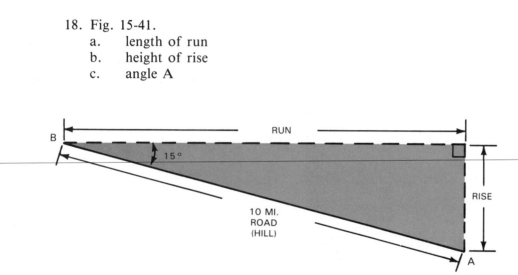

Fig. 15-41. Find the requested unknowns.

19. Fig. 15-42.
    a. length of run
    b. height of rise
    c. angle B

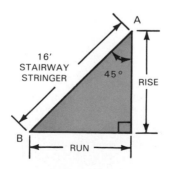

Fig. 15-42. Find the requested unknowns.

20. Fig. 15-43.
    a. height of rise
    b. length of long web
    c. distance of 1/3 point

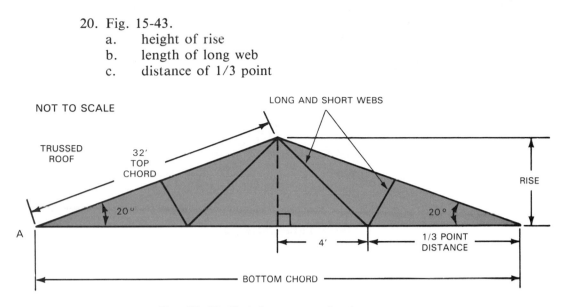

Fig. 15-43. Find the requested unknowns.

SOLVING OBLIQUE TRIANGLES

Use the Law of Sines or the Law of Cosines, as needed, to solve the following triangles (find the unknown dimensions requested). Note that the order of solving does not necessarily follow the order in which the unknowns are listed. Round all answers to two decimal places, where necessary.

21. Fig. 15-44:
    • length of westerly boundary
    • length of northerly boundary
    • angle B

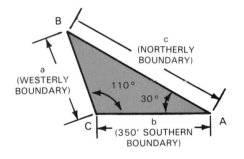

Fig. 15-44. Find the requested unknowns.

22. Fig. 15-45:
    • length of rafter a
    • length of rafter b
    • angle B

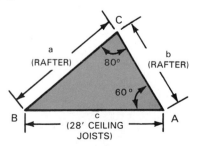

Fig. 15-45. Find the requested unknowns.

23. Fig. 15-46:
    - length of side c
    - angle B
    - angle C

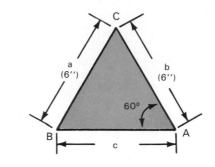

Fig. 15-46. Find the requested unknowns.

24. Fig. 15-47:
    - angle B
    - length of daredevil ramp

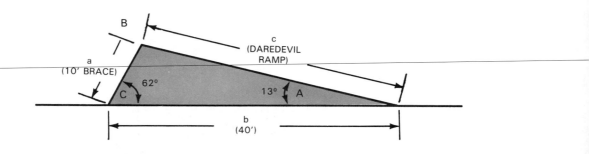

Fig. 15-47. Find the requested unknowns.

25. Fig. 15-48:
    - length of porch roof
    - angle A
    - angle B

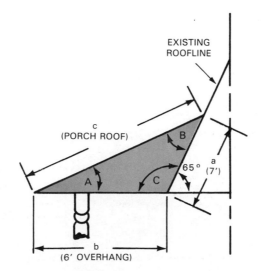

Fig. 15-48. Find the requested unknowns.

26. Fig. 15-49:
- length of side a
- angle B
- angle C

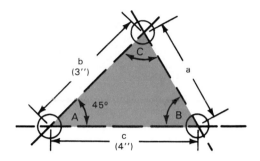

Fig. 15-49. Find the requested unknowns.

27. Fig. 15-50:
- angle A
- angle B
- angle C

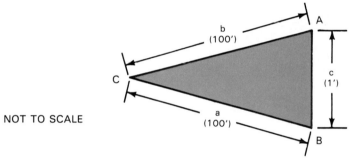

NOT TO SCALE

Fig. 15-50. Find the requested unknowns.

28. Fig. 15-51:
- angle A
- angle B
- angle C

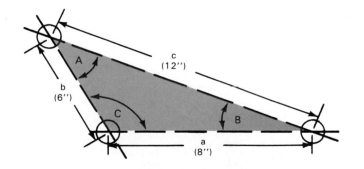

Fig. 15-51. Find the requested unknowns.

PROBLEM-SOLVING ACTIVITIES

Activity 15-1.
Layout of a large rectangle

Objective: To build a large rectangle with 2 x 4 framing lumber and to use right triangles to make sure the corners of the rectangle are exactly 90°.

Instructions:
1. Obtain enough 2" × 4" framing lumber to build the rectangle shown in Fig. 15-52. If lumber is not available, string can be used. If string is used, use masking tape to hold the corners to the floor.
2. Referring to Fig. 15-52, measure and cut side A. Make certain all measurements are accurate.
3. Measure and cut sides B, C, and D.
4. Assemble the rectangle as shown in Fig. 15-52, nailing the sides together. A carpenter's large framing square can be used to make the corners square (at a right angle).

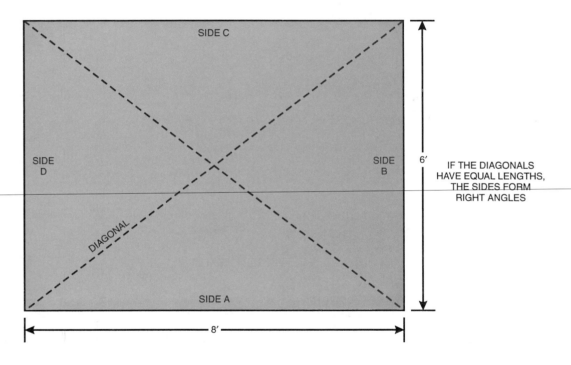

Fig. 15-52. Rectangle for Activity 15-1.

5. After the rectangle is complete, measure the lengths of the diagonals. If the rectangle is constructed properly, the diagonals will be equal.
6. If the corners of a rectangle do not form right angles, the diagonals will not be equal. See Fig. 15-53.
7. Calculate the length of the diagonals of the rectangle using the Pythagorean theorem. Compare your calculations to the measured length.

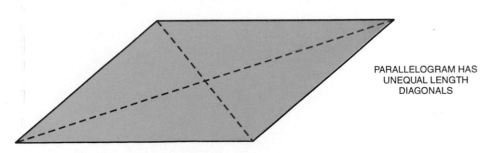

Fig. 15-53. This exaggerated illustration shows the effect when the sides of a rectangle do not form right angles.

Activity 15-2.
Estimating the height of a building

Objective: To use trigonometry to estimate the height of a building. Note: To measure angles, a professional surveyor uses a transit, which is an extremely accurate instrument that looks like a telescope on a tripod. For this activity, a drawing protractor will be used.

Instructions:

1. Tape a protractor to a straightedge and mount the straightedge on a tripod. The straightedge should be as level as possible. See Fig. 15-54. It is important to note that this activity will be accurate only if the ground between the building and the tripod is relatively level.
2. Measure the distance from the ground to the top of the straightedge (dimension "x" in Fig. 15-54). Use masking tape to mark this distance on the side of the building being measured.
3. Sight along the top of the straightedge. It should line up with the tape on the building.
4. Measure the distance from the base of the building to the center of the tripod and record this dimension.
5. Place another straightedge at the center mark on the protractor. Sight along this straightedge to the top of the building. Record the angle indicated by the second straightedge (angle "A").
6. Use the Law of Tangents to calculate dimension "a" shown in the Fig. 15-54.
7. Add dimension "a" to dimension "x" to determine the approximate height of the building.

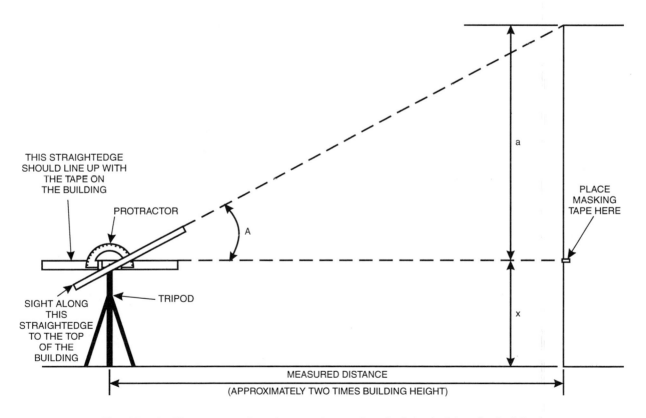

Fig. 15-54. The tangent function can be used to find the height of a building.

Activity 15-3.
Rise and run of a stairway

Objective: To determine the length of the stair stringer by measuring the rise and run of a stairway.

Instructions:

1. Determine the rise of a stairway by measuring its total height. Refer to Fig. 15-55.
2. Determine the run of a stairway by measuring its total length.
3. Use the Pythagorean theorem to find the length of the stair stinger, the board used to carry the individual steps.
4. Challenge: Obtain a large piece of heavy paper that is a little longer than the stair stringer. If necessary, tape together several smaller sheets of paper. Lay out the pattern of the stair stringer, trying to duplicate the pattern that was actually used to build the stairway.

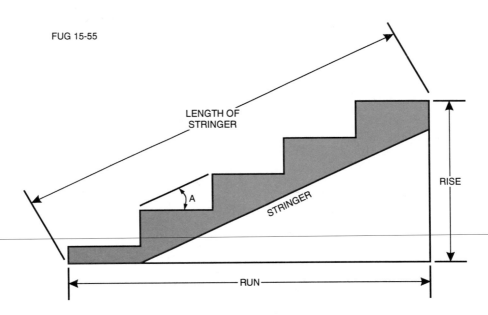

FUG 15-55

Fig. 15-55. The length of a stringer can be found using the Pythagorean theorem.

Activity 15-4.
Compass course

Objective: To follow directions using a magnetic compass.

Note: For this activity, you must have a compass with a movable outer ring. Whenever possible, this activity should be performed outdoors.

Instructions:

1. Before starting this activity, turn the outside ring of the compass until the "0" mark on the ring is in line with the "N" on the compass face.
2. Make two teams of two or three students.
3. Team #1 designs the course by placing six books on the ground in a selected pattern. These books will serve as course markers. Team #1 should lay out the course as follows:
   a) Choose the location of the first marker.
   b) While standing at the first marker and holding the compass so that the needle points to the "0" mark on the ring, specify the direction to the second marker. Write this direction on a piece of paper and attach it to the first marker.
   c) Walk in the chosen direction and place the second marker. The distance between the markers is up to the team members.

d) Continue to place the markers in this way until all the markers are placed. All directions should be given in degrees from the "0" reading on the compass. In Fig. 15-56, for example, the third marker is 95 from the second marker. Any pattern can be designed, and the course can be changed as often as desired.

4. Starting at the first marker, team #2 follows the directions on each note. If the directions are followed correctly, team #2 will move from one marker to another in the proper order. If team #2 moves to the wrong marker, the note giving the next direction might not make sense.

5. For team #1, the object of this activity is to design a challenging course that forces team #2 to use the compass. For team #2, the object is to move through the course without making mistakes.

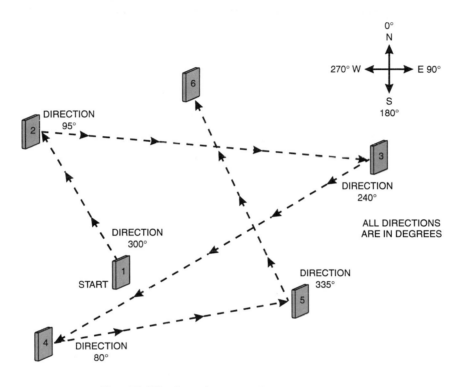

Fig. 15-56. Sample magnetic compass course.

This engineer is calculating the amount of current in an electrical circuit. When the amount of current is known, scientific notation can be used to determine the number of electrons flowing through the circuit. One ampere of current is equal to 6.25×10^{18} electrons flowing past a given point in one second. (Jack Klasey)

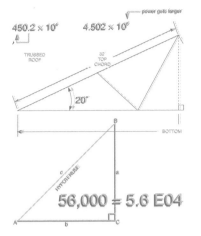

Chapter 16

Scientific and Engineering Notation

OBJECTIVES

After studying this chapter, you will be able to:
- Convert ordinary notation to scientific notation.
- Change scientific notation to ordinary notation.
- Change numbers from one form of scientific notation to another.
- Add, subtract, multiply, and divide power of 10 values.

Scientific notation uses powers of 10 to move a decimal point in a number. This provides an easier means of performing arithmetic on numbers that are either very large or very small. *Engineering notation* is a form of scientific notation, with the powers of 10 being replaced with a multiplier name.

SCIENTIFIC NOTATION

When a number is written in **scientific notation**, the decimal point is moved to follow the first significant figure. Then, to maintain the place value of the number, it is multiplied by some power of 10. This way, the number has the same value.

Powers, place values, and significant figures

Reference has been made to powers, place values, and significant figures. These have been discussed in previous chapters but are reviewed here. In regard to powers:
- A number written without an exponent has one for its exponent.
      Example: $10 = 10^1$
- A number (excluding zero) with an exponent of 0 is equal to 1.
      Example: $10^0 = 1$
- The power indicates how many times a factor is repeated.
      Example: $10^3 = 10 \times 10 \times 10$

In regard to place values, the location of a digit in a number determines the complete value of the digit. Refer to Fig. B-8 in the Technical Section to examine the relationship of the location of a digit and its power of 10. Notice, numbers greater than 1 (in magnitude) are a *positive power of 10* and numbers less than 1 (in magnitude) are a *negative power of 10*. Each place value is some power of 10, starting from the decimal point, in both directions. To determine the value of a digit, multiply by the place value. For example, in the number 462:
- The digit 4 has a value of $4 \times 10^2 = 400$.
- The digit 6 has a value of $6 \times 10^1 = 60$.
- The digit 2 has a value of $2 \times 10^0 = 2$.

In regard to significant figures:
- All nonzero digits are significant.

- Zeros used as place holders *between* two nonzero numbers are significant. Zeros at the end of a number and to the right of a decimal point are significant.
- Zeros are *not* significant if appearing in front of nonzero numbers (except as stated above). Zeros are *not* significant if appearing at the end of nonzero numbers and ending to the left of a decimal point.

Changing ordinary notation to scientific notation

A number written in scientific notation is expressed properly as the product of a number between 1 an 10 and a power of 10. This will become clear as you study Sample Problem 16-1. When changing to scientific notation, follow the applicable rules presented.

RULES FOR WRITING NUMBERS IN SCIENTIFIC NOTATION

1. Move the decimal point to follow the first significant figure. This will yield a number between 1 and 10.
2. Count the number of places the decimal is moved; this is the power of 10.
   a. If the decimal is moved to the *left*, it is a *positive* power of 10.
   b. If the decimal is moved to the *right*, it is a *negative* power of 10.
3. Drop the zeros that are no longer used as place holders.
4. Write the number with the adjusted decimal place and show multiplication by 10 raised to the power.

Sample Problem 16-1.

Write the given numbers in scientific notation.

Number Given	Scientific Notation
(a) 540,000 (move 5 places left)	drop zeros — power is +5 — 5.4×10^5
(b) 320	3.2×10^2
(c) 0.00067	6.7×10^{-4}
(d) 0.045	4.5×10^{-2}
(e) 6	6.0×10^0
(f) 90	9.0×10^1
(g) 10,300	1.03×10^4
(h) 2,090,000	2.09×10^6
(i) 0.0000000507	5.07×10^{-8}
(j) 0.4	4.0×10^{-1}

Changing improper scientific notation

If a number written in scientific notation does not have the decimal point following the first significant figure, it can be adjusted. (Although this form of scientific notation is improper, at times it is used for convenience.) When moving the decimal, count the decimal places using signed numbers. Keep in mind, positive is to the left, and negative is to the right. (Note that this is opposite the number line, discussed in Chapter 10.) Remember, if no decimal point is written, it is understood to follow the number. When adjusting a number to *proper* scientific notation, follow the rules presented.

> **RULES OF ADJUSTING IMPROPER SCIENTIFIC NOTATION**
> - Starting at the decimal point, the count begins with the given exponent.
> - Move the decimal point to follow the first significant figure.
> - Adjust the exponent to correspond with the movement of the decimal.
> - Moving to the left is toward positive (larger) numbers, and to the right is toward negative (smaller) numbers.
> - A helpful hint:
>   - If decimal number gets *smaller* by moving decimal point, exponent gets *larger* by the number of decimal places moved.
>   - If decimal number gets *larger* by moving decimal point, exponent gets *smaller* by the number of decimal places moved.

Sample Problem 16-2.

Write the given numbers in the *proper* form of scientific notation.

Number Given	Scientific Notation
(a) 450.2 $\times$ 10^4	4.502 $\times$ 10^6
move 2 places left	
(b) 920 $\times$ 10^{-3}	9.2 $\times$ 10^{-1}
(c) 78,000 $\times$ 10^6	7.8 $\times$ 10^{10}
(d) 106 $\times$ 10^{-7}	1.06 $\times$ 10^{-5}
(e) 3,000.0 $\times$ 10^{-2}	3.0 $\times$ 10^1
(f) 0.0035 $\times$ 10^4	3.5 $\times$ 10^1
(g) 0.092 $\times$ 10^{-6}	9.2 $\times$ 10^{-8}
(h) 0.5 $\times$ 10^1	5.0 $\times$ 10^0
(i) 0.000021 $\times$ 10^3	2.1 $\times$ 10^{-2}
(j) 0.089 $\times$ 10	8.9 $\times$ 10^{-1}

(The label "— power gets larger" points to the exponent in (a); for (a) the given 450.2 × 10^4 shows "move 2 places left".)

Changing scientific notation to ordinary notation

When a number is expressed in scientific notation, it can be changed to ordinary notation. This is done by moving the decimal point according to the power of 10. If the power of 10 is positive, remove it by moving the decimal point in the negative direction (right). If the power of 10 is negative, remove it by moving the decimal point in the positive direction (left). Add zeros as place holders if necessary.

When the decimal point is moved the number of places indicated by the exponent, the exponent becomes zero. The value of this factor then, is one; thereby, it can be removed. Remember, if the power was *decreased* to zero, the decimal number increases. If the power was *increased* to zero, the decimal number decreases.

Sample Problem 16-3.

Write the given numbers in ordinary notation.

Number Given	Ordinary Notations
(a) 1.5 $\times$ 10^4	15,000
(b) 3.6 $\times$ 10^{-2}	0.036
(c) 290 $\times$ 10^3	290,000
(d) 4750 $\times$ 10^{-6}	0.00475
(e) 90 $\times$ 10^0	90
(f) 0.0035 $\times$ 10^2	0.35
(g) 0.00067 $\times$ 10^{-3}	0.00000067
(h) 0.809 $\times$ 10^5	80,900
(i) 0.115 $\times$ 10^{-4}	0.0000115
(j) 0.00027 $\times$ 10^8	27,000

E notation

E notation is an abbreviated form of scientific notation. It is used with calculators and computers. The E notation replaces the written multiplication by 10 to a power. When a calculator displays 10 to a power, it displays the power using two digits. When writing numbers on paper, an E is shown preceding the two digits indicating the power. All the rules of scientific notation apply to E notation. This includes writing the answer in proper form, with the decimal following the first significant figure.

Sample Problem 16-4.

Write the given numbers in scientific notation and E notation.

Number Given	Scientific Notation	E Notation
(a) 450	4.5×10^2	4.5 E02
(b) 780,000	7.8×10^5	7.8 E05
(c) 0.00063	6.3×10^{-4}	6.3 E−04
(d) 0.00125	1.25×10^{-3}	1.25 E−03
(e) 38.5×10^9	3.85×10^{10}	3.85 E10

Changing E notation to ordinary notation

To remove the E notation, move the decimal point the same number of places as indicated by the exponent. The process is the same as for scientific notation. Remember, if the exponent is positive, it is removed by going in the negative direction. If the exponent is negative, remove it by going in the positive direction.

Sample Problem 16-5.

Write the given numbers in ordinary notation.

Number Given	Ordinary Notation
(a) 25 E04	250,000
(b) 360 E−05	0.00360
(c) 0.47 E08	47,000,000
(d) 0.0509 E−01	0.00509
(e) 801 E00	801

Practice Problems

Complete the following problems on a separate sheet of paper.

Write the following numbers in scientific notation. The decimal point should follow the first significant figure.

1. 680,000,000
2. 2,100
3. 15,700
4. 75,000,000
5. 0.000125
6. 0.018
7. 0.25
8. 0.000034
9. 0.006
10. 25

Write the following numbers in the proper form of scientific notation. The decimal should follow the first significant figure.

11. 500×10^3
12. 1200×10^5
13. $830,000 \times 10^1$
14. 21×10^6
15. 0.0025×10^5
16. 0.000304×10^3
17. 0.75×10
18. 0.000017×10^5
19. 670×10^{-4}
20. $5,600 \times 10^{-6}$
21. $14,000 \times 10^{-2}$
22. $125,000 \times 10^{-4}$
23. 0.01×10^{-3}
24. 0.00046×10^{-3}
25. 0.00203×10^{-5}
26. 0.250×10^{-7}

Remove the scientific notation to write the following numbers in ordinary notation.

27. 3.5×10^4
28. 2.8×10^6
29. 50.10×10^{-3}
30. 120.3×10^{-5}
31. 0.012×10^7
32. 0.0043×10^3
33. 0.0035×10^{-2}
34. 0.00068×10^{-6}
35. 120×10
36. 500×10^0

Write the following numbers in E notation. The decimal point should follow the first significant figure.

37. 350,000
38. 12,000
39. 5,600
40. 50
41. 0.0035
42. 0.290
43. 0.000049
44. 0.00016
45. 250×10^6
46. 1200×10^{-4}

Remove the E notation to write the following numbers in ordinary notation.

47. 5.7 E03
48. 8.2 E05
49. 4.8 E-02
50. 1.2 E-04
51. 630 E04
52. 1,200 E01
53. 25,000 E-02
54. 10 E-04
55. 0.0042 E03

ENGINEERING NOTATION

Engineering notation is also a form of scientific notation. The power of 10 is replaced by a multiplier name or letter prefix. Fig. B-9 of Technical Section B gives multiplier names and letter prefixes with their corresponding power of 10. (These prefixes belong to the SI Metric system of units.) Engineering notation names are given for exponents in multiples of 3, for example, 10^3, 10^6, 10^{-3}, 10^{-6}. The decimal point is adjusted so that the exponent can match one of these powers.

Units of measure

Units of measure — volts, watts, etc. — can be attached to a multiplier name. In this way, the multiplier name shows that the unit is multiplied by that amount. See Fig. 16-1. For example, a milliamp is a unit of measure of electrical current. One milliamp has a value of 0.001 amps. It is abbreviated as "mA." When a unit does not have a multiplier attached, it is classified as a basic unit. The basic unit has an exponent of zero.

Engineering notation is commonly used in electronics. In the sample and practice problems that follow, most of the units given are from this source. Notice that in addition to the multiplier abbreviations, each unit of measure has its own abbreviation.

As mentioned previously, when using engineering notation, an adjustment may be necessary to the decimal point. Although it can be adjusted to match any multiplier name, it is normally adjusted to allow the number to be between 0.1 and 1000.

Multiplier Name	Symbol	Power of 10	Multiply By
tera-	T	10^{12}	1,000,000,000,000
giga-	G	10^9	1,000,000,000
mega-	M	10^6	1,000,000
kilo-	k	10^3	1,000
(basic units)		10^0	1
milli-	m	10^{-3}	.001
micro-	μ	10^{-6}	.000001
nano-	n	10^{-9}	.000000001
pico-	p	10^{-12}	.000000000001

Fig. 16-1. Multiplier names, symbols, and values.

Sample Problem 16-6.

Change the given numbers to the units shown.

Number Given	Change To	New Number
(a) 1,320,000 Hz	megahertz	1.32 MHz
(b) 290 W	milliwatts	290,000 mW
(c) 0.0059 A	milliamps	5.9 mA
(d) 750 kV	volts	750,000 V
(e) 0.38 ms	microseconds	380 μs

Practice Problems

On a separate sheet of paper, change the given number to the units shown.

1. 1500 W = _____ kilowatts
2. 250 V = _____ millivolts
3. 0.0125 s = _____ milliseconds
4. 0.005 A = _____ microamps
5. 120 kHz = _____ megahertz
6. 20 MV = _____ kilovolts
7. 25 mV = _____ volts
8. 530 ns = _____ milliseconds
9. 0.05 mA = _____ amps
10. 0.0040 mW = _____ microwatts

ARITHMETIC WITH POWERS OF 10

Using scientific or engineering notation can ease the arithmetic processes. Operations performed on numbers that are very large or very small frequently turn out wrong due to the placement of the decimal in the final answer. Many of these errors can be avoided by using scientific or engineering notation. To add or subtract, multiply or divide numbers with powers of 10, follow the rules given below.

RULES FOR ADDITION AND SUBTRACTION WITH POWERS OF 10

1. To add (or subtract) numbers in scientific or engineering notation, add (or subtract) the decimal numbers. In order to do so, the exponents in the powers of 10 must be the same. If they are not, adjust a decimal point and corresponding exponent so that they are.
2. When the exponents are the same, align the decimals in a column and perform the arithmetic.
3. The powers of 10 remain the same in both addition and subtraction.
4. When finished with the arithmetic, adjust the answer to proper form.

Note: These rules are valid by the principle of common monomial factoring. See Chapter 11.

Example:
$$(a \times 10^b + c \times 10^b) = (a + c) \times 10^b$$

Sample Problem 16-7.

Add 25.4×10^3 and 0.0395×10^5 using scientific notation.

Step 1. Adjust the decimal and change the exponent of either number so the exponents are the same.

$$0.0395 \times 10^5 = 3.95 \times 10^3$$

Step 2. Align the decimal points and perform the addition. Keep the same power of 10.

$$
\begin{array}{r}
25.4 \ \times 10^3 \\
+ \ \ 3.95 \times 10^3 \\
\hline
29.35 \times 10^3
\end{array}
$$

Step 3. Adjust the decimal and exponent to the proper form of scientific notation.

$$29.35 \times 10^3 = 2.935 \times 10^4$$

Sample Problem 16-8.

Subtract $100.5 \ \text{E}-03$ and $0.22 \ \text{E}-01$ using E notation.

Step 1. Adjust the decimal point and exponent of either number.

$$0.22 \ \text{E}-01 = 22.0 \ \text{E}-03$$

Step 2. Align decimal points and perform the subtraction. Keep the same power of 10.

$$
\begin{array}{r}
100.5 \ \text{E}-03 \\
- \ \ 22.0 \ \text{E}-03 \\
\hline
78.5 \ \text{E}-03
\end{array}
$$

Step 3. Adjust the decimal to proper form.

$$78.5 \ \text{E}-03 = 7.85 \ \text{E}-02$$

Sample Problem 16-9.

Add 25 kilo and 1.3 mega using engineering notation.

Step 1. Identify the exponent values of the multiplier names.

$$kilo = 10^3$$
$$mega = 10^6$$

Step 2. Adjust the decimal and change either multiplier name so they are both the same.

$$25 \ kilo = 0.025 \ mega$$

Step 3. Align the decimals and perform the addition. The multiplier names (exponents) do not change during addition.

$$\begin{array}{r} 0.025 \ mega \\ + \ 1.3 \quad mega \\ \hline 1.325 \ mega \end{array}$$

Answer is in the proper form as shown.

RULES FOR MULTIPLICATION WITH POWERS OF 10

1. To multiply numbers in scientific or engineering notation, multiply the decimal numbers together.
2. Multiply the powers of 10 together by adding the exponents.

Note: You will recall, when multiplying exponential numbers together, if the bases are the same, the exponents are added.

Example: Multiply $a \times 10^b$ and $c \times 10^d$.
$$(a \times 10^b) \ (c \times 10^d) =$$
$$a \times 10^b \times c \times 10^d =$$
$$a \times c \times 10^b \times 10^d =$$
$$ac \times 10^{b+d}$$

3. Adjust the answer to proper form.

Sample Problem 16-10.

Multiply 4.5×10^3 and 0.012×10^2 using scientific notation.

Step 1. Multiply the decimal numbers.

$$\begin{array}{r} 4.5 \\ \times \ 0.012 \\ \hline 0.054 \end{array}$$

Step 2. Multiply the powers of 10. To do so, *add* the exponents.

$$10^3 \cdot 10^2 = 10^{3+2} = 10^5$$

Step 3. Combine the decimal number with the power of 10. Adjust decimal and exponent for the proper form of scientific notation.

$$0.0540 \times 10^5 = 5.4 \times 10^3$$

Sample Problem 16-11.

Multiply 2.5 E−05 and 50.3 E−01 using E notation.

Step 1. Multiply the decimal numbers.

$$\begin{array}{r} 2.5 \\ \times \ 50.3 \\ \hline 125.75 \end{array}$$

Step 2. Multiply the powers of 10 by *adding* the exponents.

$$E-05 + E-01 = E-06$$

Step 3. Combine the decimal number with the exponent. Adjust the decimal for proper form.

$$125.75 \ E-06 = 1.2575 \ E-04$$

Sample Problem 16-12.

Multiply 35 kilo and 1.02 micro using engineering notation.

Step 1. Multiply the decimal numbers.

$$\begin{array}{r} 35 \\ \times\ \ 1.02 \\ \hline 35.70 \end{array}$$

Step 2. Identify the exponent values of the multiplier names.

$$kilo\ =\ 10^3$$
$$micro\ =\ 10^{-6}$$

Step 3. Multiply the powers of 10 by *adding* the exponents.

$$10^3 \cdot 10^{-6}\ =\ 10^{3+(-6)}\ =\ 10^{-3}\ =\ milli$$

Step 4. Write the power of 10 as a multiplier name. Combine with the decimal number.

35.70 milli

Answer is in the proper form as shown.

RULES FOR DIVISION WITH POWERS OF 10

1. To divide numbers in scientific or engineering notation, divide the decimal numbers.
2. Divide the powers of 10 by subtracting the bottom (divisor) exponent from the top (dividend).
Note: You will recall, when dividing exponential numbers, if the bases are the same, the bottom exponent is subtracted from the top.

Example: Divide $a \times 10^b$ by $c \times 10^d$.

$$\frac{a \times 10^b}{c \times 10^d}\ =$$

$$\frac{a}{c} \times \frac{10^b}{10^d}\ =$$

$$\frac{a}{c} \times 10^{b-d}$$

3. Adjust the answer to proper form.

Sample Problem 16-13.

Divide 14.4×10^3 by 0.72×10^6 using scientific notation.

Step 1. Divide the decimal numbers.

$$\frac{14.4}{0.72}\ =\ 20.0$$

Step 2. Divide the powers of 10 by subtracting the exponents.

$$\frac{10^3}{10^6}\ =\ 10^{3-6}\ =\ 10^{-3}$$

Step 3. Combine the number with the power of 10. Adjust the decimal and exponent to the proper form of scientific notation.

$$20.0 \times 10^{-3}\ =\ 2.0 \times 10^{-2}$$

Sample Problem 16-14.

Divide 0.864 E−06 by 43.2 E−02 using E notation.

Step 1. Divide the decimal numbers.

$$\frac{0.864}{43.2}\ =\ 0.02$$

Step 2. Divide the powers of 10 by *subtracting* the exponents.

$$E-06 - E-02 = E-04$$

Step 3. Combine the number with the exponent. Adjust the decimal and exponent for proper form.

$$0.02\ E-04 = 2.0\ E-06$$

Sample Problem 16-15.

Divide 50.2 milli by 100.4 mega using engineering notation.

Step 1. Divide the decimal numbers.

$$\frac{50.2}{100.4} = 0.5$$

Step 2. Identify the exponent values of the multiplier names.

$$milli = 10^{-3}$$
$$mega = 10^{6}$$

Step 3. Divide the powers of 10 by *subtracting* the exponents.

$$\frac{10^{-3}}{10^{6}} = 10^{-3-6} = 10^{-9}$$

Step 4. Write the power of 10 as a multiplier name. Combine with the decimal number.

$$0.5\ nano$$

Answer is in the proper form as shown.

Practice Problems

With each of the following sets of numbers, perform the arithmetic indicated. Write the answer in the proper form of notation used in the problem. Do not round answers.

1. Add: $3.5 \times 10^4 + 62.1 \times 10^3$
2. Add: $900 \times 10^{-6} + 450 \times 10^{-5}$
3. Add: $0.046\ E05 + 0.00055\ E08 + 0.127\ E04$
4. Add: $48\ E-02 + 0.065\ E01 + 150\ E-03$
5. Add: 29 kilo $+ 0.050$ mega
6. Add: 0.0075 milli $+ 10.5$ micro
7. Subtract: $75 \times 10^3 - 0.32 \times 10^4$
8. Subtract: $0.68 \times 10^{-2} - 24 \times 10^{-4}$
9. Subtract: $1.5\ E06 - 48\ E04$
10. Subtract: $20\ E-05 - 50\ E-06$
11. Subtract: 4.5 mega $- 500$ kilo
12. Subtract: 2500 micro $- 0.5$ milli
13. Multiply: $3 \times 10^5 \times 0.4 \times 10^2$
14. Multiply: $0.06 \times 10^{-2} \times 50 \times 10^4$
15. Multiply: $200\ E-03 \times 0.7\ E-06$
16. Multiply: $0.05\ E01 \times 10\ E-01$
17. Multiply: 15 kilo $\times 20$ milli
18. Multiply: 30 micro $\times 2$ kilo
19. Divide: $45 \times 10^4 \div 15 \times 10^6$
20. Divide: $0.06 \times 10^{-2} \div 30 \times 10^4$
21. Divide: $4.8 \times 10^7 \div 0.06 \times 10^{-2}$
22. Divide: $0.280\ E06 \div 14\ E02$
23. Divide: $50\ E-04 \div 0.25\ E-02$
24. Divide: $1\ E00 \div 0.4\ E-03$

25. Divide: $\dfrac{10}{2\ \text{kilo}}$

26. Divide: $\dfrac{250\ \text{milli}}{100\ \text{kilo}}$

27. Divide: $\dfrac{15 \text{ kilo}}{3 \text{ milli}}$

28. Divide: $\dfrac{6.8 \times 10^4}{34 \times 10^8}$

29. Divide: $\dfrac{1000 \times 10^3}{250}$

30. Divide: $\dfrac{30 \text{ E}02 + 1.5 \text{ E}03}{0.5 \text{ E}-06 + 100 \text{ E}-8}$

TEST YOUR SKILLS

Do *not* write in this book. Use a separate sheet of paper to complete the following problems. Show your work and your final answer.

SCIENTIFIC NOTATION

Write the following numbers in the proper form of scientific notation. The decimal point should follow the first significant figure.

Example: $56,000 = 5.6 \times 10^4$

1. 890,000
2. 1,020
3. 325
4. 0.0067
5. 0.0125
6. 0.92
7. 35.6
8. 298.0
9. 1.38
10. 2,700,000

CHANGING IMPROPER SCIENTIFIC NOTATION

Adjust the decimal point and exponent to write the number in the proper form of scientific notation. The decimal should follow the first significant figure.

Example: $56.0 \times 10^3 = 5.6 \times 10^4$

11. 890×10^5
12. 0.00356×10^4
13. 12.5×10^3
14. 0.097×10^2
15. 6750×10^0
16. 320×10^{-3}
17. 0.0258×10^{-4}
18. 1500×10^{-2}
19. 0.601×10^{-5}
20. 47.5×10^{-1}

CHANGING SCIENTIFIC NOTATION TO ORDINARY NOTATION

Remove the power of 10 from the following numbers.

Example: $5.6 \times 10^4 = 56,000$

21. 7.6×10^3
22. 9.2×10^{-3}
23. 25.8×10^6
24. 0.046×10^{-5}
25. 607×10^{-4}
26. 0.00013×10^2
27. 6.53×10^2

28. 2.11×10^{-5}
29. 4.25×10^{0}
30. 1.09×10^{1}

E NOTATION

Write the following numbers in the proper form of E notation. The decimal point should follow the first significant figure and the exponent should have two digits.
    Example: 56,000 = 5.6 E04

31. 2500
32. 0.0035
33. 167 E03
34. 0.0809
35. 10
36. 0.0000015
37. 1.25×10^{3}
38. 0.303 E−02
39. 670,000,000
40. 0.050×10^{-1}

CHANGING E NOTATION TO ORDINARY NOTATION

Remove the E notation from the following numbers.
    Example: 5.6 E04 = 56,000

41. 35.6 E05
42. 2.1 E−05
43. 0.067 E−08
44. 0.0049 E04
45. 12.2 E00
46. 0.00045 E−01
47. 8070 E−07
48. 250 E03
49. 0.65 E−02
50. 0.048 E02

ENGINEERING NOTATION

Convert the number on the left to the multiplier shown with the blank lines. If no multiplier is included, it is the basic unit with an exponent of E00. If necessary, refer to Technical Section B, Fig. B-9 for the exponent value of each of the multiplier names.
    Example: 56 kW = 56,000 W (watts)

51. 3,500,000 V = _____ kV = _____ MV (volts)
52. 6,780 W = _____ MW = _____ kW (watts)
53. 4 Hz = _____ kHz = _____ mHz (hertz)
54. 0.0058 A = _____ mA = _____ μA (amps)
55. 0.000009 F = _____ μF = _____ nF (farads)
56. 250 kW = _____ W = _____ MW (watts)
57. 108 mV = _____ μV = _____ V (volts)
58. 0.0075 MHz = _____ kHz = _____ GHz (hertz)
59. 0.013 ms = _____ μs = _____ s (seconds)
60. 0.000045 μH = _____ nH = _____ pH (henry)

ARITHMETIC WITH POWERS OF 10

Perform the indicated arithmetic with the following sets of numbers. Write the answer in the proper form of notation used in the problem. Do not round answers.

61. Add: $3.56 \times 10^{3} + 0.401 \times 10^{4}$
62. Add: $25.7 \times 10^{-2} + 109 \times 10^{-4}$
63. Add: 0.0125 E06 + 6.8 E03
64. Add: 0.00090 E04 + 0.45 E−01

65. Add: 0.750 milli + 325 micro
66. Add: 92 mega + 640 kilo
67. Subtract: $9.04 \times 10^0 - 0.34 \times 10^1$
68. Subtract: $0.0245 \times 10^2 - 6.03 \times 10^{-1}$
69. Subtract: 0.318 E−04 − 0.00021 E−02
70. Subtract: 250 E03 − 1.25 E05
71. Subtract: 98 mega − 0.0102 giga
72. Subtract: 0.468 pico − 0.000067 nano
73. Multiply: $7.2 \times 10^2 \times 13 \times 10^4$
74. Multiply: $30 \times 10^{-4} \times 1.5 \times 10^{-5}$
75. Multiply: 26 E03 × 1.5 E−03
76. Multiply: 0.047 E−01 × 0.005 E08
77. Multiply: 6.1 kilo × 0.23 mega
78. Multiply: 1000 milli × 0.529 micro

79. Divide: $\dfrac{360 \times 10^3}{.09 \times 10^6}$

80. Divide: $\dfrac{0.12 \times 10^3}{4.0 \times 10^{-1}}$

81. Divide: $\dfrac{0.0093\ E-04}{30.0\ E-02}$

82. Divide: $\dfrac{6.85\ E-07}{0.005\ E-02}$

83. Divide: $\dfrac{0.072\ \text{milli}}{6.0\ \text{micro}}$

84. Divide: $\dfrac{5.6\ \text{kilo}}{0.07\ \text{mega}}$

WORD PROBLEMS WITH SCIENTIFIC NOTATION

Unless otherwise specified, give answers in scientific notation.

85. If a computer can perform an addition operation in 1.5×10^{-6} seconds, in how much time can it perform 100 addition operations?
86. The diameter of the earth is about 99.09 percent smaller than the diameter of the sun. The diameter of the sun is 8.64×10^5 miles. What is the earth's diameter? Round answer to two significant digits.
87. One foot of a steel pipe increases about 1.1×10^{-5} feet in length for a 1-degree-celsius rise in temperature. If a pipe of the same material measures 10 feet long at 32°C, what does it measure at 42°C? Give answer in ordinary notation.
88. A micron is a unit of length equal to 0.001 millimeters. What size, in inches, is a particle that has a diameter of 8 microns?
89. The volume of an underground gasoline storage tank is 4.62×10^6 cubic inches. How many gallons will it hold? Give answer in ordinary notation.
90. An aboveground storage tank has a capacity of 2.5×10^5 barrels of crude oil. If the height of the tank is 40 feet, what is its diameter, in feet? Round answer to nearest whole unit. (Conversion factor: 1 barrel = 42 gallons)
91. In 1974, the Sears Tower, measuring 1454 feet, replaced the World Trade Center as the highest building in the world. Express this figure using units of miles.
92. By area, the USSR is the largest country, having 8.649×10^6 square miles of land. Canada is second largest. It has 4.797×10^6 square miles less than the USSR. How big is Canada in terms of land area? Give answer in ordinary notation.
93. Find the distance between the sun and the moon during (a) a lunar eclipse and (b) a solar eclipse. See Fig. 16-2. (The moon is approximately 2.4×10^5 miles from the earth. The sun is about 9.3×10^7 miles from earth. Disregard the moon and earth diameters as they are so small in comparison.)

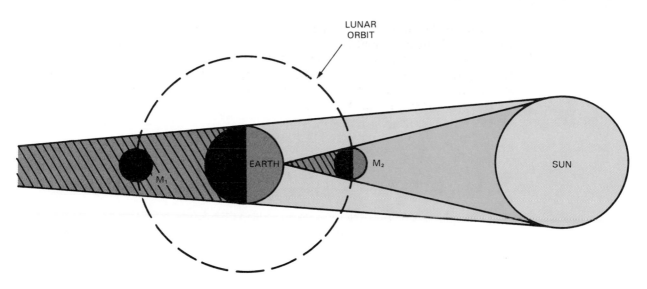

Fig. 16-2. An eclipse is the total or partial obscuring of one celestial body by another. M_1 is the moon in lunar eclipse, in which the moon is obscured. M_2 is the moon in solar eclipse, in which the sun is obscured from view.

94. In North America, Mt. McKinley is the highest mountain at 2.032×10^4 feet. What is this in meters?

95. One ampere represents a flow of 1 coulomb per second, or 6.24×10^{18} electrons past a given point per second. A current of 2 milliamperes, then, is how many electrons per second past a given point?

PROBLEM-SOLVING ACTIVITIES

Activity 16-1.
Poster of Engineering notation

Objective: To make a large poster-size chart showing engineering notation.
Instructions:

1. Use the information in this chapter to make a poster-size chart showing engineering notation. See Fig. 16-3. If desired, you can also make a chart showing complete metric notation.
2. Show 3 samples of converting from one multiplier to another on the chart.
3. Ask each of your classmates to add one example of converting multipliers to the chart.

Example:

Given Number	New Multiplier	New Multipler	Unit
3,500,000V	3500 kV	3.5 MV	volts

Fig. 16-3. Sample chart showing engineering notation.

These employees are estimating the number of hangers needed to install ductwork in a large office building.
(Jack Klasey)

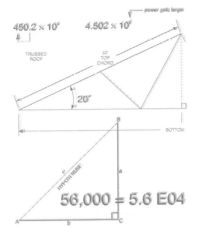

Chapter 17

Introduction to Estimating

OBJECTIVES

After studying this chapter, you will be able to:
- Calculate board feet given board length in inches or feet.
- Estimate wood needed for foundation sills.
- Estimate wood needed for wall plates.
- Estimate wood needed for door and window trim.
- Estimate wood needed for framing members.
- Estimate material requirements for sheet products.
- Estimate concrete material requirements.

Estimating the material requirements for a project requires an in-depth study of the construction techniques that will be used. An estimate of a project serves two main purposes. First, it states the amount of material needed. Second, it states the cost of a project. The accuracy of an estimate will often make the difference between profit and loss.

This chapter is an introduction to the math that is used in estimating material requirements. Estimates for a variety of materials are explored. However, these are but a sampling of the materials encountered on construction projects that require estimating. The information provided here is not intended to be all inclusive, nor is it to be used for construction techniques. It is intended to show the mathematical steps involved in estimating. Also, the focus here is on estimating material quantity and not on material cost. Estimating material quantity is only the first step in preparing a project cost estimate.

BOARD MEASURE

Dimensional lumber is lumber cut to certain standard dimensions, such as 2 x 4s or 2 x 6s. It is sold by either of two methods—*linear measure* or *board measure*. Linear measure is a measure of length. Board measure is a measure of volume of a board, given in board feet (bd. ft.). The volume of 1 board foot is derived from a board 12 inches long, 12 inches wide, and 1 inch thick. It is equal to 144 cubic inches of lumber.

Dimensional lumber is classified by its *nominal size*—the size for which it is named. With finished lumber, the nominal size is not the actual size. For example, the nominal size of a 2 x 4 is 2 inches by 4 inches. Its actual size is 1 1/2 inches by 3 1/2 inches. The nominal size of lumber describes the width and thickness.

FORMULAS TO CALCULATE BOARD FEET (BF)

Length in inches to board feet:

$$BF = \text{nominal size} \times \frac{\text{length (in inches)}}{144}$$

Length in feet to board feet:

$$BF = \text{nominal size} \times \frac{\text{length (in feet)}}{12}$$

Board feet to length in inches:

$$\text{Length (in inches)} = \frac{BF \times 144}{\text{nominal size}}$$

Board feet to length in feet:

$$\text{Length (in feet)} = \frac{BF \times 12}{\text{nominal size}}$$

Note: Substitute values in these equations, but do not try to cancel out units. It is not necessary to do so because the step is "built in" to the unitless numbers.

Sample Problem 17-1.

How many board feet are in a piece of 2 x 4 lumber, 96 inches long?

Step 1. Select the suitable formula.

$$BF = \text{nominal size} \times \frac{\text{length (in inches)}}{144}$$

Step 2. Substitute in values and solve.

$$BF = 2" \times 4" \times \frac{96"}{144}$$

$$BF = 5.33 \text{ bd. ft.}$$

Sample Problem 17-2.

Determine, in board feet, the measure of a 12-foot-long 2 x 6.

Step 1. Select the suitable formula.

$$BF = \text{nominal size} \times \frac{\text{length (in feet)}}{12}$$

Step 2. Substitute in values and solve.

$$BF = 2" \times 6" \times \frac{12'}{12}$$

$$BF = 12 \text{ bd. ft.}$$

Sample Problem 17-3.

Determine how many inches long a 1 x 8 with a board-foot measurement of 200 bd. ft. is.

Step 1. Select the suitable formula.

$$\text{Length (in inches)} = \frac{BF \times 144}{\text{nominal size}}$$

Step 2. Substitute in values and solve.

$$\text{Length (in inches)} = \frac{200 \text{ bd. ft.} \times 144}{1" \times 8"}$$

$$\text{Length} = 3600 \text{ in.}$$

Sample Problem 17-4.

How many feet long is a 1 x 10 with a board-foot measurement of 360 bd. ft.?

Step 1. Select the suitable formula.

$$\text{Length (in feet)} = \frac{\text{BF} \times 12}{\text{nominal size}}$$

Step 2. Substitute in values and solve.

$$\text{Length (in feet)} = \frac{360 \text{ bd. ft.} \times 12}{1\text{"} \times 10\text{"}}$$

Length: = 432 ft.

Practice Problems

Complete the following problems on a separate sheet of paper.

Convert the linear measurements to board feet.

	Nominal Size	Linear Length
1.	2 × 4	136 inches
2.	2 × 6	14 feet
3.	2 × 8	16 feet
4.	2 × 10	200 inches
5.	2 × 12	8 feet
6.	1 × 4	6 feet
7.	1 × 6	96 inches
8.	1 × 8	50 feet
9.	1 × 10	40 feet
10.	1 × 12	480 inches

Change the board feet measurements to linear feet.

	Nominal Size	Board Feet
11.	1 × 4	20 bd. ft.
12.	1 × 6	30 bd. ft.
13.	1 × 8	800 bd. ft.
14.	1 × 10	1200 bd. ft.
15.	1 × 12	560 bd. ft.
16.	2 × 4	300 bd. ft.
17.	2 × 6	120 bd. ft.
18.	2 × 8	240 bd. ft.
19.	2 × 10	680 bd. ft.
20.	2 × 12	1400 bd. ft.

LINEAR MEASURE

Linear measure is taken in a straight line. Standard units are linear feet (lin. ft.). Although straight-line measurements appear simple, caution must be exercised to take into account the width of the boards, especially at joints.

Standard lumber lengths

Lumber is available in standard lengths of 8, 10, 12, 14, and 16 feet. When making calculations for linear measurements, do not ignore the standard lengths. If the length needed is longer than the available lumber, it is usually best to place the joint somewhere near the middle to avoid having one very short piece. For example, if the length needed is 18 feet, a 16-foot and a 2-foot length could be used, but a 10-foot and an 8-foot would be better.

Foundation sill

The *foundation sill* lays flat on the top of concrete foundation walls. A typical sill is made from pressure-treated, 2 x 6 lumber. The sill is flush with the outside edge of the wall. To determine the total length of 2 x 6 required, calculate the perimeter of the foundation. Then, subtract the widths of the sill as necessary. A sketch helps to determine what widths to subtract.

Sample Problem 17-5.

Using Fig. 17-1, calculate the total length of 2 x 6 sill, and determine the standard lengths desired for the job. Use actual board width — 5 1/2 inches.

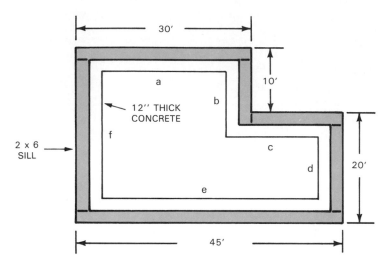

Fig. 17-1. A concrete foundation plan showing foundation sill.

Step 1. Find total length required. Study drawing. Notice that horizontal boards run the full length of the foundation. Notice that the length of the vertical boards is reduced by the width of the horizontal boards. (Boards in this problem are laid out this way only to simplify the calculations presented.)

Sum length of horizontal pieces:

$$
\begin{array}{rl}
a = & 30' \\
c = & 15' \\
e = & 45' \\
\hline
\text{Horizontal} = & 90'
\end{array}
$$

Sum length of vertical pieces:

$$
\begin{array}{rll}
b = (10' + 5\,1/2") - 5\,1/2" & = 10' \\
d = 20' - (5\,1/2" \times 2) & = 19'1" \\
f = 30' - (5\,1/2" \times 2) & = 29'1" \\
\hline
\text{Vertical} & = 58'2"
\end{array}
$$

Sum length of horizontal and vertical pieces:

Horizontal	90'
Vertical	58'2"
Length required	148'2"

Note: To verify that this is indeed perimeter of foundation minus sill widths:

$$
\begin{array}{ll}
\text{Perimeter} = 30' + 15' + 45' + 10' + 20' + 30' = 150' \\
\text{Sill widths deducted} = (5\,1/2" \times 4) = 22" \qquad\qquad 1'10" \\
\hline
\text{Length required} \qquad\qquad\qquad\qquad\qquad\qquad\qquad 148'2"
\end{array}
$$

Step 2. Examine drawing for best combination of standard lengths. Standard 10-foot lengths should be ordered as the best combination for all sides.

$$
\begin{array}{rl}
\text{a:} & 3 \text{ -- } 10' = 30' \\
\text{b:} & 1 \text{ -- } 10' = 10' \\
\text{c:} & 1.5 \text{ -- } 10' = 15' \\
\text{d:} & 2 \text{ -- } 10' = 20' \\
\text{e:} & 4.5 \text{ -- } 10' = 45' \\
\text{f:} & 3 \text{ -- } 10' = 30' \\
\hline
\text{Total:} & 15 \text{ -- } 10' = 150'
\end{array}
$$

Wall plates

The 2 x 4s used to frame a wall are attached at the top and bottom with 2 x 4s called *wall plates*. Refer to Fig. 17-2. The bottom plate has a single 2 x 4. The top uses two 2 x 4s nailed together. To calculate the total length of 2 x 4s needed, determine the total length of the walls and multiply by 3.

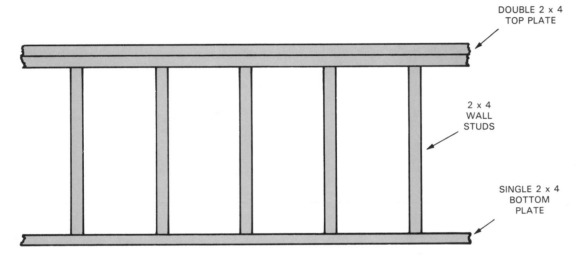

DOUBLE 2 x 4 TOP PLATE

2 x 4 WALL STUDS

SINGLE 2 x 4 BOTTOM PLATE

Fig. 17-2. Standard wall framing showing top and bottom plates.

Sample Problem 17-6.

Using the foundation drawing in Sample Problem 17-5, determine the total length of 2 x 4s needed for top and bottom plates of outside walls and standard lengths for the job.

Step 1. Total length of outside walls is 150 feet (from calculations for Sample Problem 17-5).

Step 2. Determine the total length required for the plates.

Total plates = 150 ft. × 3 = 450 ft.

Step 3. Determine quantity of standard lengths. The best combination of standard lengths is same for Sample Problem 17-5, times 3.

Total: (15 × 3) -- 10 ft.
45 -- 10 ft. = 450 ft.

Door and window trim

Dimensions of door and window casings (or any other such trim that encloses something) are given for the inside edge of the trim. If trim is applied to the *exterior* of something, a picture, for example, the length needed is the length of each side to be trimmed *plus* the width of the trim board for each mitered end. (If trim is applied to the *interior* of something, this does not apply. With mitered corners, in this case, the length of trim equals the length of each side being trimmed.)

Sample Problem 17-7.

Determine the total length of trim needed to frame the window shown in Fig. 17-3 and standard lengths for the job.

Step 1. Find length of top and bottom pieces.

Window width	=	2'
Double trim width	=	6" (3" for each mitered end)
Total one end	=	2'6"
	×	2
Total top and bottom	=	5'

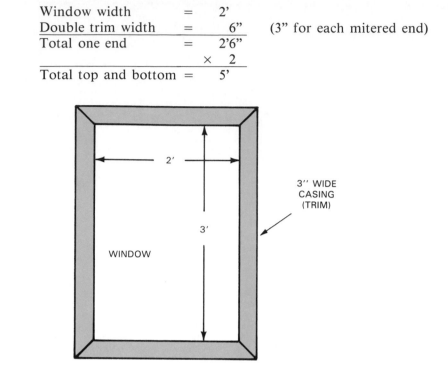

Fig. 17-3. Window casing.

Step 2. Find length of side pieces.

Window height	=	3'
Double trim width	=	6"
Total one side	=	3'6"
	×	2
Total both sides	=	7'

Step 3. Total all sides.

$$5' + 7' = 12'$$

Step 4. Best standard length:

1 -- 12'

Sample Problem 17-8.

Find the length of trim needed to frame the door shown in Fig. 17-4 and standard lengths for the job.

Step 1. Find length of top piece.

Door width	= 3'
Double trim width	= 7"
Total top piece	= 3'7"

Step 2. Find length of sidepieces.

Door height	=	6'8"
Trim width	=	3 1/2"
Total one side	=	6'11 1/2"
	×	2
Total both sides	=	13'11"

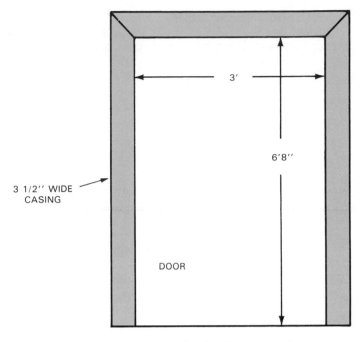

Fig. 17-4. Door casing.

Step 3. Total three sides.

3'7" + 13'11" = 17'6"

Step 4. Best standard lengths:

Choice #1:
Top: 1 -- 8' (waste = 4'5")
Sides: 1 -- 14' (waste = 1")

Choice #2:
Top + one side: 1 -- 12' (waste = 1'5 1/2")
Other side: 1 -- 8' (waste = 1'1/2")

Practice Problems

Complete the following problems on a separate sheet of paper.

1. The overall length of the foundation shown in Fig. 17-1 was changed from 45' to 60'. If all the other dimensions given remain the same, what is the new length of 2 × 6 sill needed?
2. What is the total length of 2 × 6 needed for the double top plate and single bottom plate of a 60' long wall?
3. What is the total length of window casing needed if the window dimensions in Fig. 17-3 are changed to 4'6" × 6'?
4. What is the total length of door casing needed if the size of the door shown in Fig. 17-4 is changed to 14' wide and 7' tall?

FRAMING MEMBERS

Framing is the timber structure of a building that gives it its shape; including interior and exterior walls, floor, roof, and ceilings. The *framing members* are the individual boards. The number of framing members needed depends primarily on variations in construction techniques and:
• Spacing of framing.
• Corners and wall intersections.
• Openings in framing.

Spacing of framing. Framing members, wherever they are used, have a standard spacing of either 16 or 24 inches between centers. See Fig. 17-5. On drawings, this spacing would be noted as "16″ O.C." or "24″ O.C." The O.C. stands for *on center.* The standard spacing has been determined by the dimensions of sheet products, which come in 4-foot widths and in length in multiples of 4 feet. When calculating the number of framing members required, if the result is not a whole number, round up to the next highest whole number.

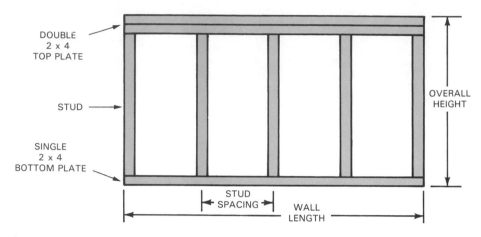

Fig. 17-5. Wall framing.

FORMULA TO CALCULATE NUMBER OF FRAMING MEMBERS

$$\# = \frac{\text{length (in inches)}}{\text{spacing (in inches)}} + 1$$

Notes on formula:
• Length is *framing* length. It runs in direction of the plates.
• The additional one is for closing off the last space.
• Works for walls, joists, and rafters.
• For estimating connected, exterior walls, length is the building perimeter. Do not figure each wall individually. There will be only one additional stud.

Sample Problem 17-9.

If the 2 x 4 studs are spaced 16' O.C., how many are required for a wall 16 feet long? Note: Conversion factors are found in Chapter 7 if review is necessary.

Step 1. Change the measurement of the length from feet to inches.

Conversion factor: 12 inches = 1 foot

Conversion bar: $\left(\dfrac{16 \text{ ft.}}{}\right) \left(\dfrac{12 \text{ in.}}{1 \text{ ft.}}\right)$

Length = 192 in.

Step 2. Use the formula to find the number of studs needed.

Formula: $\# = \dfrac{\text{length (in inches)}}{\text{spacing (in inches)}} + 1$

Substitution: $\# = \dfrac{192''}{16''} + 1$

Solution: $\# = 13$

Corners and wall intersections. Corners and wall intersections require extra studs. Since a corner (using 2 x 4s) is usually made up of 3 studs, add 2 extra for each corner. The same is true for wall intersections. For each intersection, add 2 more studs.

Openings in framing. Framing an opening, such as a window or door, alters the number and type of framing members used. Fig. 17-6 depicts one type of framing used for a door opening. Framing details can be complex and have many variations. For the sake of simplicity, account for openings in framing by adding 3 to the number of framing members for each opening encountered. Ignore the 2 x 12 used as a header. For example, if the wall of Sample Problem 17-9 had one window and one door, the number of studs required would be 19—the original 13 for the wall *plus* three for each opening.

Practice Problems

On a separate sheet of paper, calculate the number of framing members needed. If the result is not a whole number, round *up* to the next whole number. Be sure to add 1 for the beginning stud.

	Stud spacing	Wall length
1.	16" O.C.	25'
2.	16" O.C.	30'
3.	16" O.C.	38'
4.	16" O.C.	120"
5.	16" O.C.	150"
6.	24" O.C.	26'
7.	24" O.C.	35'
8.	24" O.C.	50'
9.	24" O.C.	136"
10.	24" O.C.	256"

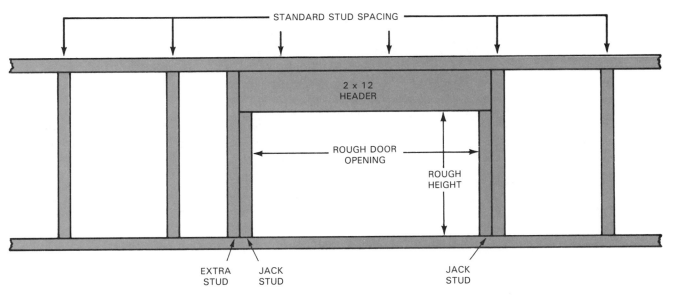

Fig. 17-6. Framing of a door opening.

SQUARE MEASURE

Square measure is the measure for surface area. Surface area calculations determine the material requirements for sheet products. Sheet products include such items as flooring, wall covering, roofing, plywood, drywall, paneling, etc. Determine the surface area as discussed in Chapter 9. (Remember that surface area is the *total* area of all concerned surfaces.) Then, figure the amount of material needed.

Sheet products

Sheet products are available in certain standard sizes. Examples of standard sizes:
- Plywood: 4 feet × 8 feet = 32 square feet
- Drywall: 4 feet × 8 feet = 32 square feet

- Drywall: 4 feet × 12 feet = 48 square feet
- Ceiling tiles: 2 feet × 4 feet = 8 square feet
- Roofing shingles: 1 square = 100 square feet
  Note: Roofing materials are sold by *the square*. This is the amount of material needed to provide 100 square feet of finished roof surface.

To calculate the number of sheets required, divide the surface area to be covered by the area covered per sheet.

FORMULA TO CALCULATE NUMBER OF SHEETS

$$\# = \frac{\text{area to be covered}}{\text{area of sheet}}$$

Notes on formula:
- Always round to the next highest whole number.
- Be sure units are the same.

Sample Problem 17-10.

Determine the number of sheets of 4' x 8' plywood needed for the side of the house shown in Fig. 17-7. Subtract for the window and door openings.

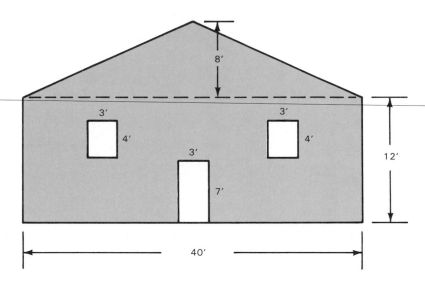
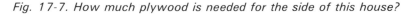

Fig. 17-7. How much plywood is needed for the side of this house?

Step 1. Find area of rectangular portion.

Formula: $A_1 = l \times w$
Substitution: $A_1 = 12 \text{ ft.} \times 40 \text{ ft.}$
Solution: $A_1 = 480 \text{ ft.}^2$

Step 2. Find area of triangular portion.

Formula: $A_2 = 1/2 \times b \times h$
Substitution: $A_2 = 1/2 \times 40 \text{ ft.} \times 8 \text{ ft.}$
Solution: $A_2 = 160 \text{ ft.}^2$

Step 3. Find area of windows.

Formula: $A_3 = l \times w$
Substitution: $A_3 = 4$ ft. $\times 3$ ft.
Solution: $A_3 = 12$ ft.^{2}
Two windows: $A_3 = 24$ ft.^{2}

Step 4. Find area of door.

Formula: $A_4 = l \times w$
Substitution: $A_4 = 3$ ft. $\times 7$ ft.
Solution: $A_4 = 21$ ft.^{2}

Step 5. Find combined area.

Formula: $A_T = A_1 + A_2 - (A_3 + A_4)$
Substitution: $A_T = 480$ ft.$^2 + 160$ ft.$^2 - (24$ ft.$^2 + 21$ ft.$^2)$
Solution: $A_T - 595$ ft.^{2}

Step 6. Find number of sheets of 4-foot by 8-foot plywood.

Area of plywood $= 32$ ft.^{2} per sheet

Formula: $\# = \dfrac{\text{area to be covered}}{\text{area of sheet}}$

Substitution: $\# = \dfrac{595 \text{ ft.}^2}{32 \text{ ft.}^2}$

Solution: $\# = 18.6$ sheets $\Rightarrow 19$ sheets (rounded)

Practice Problems

Complete the following problems on a separate sheet of paper.

1. How many 8" square tiles are needed to cover a 32' × 48' floor?
2. How many 4" square tiles are needed to cover a wall that is 8' high and 12' long?
3. If the measurements of the side of the house shown in Fig. 17-7 are changed so that the wall is 32' wide and 16' high, how many 4' × 8' sheets of plywood will be needed?
4. How many 2' × 4' tiles are needed to cover a 60' × 120' ceiling?
5. How many 1" square tiles are needed to cover a 12' × 14' bathroom floor?

CUBIC MEASURE

Cubic measure is the measure for volume. Concrete is sold by the cubic yard. Volume calculations determine the amount of concrete required for a job. Refer to Chapter 8 for a review of volume calculations. Since concrete is bought by the cubic yard, volume in cubic feet must be converted. A conversion bar will do the job; however, for convenience, a formula is provided. In the formula, the numerator produces a calculation of cubic feet. Dividing by 27 converts cubic feet to cubic yards (27 ft.^{3} = 1 yd.^{3}).

FORMULA TO CALCULATE CUBIC YARDS OF CONCRETE

Volume (in cubic yards) = $\dfrac{\text{length} \times \text{width} \times \text{depth (all in feet)}}{27}$

Note: For convenience, decimal-foot equivalents of 1 through 12 inches are provided.

INCHES	FEET
1	0.08
2	0.17
3	0.25
4	0.33
5	0.42
6	0.50
7	0.58
8	0.67
9	0.75
10	0.83
11	0.92
12	1.00

Sample Problem 17-11.

Fig. 17-8 shows a concrete footing. Calculate the amount of concrete needed.

Step 1. Since width and depth are the same for both Section A and Section B, their lengths may be combined and used in the formula as one length. Subtract the width of Section A from 6 feet to get the length of Section B.

Length of Section B = 6 ft. − 2 ft. = 4 ft.

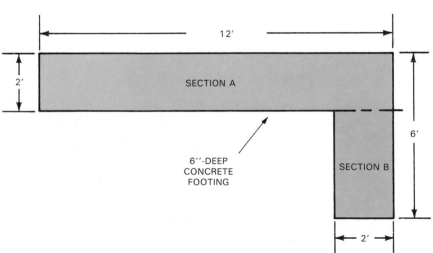

Fig. 17-8. Plan view of a concrete footing.

Step 2. Sum the lengths.

Total length = Section A length + Section B length
Total length = 12 ft. + 4 ft. = 16 ft.

Step 3. Use formula to calculate volume.

Formula: Volume (in cubic yards) =
$$\frac{\text{length} \times \text{width} \times \text{depth (all in feet)}}{27}$$

Substitution: Volume (in cubic yards) =
$$\frac{16 \text{ ft.} \times 2 \text{ ft.} \times 0.5 \text{ ft.}}{27}$$

Solution: Volume = 0.6 yd.^{3}

Note: This method of finding cubic yards of concrete works for footings because the dimensions of width and depth are generally the same throughout. Any time they are not, cubic yards should be found by figuring volume of each section separately and then totaling the values.

Sample Problem 17-12.

Find the concrete needed for the walls and floor in Fig. 17-9.

Step 1. Subtract double the wall thickness from the length of one side wall for the length of Side a.

Length of Side a = 16' − (8" × 2) = 14'8"

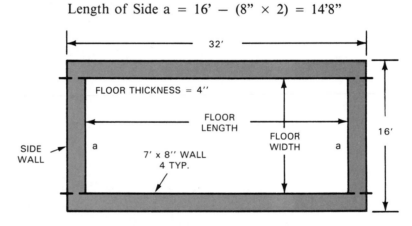

Fig. 17-9. Plan view of concrete "room."

Step 2. Sum the lengths of the walls.

Total length = 2(32') + 2(14'8")
Total length = 64' + 29'4" = 93'4" = 93.33'

Step 3. Calculate volume of the walls.

Formula: Volume (in cubic yards) =
$$\frac{\text{length} \times \text{width} \times \text{depth (all in feet)}}{27}$$

Substitution: Volume (in cubic yards) =
$$\frac{93.33' \times 7' \times 0.67'}{27}$$

Solution: Volume = 16.2 yd.^{3}

Step 4. Find the dimensions of the floor by subtracting double the wall thickness in each direction.

Length of floor = 32' − (8" × 2) = 30'8" = 30.67'
Width of floor = 16' − (8" × 2) = 14'8" = 14.67'

Step 5. Calculate volume of the floor.

Formula: Volume (in cubic yards) =
$$\frac{\text{length} \times \text{width} \times \text{depth (all in feet)}}{27}$$

Substitution: Volume (in cubic yards) =
$$\frac{30.67' \times 14.67' \times 0.33'}{27}$$

Solution: Volume = 5.5 yd.³

Step 6. Find total volume of concrete needed.

Total volume = volume of walls + volume of floor
Total volume = 16.2 yd.³ + 5.5 yd.³
Total volume = 21.7 yd.³

Practice Problems

Find the cubic measure for the following problems. If necessary, round your answers to 1 decimal place.

1. How many cubic yards of concrete are needed for the footing shown in Fig. 17-8 if its length is changed from 12' to 20' and its depth is changed from 6" to 8".
2. Determine the concrete needed for the walls of the room in Fig. 17-9 if the width is changed from 16' to 24'.
3. Calculate the cubic yards of concrete needed for a 4" thick floor of a storage building measuring 8' wide and 24' long. If the concrete costs $50 per cubic yard, how much will the concrete for the floor cost?
4. How many cubic yards of dirt are needed to fill a trench that is 50' long, 10' wide, and 6' deep?
5. Determine the cubic yards of topsoil needed for a 3" cover over a tract of land that is 150' long and 100' wide.

TEST YOUR SKILLS

Do *not* write in this book. Use a separate sheet of paper to complete the following problems. Show your work and your final answer. Include units where necessary.

BOARD MEASURE

1. Convert the linear measurements to board feet.

	Nominal	Linear Length
a.	2 x 4	120 inches
b.	2 x 4	16 feet
c.	2 x 6	36 inches
d.	2 x 6	12 feet
e.	1 x 8	90 inches
f.	1 x 8	15 feet
g.	1 x 12	144 inches
h.	1 x 12	4 feet
i.	2 x 10	70 inches
j.	2 x 10	256 feet

2. Convert the measurements of board feet to linear measurements.

Nominal	Board Feet	Linear
a. 2 x 4	35 bd. ft.	inches
b. 2 x 6	120 bd. ft.	feet
c. 2 x 8	48 bd. ft.	inches
d. 2 x 10	580 bd. ft.	feet
e. 2 x 12	360 bd. ft.	inches
f. 1 x 4	10 bd. ft.	feet
g. 1 x 6	40 bd. ft.	inches
h. 1 x 8	260 bd. ft.	feet
i. 1 x 10	150 bd. ft.	inches
j. 1 x 12	12 bd. ft.	feet

LINEAR MEASURE

3. a. Determine the exact amount of pressure-treated, 2 x 6 nominal (1 1/2" x 5 1/2" actual) lumber needed as a sill for the concrete foundation walls having outside wall dimensions shown in Fig. 17-10. Note: Sill will be used above the door also.
   b. Make a material list using standard lumber lengths (8', 10', 12', 14', 16').

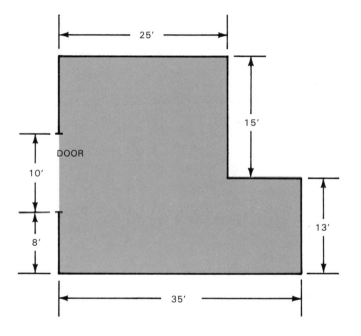

Fig. 17-10. Outside-wall dimensions of concrete foundation.

4. Determine the total length of 2 x 4 needed for a double top plate and a single bottom plate on a wall 56 feet long.
5. What length of 1 x 4 board is needed to trim around a 4-foot by 6-foot window? (The board measures 3/4 inches thick by 3 1/2 inches wide.)
6. Use Fig. 17-11 to determine the total length of trim needed for the sliding-glass door.

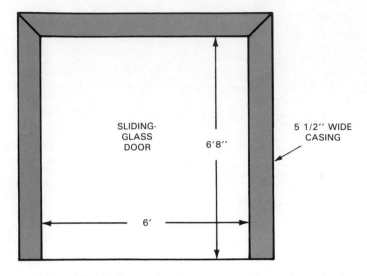

SLIDING-
GLASS
DOOR

6'8''

5 1/2'' WIDE
CASING

6'

Fig. 17-11. Determine the amount of wood required.

7. Calculate the number of framing members needed.

Stud Spacing	Wall Length
a. 16" O.C.	24'
b. 16" O.C.	40'
c. 16" O.C.	18'9"
d. 16" O.C.	10'6"
e. 16" O.C.	110"
f. 24" O.C.	15'
g. 24" O.C.	38'
h. 24" O.C.	86"
i. 24" O.C.	20'4"
j. 24" O.C.	14'8"

SQUARE MEASURE

Answer questions and round to the next largest whole number, if necessary.

8. How many 4-foot by 8-foot sheets of plywood are needed to cover a roof measuring 24 feet wide and 36 feet long?
9. Determine how many 2-foot by 4-foot tiles are needed to cover a ceiling that measures 12 feet by 16 feet.
10. How many 4-foot by 8-foot sheets of drywall are needed for the 8-foot high-walls and the ceiling of a 14-foot by 18-foot room? Ignore waste for doors and windows.
11. How many 4-foot by 8-foot sheets of plywood will be needed for the house shown in the elevational drawings of Fig. 17-12? Give answer in terms of the following list:
    a. Floor.
    b. Front and rear.
    c. Left and right sides.
    d. Roof.

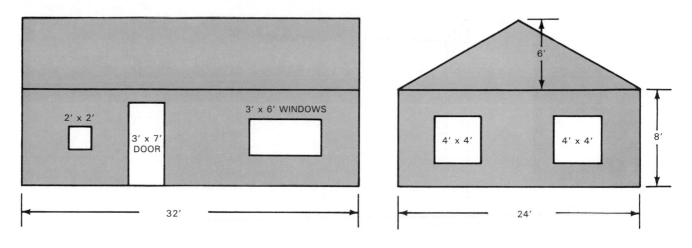

FRONT AND REAR ELEVATIONS   LEFT AND RIGHT SIDE ELEVATIONS

Fig. 17-12. A multiview drawing of a house.

CUBIC MEASURE

All answers for concrete must be given in cubic yards. Round answer to one decimal place, if necessary.

12. What volume of concrete is needed to pour a small pad 8 feet by 8 feet and 4 inches thick?
13. A 5-inch-thick concrete pad will be used for a parking area 32 feet by 20 feet.
    a. How much concrete will be needed?
    b. If the concrete costs $60 per yard, what will be the cost for material?

BUILDING ESTIMATE

14. Make an estimate of materials needed for a 24-foot by 24-foot garage. There is one opening for an 8-foot-wide by 7-foot-high door. Walls rise 8 feet above the foundation. The roof is flat, and a center beam runs the length of the garage. The foundation is concrete 4 feet deep and 6 inches thick. The floor is poured concrete, 4 inches thick. Draw diagrams and give answers in terms of the following:
    a. Concrete for foundation.
    b. Concrete for floor.
    c. 2 x 4 wall studs spaced 16 inches on center.
    d. 2 x 4 for double top and single bottom plates.
    e. 2 x 10 roof rafters 14 feet long, spaced 16 inches on center.
    f. 4 x 8 plywood for walls.
    g. 4 x 8 plywood for roof.

PROBLEM-SOLVING ACTIVITY

Activity 17-1.
Bill of materials and cost estimate for a small building
Objective: Make a complete bill of materials and cost estimate for a small building, such as a storage shed.
  Instructions:
  1. Design a small building. Draw the building, showing all construction details.
  2. Using the information on your drawings, make a bill of materials for the building. See Fig. 17-14.

Sample Bill of Materials		
Section	**Material**	**Quantity**
Framing	2 x 4s	
Foundation	Concrete block	
Floor	1/2" Plywood	
Walls	3/8" Plywood siding	
Door	32 x 80 metal	
Roof	1/2" Plywood	
Roof	Fiberglass shingles	
Misc.	Nails	

Fig. 17-14. Sample bill of materials.

3. Call or visit a lumber yard to determine the cost of each item on the bill of materials.

Technical Section A

Using the Electronic Calculator

The calculator is a very useful tool when used correctly. Calculators do not make errors; people do. Any computer will only perform as instructed . . . "garbage in equals garbage out." It is best to use the calculator as an aid in solving math problems, using a pencil and paper to maintain order to the problem. This will help to reduce mistakes. This section explains how to use a standard calculator. Calculators vary in their operation. Most come with operation manuals at the time of purchase. It is recommended that you consult yours for additional information.

BASIC ARITHMETIC

As a general rule, perform calculations with addition and subtraction separately from multiplication and division. With no grouping symbols, multiply ($\times$) and divide ($\div$) first, then, add ($+$) and subtract ($-$). Refer to Chapter 10, for a complete list of mathematical operations, and for the order in which they are to be performed.

Using the equal sign ($=$)

The $=$ is used as the end of a problem. Using it clears the calculator, making it ready for the next operation.

If a string of numbers is being added or subtracted, end the string with $=$ before proceeding. You will notice that the calculator will perform the arithmetic every time a basic function key is depressed. The problem may be solved, but this number remains in the calculator and must be cleared before proceeding.

To add or subtract several numbers, then multiply or divide, the $=$ must be pressed between $+/-$ and $\times/\div$ steps. If not, the multiplication or division will only be performed on the last number entered, rather than on the total. For example, if taking an average, perform all addition, then press $=$ before performing the division.

FRACTIONS

Some calculators accept fractions, however, most work only with fractions in decimal form. When fractions are to remain in fraction form, for whatever reason, the calculator can be used to perform the arithmetic. This will help to reduce errors. However, a pencil and paper will be needed for keeping track.

Mixed numbers and improper fractions

Change to an improper fraction:
Rule: Multiply the denominator times the whole number and add to the numerator. Keep the same denominator.

$$4 \frac{2}{3} = \frac{3 \times 4 + 2}{3} = \frac{14}{3}$$

Change to a mixed number:
Rule: Divide the numerator by the denominator. The remainder becomes the fractional part; it is written over the denominator.

$$\frac{30}{4} = 30 \div 4 = 7 \text{ remainder } 2 = 7 \frac{2}{4} = 7 \frac{1}{2}$$

$$\frac{30}{4} = 7 \frac{2}{4} = 7 \frac{1}{2}$$

Using a calculator: Perform the division. This will result in a decimal number, consisting of a whole number and a decimal fraction. Multiply the whole number by the denominator; then subtract from the numerator. This will give the numerator of the fractional part; it is written over the denominator.

$$\frac{30}{4} = 7.5 = 7 + \frac{30 - (7 \times 4)}{4} = 7\frac{2}{4} = 7\frac{1}{2}$$

SCIENTIFIC NOTATION

Many calculators have a special key for entering powers of 10. Arithmetic can be performed in scientific notation. The answer will be displayed in E notation. Refer to Chapter 16 for details on dealing with exponents.

The $+/-$ key is used to change the sign of a number. It is also used to change the sign of an exponent.

TRIGONOMETRY

Refer to Chapter 15 for review of the trigonometric functions. The keys for the basic functions of sine, cosine, and tangent are used like the keys for the basic arithmetic operations. The function keys can be contained within complex problems. Be aware that the function will be taken of the last number entered. If it is desired to find the trigonometric function of an entire expression, be sure to first use the $=$ key.

• To find the value of a trigonometric function, such as:

tan 53°
   1. Enter the angle (53°).
   2. Press tan. (Calculator is in degree mode.)
   3. Answer is displayed (1.33).

$$\cos\left(\frac{\pi}{3} + \frac{\pi}{3}\right)$$

   1. Perform the arithmetic inside parentheses. Convert to decimal number.

$$\cos \frac{2\pi}{3} = \cos 2.094$$

   2. Press cos. (Calculator is in radian mode.)
   3. Answer is displayed (-0.5).

• To find the angle when given the value of the function:

sin A = 0.707
   1. Rewritten: a = $\sin^{-1} 0.707$.
   2. Press INV key. May be called 2nd F key, SHIFT key, or other, depending on calculator.
   3. Press sin. (Calculator is in degree mode.)
   4. Answer displayed is the angle (45°).

SQUARE AND SQUARE ROOT

The x^2 key is used by first entering a number. Pressing this key, then, the number is squared. The $\sqrt{\ }$ key is used to take the square root of any number. Enter the number, then, press $\sqrt{\ }$. The number displayed is the square root. If it is desired to find the square or square root of an entire expression, be sure to first use the $=$ key.

On many calculators, x^2 and $\sqrt{\ }$ are contained on the same key. One is written on the key. The other is written on the body of the calculator, which is again used by first pressing INV.

Example: A triangle has sides of 3 feet and 4 feet. Find the hypotenuse.

Formula: c = $\sqrt{a^2 + b^2}$

Substitution: c = $\sqrt{3^2 + 4^2}$

Calculator operations:
enter 3/press x^2/press $+$/enter 4/press x^2/press $=$/press $\sqrt{\ }$
Answer on display: 5 feet

Technical Section B

Tables

Length—smallest unit: inch (in.)
1 foot (ft.)   = 12 inches
1 yard (yd.)   = 3 feet
                 = 36 inches
1 mile (mi.) (also called statute mile)
                 = 1760 yards
                 = 5280 feet
1 nautical mile = 1.15 statute miles
                 = 6076 feet

Area—smallest unit: square inches (sq. in.) (in.²)
1 square foot (sq. ft.) (ft.²) = 144 square inches
1 square yard (sq. yd.) (yd.²) = 9 square feet
                 = 1296 square inches
1 acre (a.) = 4840 square yards
            = 43,560 square feet

Volume—smallest unit: cubic inches (cu. in.) (in.³)
1 cubic foot (cu. ft.) (ft.³) = 1728 cubic inches
1 cubic yard (cu. yd.) (yd.³) = 27 cubic feet
                 = 46,656 cubic inches

Capacity—smallest unit: fluid ounces (fl. oz.)
1 cup (c.)   = 8 fluid ounces
1 pint (pt.)   = 2 cups
                = 16 fluid ounces
1 quart (qt.)   = 2 pints
                = 4 cups
                = 32 fluid ounces
1 gallon (gal.) = 4 quarts
                = 16 cups
                = 128 fluid ounces
                = 0.1337 cubic feet
                = 231 cubic inches
1 barrel (bbl.) = 42 gallons

Weight—smallest unit: ounces (oz.) (dry weight)
1 pound (lb.) = 16 ounces
1 ton (tn.)   = 2000 pounds

Miscellaneous
1 mil = 0.001 inches
1 cu. ft. of water = 62.425 lb. (weight)
1 lb. per sq. in. = 27.70 in. of water (pressure)
                   = 2.31 ft. of water
                   = 2.036 in. of Hg
1 atmosphere   = 14.696 lb. per sq. in. (pressure)
                   = 33.899 ft. of water
                   = 29.921 in. of Hg
1 cu. ft. per sec. = 448.83 gal. per min. (flow)
1 horsepower = 33,000 ft. lb. per min. (power)
             = 550 ft. lb. per sec.
             = 2546.5 Btu per hr.
             = 745.7 watts
1 Btu = 777.97 ft. lb. (energy)
1 kilowatt hr. = 2.655 × 10⁶ ft. lb. (energy)
1 degree = 60 minutes (angle measure)
1 mi. per hr. = 88.028 ft. per min. (velocity)
              = 1.467 ft. per sec.
degrees Rankine = degrees Fahrenheit + 459.69
                 (temperature)

Fig. B-1. U.S. Conventional System of Units.

PREFIXES: The metric system uses prefixes combined with the basic unit to form a larger or smaller unit. The prefixes represent a multiplier. The common prefixes include those listed below. Other prefixes are listed in Fig. B-9 of this section.

smaller:
    milli- (m) = × 1/1000 = × 0.001
    centi- (c) = × 1/100 = × 0.01
    deci- (d) = × 1/10 = × 0.1
larger:
    deka- (da) = × 10
    hecto- (h) = × 100
    kilo- (k) = × 1000

Length—basic unit: meter (m)
1 meter = 10 decimeters (dm)
        = 100 centimeters (cm)
        = 1000 millimeters (mm)
10 meters = 1 dekameter (dam)
100 meters = 1 hectometer (hm)
1000 meters = 1 kilometer (km)

Area—basic unit: square meter (m²)
1 square meter = 100 square decimeters (dm²)
               = 10 000 square centimeters (cm²)
100 square meters = 1 are (a)
10 000 square meters = 1 hectare (ha)

Volume—basic unit: cubic meter (m³)
1 cubic meter = 1000 cubic decimeters (dm³)
              = 1 000 000 cubic centimeters (cm³)
1000 cubic meters = 1 cubic dekameter (dam³)

Capacity—basic unit: liter (L)
1 liter = 1000 milliliters (mL)
        = 0.001 cubic meters
        = 1000 cubic centimeters
1000 liters = 1 kiloliter (kL)

Mass—basic unit: gram (g)
1 gram = 10 decigrams (dg)
       = 100 centigrams (cg)
       = 1000 milligrams (mg)
1000 grams = 1 kilogram (kg)
1000 kilograms = 1 metric ton

Miscellaneous
1 millimeter = 39.37 mils
1 micron = 0.001 millimeters
degrees Kelvin = degrees Celsius + 273.16

Fig. B-2. SI Metric System of Units.

```
1 year (yr.) = 12 months (mo.)
            = 52 weeks (wk.) (52.18 wk. avg.)
            = 365 days (366 days in leap year)
            = 365.25 days avg.
1 month = 31 days in: January, March, May, July,
                      August, October, December
        = 30 days in: April, June, September,
                      November
        = 28 days in: February (29 in leap year)
        = 30.44 days avg.
        = 4.35 weeks avg.
1 week  = 7 days
        = 168 hours (hr.)
1 day   = 24 hours
        = 1440 minutes (min.)
1 hour  = 60 minutes
        = 3600 seconds (sec.)
1 minute = 60 seconds
```

Fig. B-3. Units of time.

	MULTIPLY	BY	TO OBTAIN
Length:	inches	2.54	centimeters
	feet	0.305	meters
	yards	0.914	meters
	miles	1.609	kilometers
Area:	square inches	6.452	square centimeters
	square feet	0.093	square meters
	square yards	0.836	square meters
	acres	4047	square meters
Volume:	cubic inches	16.387	cubic centimeters
	cubic feet	0.028	cubic meters
	cubic feet	28.317	liters
	cubic yards	0.7646	cubic meters
Capacity:	fluid ounces	0.029	liters
	quarts	0.940	liters
	gallons	3.785	liters
	gallons	3785.0	cubic centimeters
Weight:	ounces av.	28.35	grams
	pounds av.	0.454	kilograms
	tons	907.2	kilograms
Miscellaneous	feet per second	30.48	centimeters per second (velocity)
	degrees Fahrenheit	$(F° - 32) \times 5/9$	degrees Celsius (temperature)
	atmosphere	760	mm of Hg. (pressure)
	cu. ft. per second	1699.3	liters per minute (flow)
	gal. per minute	0.06308	liters per second (flow)
	lb. per cubic foot	16.018	kg. per cubic meter (density)
	Btu per hr. per ft.^{2} per °F	4.88	kcal. per hr. per m^2 per °C (heat transfer)

Fig. B-4. Conversion factors for U.S. Conventional to SI Metric.

	MULTIPLY	BY	TO OBTAIN
Length:	centimeters	0.394	inches
	meters	39.37	inches
	meters	3.281	feet
	meters	1.094	yards
	kilometers	3281	feet
	kilometers	0.621	miles
Area:	square centimeters	0.155	square inches
	square meters	10.76	square feet
	square kilometers	247.1	acres
Volume:	cubic centimeters	0.006	cubic inches
	cubic meters	35.31	cubic feet
	cubic centimeters	2.64×10^{-4}	gallons
Capacity:	liters	1.056	quarts
	liters	0.264	gallons
	liters	0.03531	cubic feet
	liters	61.023	cubic inches
Mass:	grams	0.035	ounces av.
	kilograms	2.205	pounds av.
Miscellaneous:	degrees Celsius	$(C° \times 9/5) + 32$	degrees Fahrenheit (temperature)
	calories	3.968×10^{-3}	British thermal units
	gram per cubic centimeter	0.03613	pound per cubic inch

Fig. B-5. Conversion factors for SI Metric to U.S. Conventional.

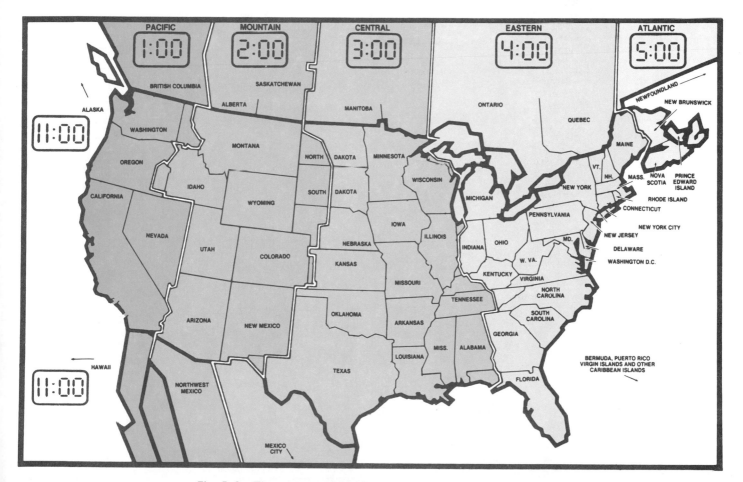

Fig. B-6. Time zones of Continental United States and Canada.
(Ameritech Industrial Yellow Pages, Inc.)

Standard	24 Hour
1 a.m.	0100 hours
2 a.m.	0200 hours
3 a.m.	0300 hours
4 a.m.	0400 hours
5 a.m.	0500 hours
6 a.m.	0600 hours
7 a.m.	0700 hours
8 a.m.	0800 hours
9 a.m.	0900 hours
10 a.m.	1000 hours
11 a.m.	1100 hours
12 noon	1200 hours

Standard	24 Hour
1 p.m.	1300 hours
2 p.m.	1400 hours
3 p.m.	1500 hours
4 p.m.	1600 hours
5 p.m.	1700 hours
6 p.m.	1800 hours
7 p.m.	1900 hours
8 p.m.	2000 hours
9 p.m.	2100 hours
10 p.m.	2200 hours
11 p.m.	2300 hours
12 midnight	2400 hours

Fig. B-7. The standard clock system compared to the 24-hour clock system.

Place Name	Place Value	Power of 10
millions	1,000,000.	10^6
hundred thousands	100,000.	10^5
ten thousands	10,000.	10^4
thousands	1,000.	10^3
hundreds	100.	10^2
tens	10.	10^1
ones or units	1.	10^0
decimal point	.	
tenths	.1	10^{-1}
hundredths	.01	10^{-2}
thousandths	.001	10^{-3}
ten-thousandths	.0001	10^{-4}
hundred-thousandths	.00001	10^{-5}
millionths	.000001	10^{-6}

Fig. B-8. Place values and powers of 10.

Multiplier Name	Symbol	Power of 10	Multiply By
tera-	T	10^{12}	1,000,000,000,000
giga-	G	10^9	1,000,000,000
mega-	M	10^6	1,000,000
kilo-	k	10^3	1,000
(basic units)		10^0	1
milli-	m	10^{-3}	.001
micro-	μ	10^{-6}	.000001
nano-	n	10^{-9}	.000000001
pico-	p	10^{-12}	.000000000001

Fig. B-9. Engineering notation.

FRACTIONAL INCH	DECIMAL INCH	MILLIMETER	FRACTIONAL INCH	DECIMAL INCH	MILLIMETER
1/64	.015625	0.397	33/64	.515625	13.097
1/32	.03125	0.794	17/32	.53125	13.494
3/64	.046875	1.191	35/64	.546875	13.891
1/16	.0625	1.588	9/16	.5625	14.288
5/64	.078125	1.984	37/64	.578125	14.684
3/32	.09375	2.381	19/32	.59375	15.081
7/64	.109375	2.778	39/64	.609375	15.478
1/8	.125	3.175	5/8	.625	15.875
9/64	.140625	3.572	41/64	.640625	16.272
5/32	.15625	3.969	21/32	.65625	16.669
11/64	.171875	4.366	43/64	.671875	17.066
3/16	.1875	4.762	11/16	.6875	17.462
13/64	.203125	5.159	45/64	.703125	17.859
7/32	.21875	5.556	23/32	.71875	18.256
15/64	.234375	5.953	47/64	.734375	18.653
1/4	.25	6.350	3/4	.75	19.05
17/64	.265625	6.747	49/64	.765625	19.447
9/32	.28125	7.144	25/32	.78125	19.844
19/64	.296875	7.541	51/64	.796875	20.241
5/16	.3125	7.938	13/16	.8125	20.638
21/64	.328125	8.334	53/64	.828125	21.034
11/32	.34375	8.731	27/32	.84375	21.431
23/64	.359375	9.128	55/64	.859375	21.828
3/8	.375	9.525	7/8	.875	22.225
25/64	.390625	9.922	57/64	.890625	22.622
13/32	.40625	10.319	29/32	.90625	23.019
27/64	.421875	10.716	59/64	.921875	23.416
7/16	.4375	11.112	15/16	.9375	23.812
29/64	.453125	11.509	61/64	.953125	24.209
15/32	.46875	11.906	31/32	.96875	24.606
31/64	.484375	12.303	63/64	.984375	25.003
1/2	.5	12.700	1	1.	25.400

Fig. B-10. Decimal and metric equivalents of common fractions of an inch.

NATURAL TRIGONOMETRIC FUNCTIONS

Angle	Sine	Cosine	Tangent	Angle	Sine	Cosine	Tangent
1°	.0175	.9998	.0175	46°	.7193	.6947	1.0355
2°	.0349	.9994	.0349	47°	.7314	.6820	1.0724
3°	.0523	.9986	.0524	48°	.7431	.6691	1.1106
4°	.0698	.9976	.0699	49°	.7547	.6561	1.1504
5°	.0872	.9962	.0875	50°	.7660	.6428	1.1918
6°	.1045	.9945	.1051	51°	.7771	.6293	1.2349
7°	.1219	.9925	.1228	52°	.7880	.6157	1.2799
8°	.1392	.9903	.1405	53°	.7986	.6018	1.3270
9°	.1564	.9877	.1584	54°	.8090	.5878	1.3764
10°	.1736	.9848	.1763	55°	.8192	.5736	1.4281
11°	.1908	.9816	.1944	56°	.8290	.5592	1.4826
12°	.2079	.9781	.2126	57°	.8387	.5446	1.5399
13°	.2250	.9744	.2309	58°	.8480	.5299	1.6003
14°	.2419	.9703	.2493	59°	.8572	.5150	1.6643
15°	.2588	.9659	.2679	60°	.8660	.5000	1.7321
16°	.2756	.9613	.2867	61°	.8746	.4848	1.8040
17°	.2924	.9563	.3057	62°	.8829	.4695	1.8807
18°	.3090	.9511	.3249	63°	.8910	.4540	1.9626
19°	.3256	.9455	.3443	64°	.8988	.4384	2.0503
20°	.3420	.9397	.3640	65°	.9063	.4226	2.1445
21°	.3584	.9336	.3839	66°	.9135	.4067	2.2460
22°	.3746	.9272	.4040	67°	.9205	.3907	2.3559
23°	.3907	.9205	.4245	68°	.9272	.3746	2.4751
24°	.4067	.9135	.4452	69°	.9336	.3584	2.6051
25°	.4226	.9063	.4663	70°	.9397	.3420	2.7475
26°	.4384	.8988	.4877	71°	.9455	.3256	2.9042
27°	.4540	.8910	.5095	72°	.9511	.3090	3.0777
28°	.4695	.8829	.5317	73°	.9563	.2924	3.2709
29°	.4848	.8746	.5543	74°	.9613	.2756	3.4874
30°	.5000	.8660	.5774	75°	.9659	.2588	3.7321
31°	.5150	.8572	.6009	76°	.9703	.2419	4.0108
32°	.5299	.8480	.6249	77°	.9744	.2250	4.3315
33°	.5446	.8387	.6494	78°	.9781	.2079	4.7046
34°	.5592	.8290	.6745	79°	.9816	.1908	5.1446
35°	.5736	.8192	.7002	80°	.9848	.1736	5.6713
36°	.5878	.8090	.7265	81°	.9877	.1564	6.3138
37°	.6018	.7986	.7536	82°	.9903	.1392	7.1154
38°	.6157	.7880	.7813	83°	.9925	.1219	8.1443
39°	.6293	.7771	.8098	84°	.9945	.1045	9.5144
40°	.6428	.7660	.8391	85°	.9962	.0872	11.4301
41°	.6561	.7547	.8693	86°	.9976	.0698	14.3006
42°	.6691	.7431	.9004	87°	.9986	.0532	19.0811
43°	.6820	.7314	.9325	88°	.9994	.0349	28.6363
44°	.6947	.7193	.9657	89°	.9998	.0175	57.2900
45°	.7071	.7071	1.0000	90°	1.0000	.0000	

Fig. B-11. The table of natural trigonometric functions.

Inches	Decimal of a Foot	Inches	Decimal of a Foot	Inches	Decimal of a Foot
1/8	.01042	3 1/8	.26042	6 1/4	.52083
1/4	.02083	3 1/4	.27083	6 1/2	.54167
3/8	.03125	3 3/8	.28125	6 3/4	.56250
1/2	.04167	3 1/2	.29167	7	.58333
5/8	.05208	3 5/8	.30208	7 1/4	.60417
3/4	.06250	3 3/4	.31250	7 1/2	.62500
7/8	.07291	3 7/8	.32292	7 3/4	.64583
1	.08333	4	.33333	8	.66666
1 1/8	.09375	4 1/8	.34375	8 1/4	.68750
1 1/4	.10417	4 1/4	.35417	8 1/2	.70833
1 3/8	.11458	4 3/8	.36458	8 3/4	.72917
1 1/2	.12500	4 1/2	.37500	9	.75000
1 5/8	.13542	4 5/8	.38542	9 1/4	.77083
1 3/4	.14583	4 3/4	.39583	9 1/2	.79167
1 7/8	.15625	4 7/8	.40625	9 3/4	.81250
2	.16666	5	.41667	10	.83333
2 1/8	.17708	5 1/8	.42708	10 1/4	.85417
2 1/4	.18750	5 1/4	.43750	10 1/2	.87500
2 3/8	.19792	5 3/8	.44792	10 3/4	.89583
2 1/2	.20833	5 1/2	.45833	11	.91667
2 5/8	.21875	5 5/8	.46875	11 1/4	.93750
2 3/4	.22917	5 3/4	.47917	11 1/2	.95833
2 7/8	.23959	5 7/8	.48958	11 3/4	.97917
3	.25000	6	.50000	12	1.00000

Fig. B-12. Inches converted to decimals of feet.

Glossary

absolute value: the numerical value of a number without regard to its sign.

abstract number: a number with no associated units.

accuracy: reference to the number of significant digits in a number.

acute angle: any angle measuring less than 90-degrees.

acute triangle: an oblique triangle that has all angles less than 90-degrees.

algebraic expression: a mathematical phrase made up of numbers, letters, and operations.

annual percentage rate (APR): interest rate given on a one-year basis equal to average annual interest divided by outstanding principal.

arc: a portion of a circle, represented by a curved line.

architect's scale: a rule having four or six different scales divided into feet, inches, and fractional parts of an inch.

area (A): a measure of the size of a two-dimensional surface, or of a region on such a surface.

arithmetic operation: one of the basic operations of addition, subtraction, multiplication, and division.

bar graph: a graph that presents relative size of data using bars.

board foot (bd. ft.): unit of volume in measuring lumber; equals 144 cubic inches, or the volume of a board 1 foot square and 1 inch thick.

board measure: measurement of lumber in board feet.

capacity: a measure of what is contained within an object.

circle: a continuous line or the plane bounded by such a line, in which every point of the line is equidistant from a center point lying on the plane.

circle graph: a circular graph that presents relative size of data using sectors, which show the percentage of each group of data to the total.

circumference (c): the perimeter, or distance around, a circle. Also, the boundary of a circle.

civil engineer's scale: a rule having four or six different scales graduated by 10, 20, 30, 40, 50, or 60 minor divisions per inch.

coefficient: *See* numerical coefficient.

common denominator: a denominator that is common to all concerned fractions.

compound interest: interest paid on principal plus accrued interest.

conditional equation: an equation whose two sides are equal only when certain values are substituted for the unknown.

cone: a solid having a circular base and another surface that tapers to a point.

constant: a value that does *not* change.

constant of proportionality (k): the constant ratio of one variable quantity to another to which it is proportional.

conversion factor: a relationship used to change from one unit to another.

coordinate: any x or y value of an ordered pair.

cross multiplication: a method used to transpose numbers in a proportion, which involves multiplying the numerator of the first ratio by the denominator of the second and vice versa.

cross-sectional area: the area measurement at the surface of a solid through which a real or imaginary cutting plane was passed.

cube: a solid of six square sides, each of which is at a right angle to each adjacent side. Also, to raise a number to the third power.

cube root: a factor of a number that when cubed gives the number.

cylinder: a solid having two equal-sized circular bases and a third side joining the bases.

decimal fraction: *See* decimal number.

decimal number: any number written in the form of an integer followed by a decimal point followed by a string of numbers; where the numbers to the right of the decimal point represent a fractional part of a whole.

decimal place: reference to one of the digits right of the decimal point in a decimal number.

decimal point: a dot (.) appearing in any decimal number to mark the point at which place values change from positive to negative powers of 10.

decimal scale: *See* civil engineer's scale.

decimal system: the system of numbers in which each digit is assigned a value that is a multiple of 10.

degree (°): a unit of angular measurement; there are 360 degrees in a circle.

denominate number: a number with units attached.

denominator: the bottom number of a fraction, which states into how many parts a whole was divided.

dependent variable: a variable whose value is dependent on the value of another variable.

diameter (d): the distance from one side of a circle to the other side, measured through the center and equal to twice the radius.

difference: the answer of a subtraction operation.

digit: any numeral, 0 through 9.

direct proportion: *See* direct variation.

direct variation: any function defined by the equation $y = kx$.

discount: the amount subtracted from retail price to arrive at a sale price.

dividend: the number that is being divided in division.

divisor: the number by which a dividend is divided.

E notation: an abbreviated form of scientific notation.

engineering notation: a form of scientific notation in which the powers of 10 are replaced by metric prefixes. (The number is in a form such that the exponent of the power of 10 is some multiple of 3.)

engineer's scale: *See* civil engineer's scale.

equation: an algebraic statement that two algebraic expressions are equal.

equilateral triangle: any triangle having three equal sides and three equal angles.

equivalent fractions: different fractions representing the same value.

exponent: a symbol written above and to the right of a number or a mathematical expression to indicate the operation of raising to a power, or how many times the number or expression will be repeated as a factor.

factoring: breaking down a number into factors, the reverse process of finding a product.

factors: all numbers or quantities that are being multiplied in the multiplication process. Also, any set of numbers that will equal a given number when multiplied together. Also, each number, letter, or quantity of a term that is formed of two or more symbols multiplied together to form a product.

formula: a mathematical statement of a real-life situation that expresses one quantity in terms of one or more other quantities.

fraction: a portion of a whole amount.

functional relationship: a mathematical relationship between two variables, in which for every value of a variable x there is only one value of a corresponding variable y.

graduation: a ruled marking on a measurement scale.

graph: a mathematical picture of data.

gross income: the money that a person earns prior to payroll deductions.

histogram: a bar graph of statistical data that depicts, by intervals, the frequency of the occurrence of the value of a variable.

hypotenuse: the side opposite the right angle of a right triangle.

identity: an equation that is true for any value of unknown, having an endless number of solutions.

improper fraction: a fraction with the numerator the same or larger than the denominator.

independent variable: the controlling factor between variables, on which the value of the other variable depends.

inequality: a mathematical statement comparing two or more numbers or algebraic expressions that are not equal.

integer: any whole number, its opposite, or zero.

interest: a percentage of a sum of money that is earned on a savings account or owed on a loan.

interest rate: a percentage applied to a principal in order to determine an interest amount.

intersect: to pass through or to cross.

inverse proportion: *See* inverse variation.

inverse variation: any function defined by the equation $y = k/x$.

invert: to reverse the positions of numerator and denominator in, or take the reciprocal of, a fraction.

isolate the unknown: rearrange an equation so that the unknown is by itself on one side of the equal sign.

isosceles triangle: any triangle having two equal sides and two equal angles.

Law of Cosines: a law stating that in any triangle, the square of any side is equal to the sum of the squares of the other two sides, less double the product of these two sides multiplied by the cosine of the angle included between them.

Law of Sines: a law stating that in any triangle, right or oblique, any side is proportional to the sine of its opposite angle.

length (l): the base quantity of distance.

like term: a term that differs from another term only in its numerical coefficient.

line graph: a graph of data that is depicted by a line.

linear equation: an equation of the first degree, having unknowns raised to the first power only, and the graph of which is a straight line.

literal number: any letter used in algebra to represent a number.

lowest common denominator (LCD): the common denominator that is smallest in value.

magnitude: the absolute value of a number.

markup: the amount added to wholesale price to arrive at retail price.

mechanical engineer's scale: a rule having four different scales divided into inches and fractional inches.

metric scale: a rule having four or six different metric scales, which are given as ratios—drawing to actual size.

micrometer: an instrument used for taking linear measurements where extreme accuracy is needed.

mill rate: *See* millage.

millage (mils): a rate, given as dollars per thousand, or mils, applied to assessed property value to compute property tax.

minuend: the number that is being deducted from in subtraction.

mixed number: a number containing both a whole number and a fraction.

monomial: a polynomial of only one term.

multimeter: an instrument used to measure electrical quantities of voltage, current, and resistance.

natural number: sometimes called the counting numbers—any whole number except zero.

negative number: a number preceded by a minus sign.

net income: the money that a person receives after payroll deductions.

nominal: designated size that may vary from actual size.

numerator: the top number of a fraction, which states how many parts are used.

numerical coefficient: the product of all factors of a term that are in number form, commonly referred to simply as "coefficient."

oblique triangle: any triangle that does not contain a 90-degree angle.

obtuse angle: any angle measuring greater than 90-degrees.

obtuse triangle: an oblique triangle that has one angle greater than 90-degrees.

ordered pair: the pair of x and y coordinates that define a point on a graph.

origin: the point of a rectangular coordinate system, designated by ordered pair (0,0), where x- and y-axes meet or intersect.

parallel: nonintersecting lines or planes, of which the perpendicular distance between is the same at any location.

percentage: a quantity expressed on a basis of 100.

perfect square: any number having an exact square root.

perimeter (p): a measure of length given by the distance around a plane shape. Also, the boundary of the plane shape.

perpendicular: being at right angles to a given line or plane.

pie chart: *See* circle graph.

place value: the value of the location of a digit in a number.

plane shape: object with two dimensions.

polygon: any plane figure bounded by straight lines.

polynomial: an algebraic expression of one or more terms in which all are in some form of $ax^n y^m$ . . ., where n and m are any whole numbers.

positive number: a number preceded by a plus sign.

power: the number of times as indicated by an exponent that a number occurs as a factor in a product; also, the product itself.

precision: reference to the decimal position of the last significant digit in a number.

prime factors: the set of all factors that will equal a given number when multiplied together, of which all numbers are prime.

prime number: a positive whole number that can be exactly divided only by itself and the number 1.

principal: the original amount of a loan or deposit on which interest is paid.

principal square root: the positive square root of a number.

product: the answer of a multiplication operation.

proper fraction: a fraction with the numerator smaller than the denominator.

proportion: a comparison of two equal ratios.

Pythagorean theorem: a theorem in geometry that states that the square of the hypotenuse of a right triangle is equal to the sum of the squares of the other two sides.

quadrant: one of the four quarters into which the plane of a rectangular coordinate system is divided.

quotient: the answer of a division problem.

radian: an angular measure defined by two radii of a circle and an arc joining them, all of the same length; there are 2π radians in a circle.

radical: an indicated root of a quantity symbolized by a radical sign.

radius (r): the line segment extending from the center of a circle or sphere to the curve or surface.

ratio: the relationship of one quantity to one or more other quantities.

rectangle: a four-sided figure that has four 90-degree angles and opposite sides equal in length and parallel to each other.

rectangular coordinate system: a standardized system of drawing a graph that employs an x-axis, a y-axis, and, if a three-dimensional plot, a z-axis; all of which are perpendicular and which pass through a common point called the origin.

retail price: the amount charged to the consumer.

right angle: an angle having a measure of 90-degrees.

right triangle: any triangle with a 90-degree angle.

root: a number that is equal to a given number when repeated as a factor an indicated number of times.

rounding: giving a close approximation of a number by cutting off the least significant digit or digits.

sale price: the price paid for an item after discount.

scale: a proportion between two sets of dimensions, as between those of a drawing and an original. Also, a measuring instrument.

scalene triangle: any triangle having no equal sides and no equal angles.

scientific notation: a system of number notation whereby a number is expressed as a product of a number between 1 and 10 and multiplied by an appropriate power of 10.

sector: a pie-shaped geometrical figure bounded by two radii and the included arc of a circle.

signed number: one of a system of numbers represented by a positive ($+$) or negative ($-$) sign.

significant figure: any of the figures of a number that begin with the first nonzero figure to the left and end with the last figure to the right that is not zero or is a zero that indicates an number to be exact and not approximate.

similar triangles: triangles with proportional sides that have the exact same angles.

simple interest: interest applied only to the principal of a savings account or loan.

slope (m): a measure of the upward or downward slant of a line or plane.

solids: objects with three dimensions.

solution: the answer to a problem. Also, any number that satisfies an equation.

sphere: a solid bounded by a curved surface of which any point on it is equidistant from a center point within.

square: a rectangle with four equal sides. Also, forming a right angle. Also, to multiply a number by itself.

square root: a factor of a number that when squared gives the number.

statistics: collections of numbers derived through records or sampling.

subtrahend: the number that is deducted in subtraction.

sum: the answer of an addition problem.

surface area: the area of a boundary that separates a shape from surrounding space.

system of simultaneous equations: two linear equations each containing the same two unknowns and treated as a system — the solution of which is any pair of values which satisfies both equations.

term: a part of an algebraic expression.

tolerance: the acceptable range of accuracy of a measurement.

transpose the equation: reversing operations of an equation in order to isolate an unknown.

trend: the general positive or negative direction depicted by a graph or a segment of a graph.

triangle: a three-sided figure with three inside angles, the measure of which adds up to 180-degrees.

trinomial: a polynomial of three terms.

unknown: a symbol in an equation representing an unknown quantity.

unlike term: any term that is not a like term — differing from another term in the literal symbol or symbols it contains or in the degree of the symbols.

variable: a literal number that represents two or more numbers.

variation: an equation of a functional relationship that relates one variable to one or more other variables by means of multiplication, division, or both.

vector: a quantity that has both magnitude and direction, commonly represented by a directed line segment whose length represents the magnitude and whose orientation in space represents the direction.

volume (V): a measure of the size of a body or definite region in three-dimensional space.

weight: a measure of how heavy something is, which varies with the force of gravity.

whole number: zero and any positive number that contains no fractional parts.

wholesale price: the price a store pays for an item.

x-axis: the major horizontal line on a graph of rectangular coordinates.

x-intercept: the x-coordinate of a point where a line, curve, or surface intersects the x-axis.

y-axis: the major vertical line on a graph of rectangular coordinates.

y-intercept: the y-coordinate of a point where a line, curve, or surface intersects the y-axis.

Index

Answers to Practice Problems and Odd-Numbered Test-Your-Skills Questions

PRACTICE PROBLEMS—CHAPTER 1
Word problems using addition, pages 10-11
1. 17 miles
2. $977
3. 50 hours
4. 16 amps
5. 43 feet
6. 2010 ohms
7. 3465 pounds
8. 1201 pints
9. 173 meters
10. 334 tiles

Word problems using subtraction, pages 13-14
1. 1205 kilowatts
2. 52 remaining
3. 105 remaining
4. 3"
5. 750 pounds
6. $677
7. 31 gallons evaporated
8. 390 ohms
9. 28 pounds for the puppy
10. 79 remaining

Word problems using multiplication, pages 16-17
1. 330 miles
2. 160 gallons
3. 224 tiles
4. $320
5. 310 mg
6. 60 ft.-lb.
7. 960 watts
8. 525 screws
9. 168 pieces
10. 350 brackets

Word problems using division, pages 18-19
1. $2 per can
2. 12 pieces
3. $13 per hour
4. 10 amps
5. 18 mpg
6. 15 studs
7. 17 studs
8. 12"
9. 35 containers
10. 160 bottles

Word problems using combination arithmetic, pages 21-22
1. ABC, $35 more
2. $215 for labor
3. It is under maximum. A 120-pound passenger is not safe.
4. four 8' boards are needed
5. 8 mpg
6. 110 cups

7. 32"
8. 50 average
9. 9 shelves
10. 241 pounds average

TEST YOUR SKILLS—CHAPTER 1
Answers to odd-numbered problems, pages 22-23
1. 1260 lin. ft.
3. 4520 lb.
5. $2 per can
7. 750 lb. per day
9. 80%

PRACTICE PROBLEMS—CHAPTER 2
Reading fractions, pages 25-26
1. seven-sixteenths
2. six-eighths
3. five-tenths
4. three-fourths
5. three and five-sixths
6. five and two-sevenths
7. one and one-ninth
8. ten and one-third
9. four and one-fourth
10. nine and two-fifths

Value of fractions, page 27
1. numerator: 3, denominator: 5
2. numerator: 2, denominator: 7
3. numerator: 5, denominator: 9
4. numerator: 1, denominator: 2
5. $\dfrac{1}{12} \quad \dfrac{1}{10} \quad \dfrac{1}{7} \quad \dfrac{1}{6} \quad \dfrac{1}{4}$
6. $\dfrac{2}{16} \quad \dfrac{3}{16} \quad \dfrac{7}{16} \quad \dfrac{15}{16} \quad \dfrac{22}{6}$
7. $\dfrac{1}{16} \quad \dfrac{1}{12} \quad \dfrac{1}{9} \quad \dfrac{1}{5} \quad \dfrac{1}{2}$
8. $\dfrac{1}{5} \quad \dfrac{2}{5} \quad \dfrac{3}{5} \quad \dfrac{4}{5} \quad \dfrac{7}{5}$
9. $\dfrac{1}{18} \quad \dfrac{1}{14} \quad \dfrac{1}{7} \quad \dfrac{1}{5} \quad \dfrac{1}{3}$
10. $\dfrac{6}{12} \quad \dfrac{7}{12} \quad \dfrac{10}{12} \quad \dfrac{12}{12} \quad \dfrac{17}{12}$

Numbers with and without fractions, pages 27-28
1. proper fraction
2. mixed number
3. whole number
4. improper fraction
5. mixed number with an improper fraction
6. whole number
7. proper fraction
8. mixed number
9. improper fraction
10. proper fraction

Changing fractions from one form to another, page 29

1. $\dfrac{16}{3}$

2. $\dfrac{19}{4}$

3. $\dfrac{7}{2}$

4. $\dfrac{6}{1}$

5. $\dfrac{47}{5}$

6. $\dfrac{7}{3}$

7. $\dfrac{10}{1}$

8. $\dfrac{18}{7}$

9. $\dfrac{9}{1}$

10. $\dfrac{77}{9}$

11. $1\dfrac{5}{7}$

12. $2\dfrac{3}{5}$

13. $2\dfrac{2}{4}$ or $2\dfrac{1}{2}$

14. 5

15. $1\dfrac{1}{2}$

16. $1\dfrac{1}{6}$

17. $5\dfrac{1}{6}$

18. $2\dfrac{4}{8}$ or $2\dfrac{1}{2}$

19. $1\dfrac{3}{12}$ or $1\dfrac{1}{4}$

20. $1\dfrac{7}{10}$

Equivalent fractions, page 33

1. $\dfrac{10}{24}$

2. $\dfrac{4}{16}$

3. $\dfrac{16}{20}$

4. $\dfrac{30}{32}$

5. $\dfrac{6}{8}$

6. $\dfrac{3}{6}$

7. $\dfrac{32}{4}$

8. $\dfrac{60}{12}$

9. $\dfrac{10}{15}$

10. $\dfrac{20}{28}$

11. $\dfrac{1}{2}$

12. $\dfrac{1}{3}$

13. $\dfrac{3}{4}$

14. $\dfrac{3}{4}$

15. $\dfrac{4}{5}$

16. $\dfrac{1}{4}$

17. 3

18. $3\dfrac{3}{4}$

19. $5\dfrac{1}{2}$

20. $3\dfrac{5}{8}$

21. LCD = 24 $\dfrac{1}{3}, \dfrac{5}{12}, \dfrac{11}{24}, \dfrac{5}{6}$

22. LCD = 32 $\dfrac{3}{4}, \dfrac{7}{8}, \dfrac{29}{32}, \dfrac{15}{16}$

23. LCD = 20 $\dfrac{1}{2}, \dfrac{3}{5}, \dfrac{13}{20}, \dfrac{7}{10}$

24. LCD = 30 $\dfrac{3}{15}, \dfrac{7}{30}, \dfrac{1}{3}, \dfrac{2}{5}$

25. LCD = 64 $1\frac{1}{64}$, $\frac{33}{32}$, $1\frac{1}{16}$, $\frac{9}{8}$

Addition of fractions, pages 36-37

1. $\frac{1}{2}$

2. $\frac{7}{8}$

3. $1\frac{5}{12}$

4. $\frac{2}{3}$

5. $\frac{73}{84}$

6. $\frac{25}{48}$

7. $8\frac{13}{30}$

8. $5\frac{23}{90}$

9. $22\frac{1}{3}$

10. $14\frac{1}{2}$

11. $24\frac{1}{10}$

12. $6\frac{77}{120}$

13. $9\frac{49}{60}$

14. 27

15. $27\frac{6}{7}$

16. 20

17. $11\frac{5}{16}$

18. $29\frac{5}{12}$

19. $5\frac{5}{16}$ inches

20. $1\frac{37}{48}$ yards

Subtraction of fractions, page 39

1. $\frac{1}{4}$

2. $\frac{1}{6}$

3. $\frac{3}{16}$

4. $\frac{1}{12}$

5. $\frac{8}{35}$

6. $\frac{17}{60}$

7. $2\frac{1}{6}$

8. $3\frac{5}{8}$

9. $3\frac{5}{6}$

10. $8\frac{7}{12}$

11. $11\frac{2}{5}$

12. $1\frac{1}{2}$

13. $2\frac{1}{6}$

14. $\frac{5}{9}$

15. $4\frac{1}{3}$

16. $10\frac{11}{16}$

17. 5

18. $9\frac{5}{6}$

19. $47\frac{13}{16}$ inches

20. $32\frac{7}{12}$ inches

Multiplication of fractions, pages 40-41

1. $\frac{2}{7}$

2. $\dfrac{9}{20}$

3. $\dfrac{49}{120}$

4. $\dfrac{8}{15}$

5. $\dfrac{37}{100}$

6. $14\dfrac{11}{14}$

7. $12\dfrac{3}{7}$

8. 0

9. 11

10. $22\dfrac{2}{7}$

11. 8 cups

12. $1059\dfrac{3}{8}$ pounds, does exceed maximum weight

13. $\dfrac{15}{32}$ inches, less than $\dfrac{1}{2}$ inch

14. Long sides: $487\dfrac{1}{2}$ inches, short sides: $213\dfrac{3}{4}$ inches

Division of fractions, page 42

1. 3

2. $\dfrac{1}{2}$

3. $2\dfrac{1}{4}$

4. $\dfrac{5}{6}$

5. $3\dfrac{1}{2}$

6. $\dfrac{5}{8}$

7. $\dfrac{29}{46}$

8. $\dfrac{71}{80}$

9. $\dfrac{261}{295}$

10. $1\dfrac{17}{33}$

11. 5 pieces

12. $28\dfrac{23}{48}$ inches

TEST YOUR SKILLS—CHAPTER 2
Answers to odd-numbered problems, pages 43-46

1. $4\dfrac{5}{6}$

3. $\dfrac{9}{1}$

5. $\dfrac{23}{9}$

7. $8\dfrac{1}{6}$

9. $\dfrac{19}{4}$

11. $\dfrac{9}{12}$

13. $\dfrac{72}{12}$

15. $3\dfrac{9}{21}$

17. $\dfrac{100}{10}$

19. $\dfrac{9}{48}$

21. $\dfrac{1}{2}$

23. $\dfrac{1}{4}$

25. $1\dfrac{1}{2}$

27. 4

29. $\dfrac{1}{10}$

31. $\dfrac{5}{6}, \dfrac{9}{12}, \dfrac{2}{3}, \dfrac{8}{24}, 12$

33. $\dfrac{3}{6}, \dfrac{35}{72}, \dfrac{7}{18}, \dfrac{11}{36}, 72$

35. $\dfrac{17}{22}, \dfrac{8}{11}, \dfrac{31}{44}, \dfrac{15}{22}, 44$

37. $1\dfrac{1}{3}$

39. $6\dfrac{3}{10}$

41. $4\dfrac{1}{8}$

43. $5\frac{4}{21}$

45. $5\frac{15}{32}$

47. $\frac{1}{12}$

49. $1\frac{1}{2}$

51. $3\frac{11}{15}$

53. $2\frac{1}{6}$

55. $\frac{23}{30}$

57. $\frac{3}{10}$

59. $\frac{1}{4}$

61. 3

63. $15\frac{13}{15}$

65. $6\frac{1}{3}$

67. $\frac{3}{8}$

69. $\frac{25}{36}$

71. $\frac{23}{30}$

73. $1\frac{31}{33}$

75. $\frac{2}{3}$

77. $8\frac{3}{4}$ hr.

79. $6\frac{9}{10}$ yr.

81. 12 in. ID

83. $50\frac{1}{4}$ ft.

85. $\frac{7}{32}$ in.

PRACTICE PROBLEMS—CHAPTER 3
Value of decimal numbers, page 51

1. 2×100
   0×10
   9×1

2. 3×1000
   5×100
   6×10
   7×1

3. $1 \times 10,000$
   0×1000
   4×100
   5×10
   1×1

4. 1×100
   5×10
   0×1

5. 8×100
   7×10
   2×1

6. 1×10
   0×1

7. 2×1000
   0×100
   8×10
   3×1

8. $2 \times 100,000$
   $3 \times 10,000$
   5×1000
   7×100
   1×10
   0×1

9. $1 \times 1,000,000$
   $0 \times 100,000$
   $4 \times 10,000$
   5×1000
   1×100
   0×10
   2×1

10. $2 \times 10,000$
    6×1000
    8×100
    2×10
    1×1

11. 3×1
    5×0.1
    2×0.01

12. 8×1
    0×0.1
    6×0.01
    8×0.001

13. 9×1
    1×0.1
    2×0.01
    5×0.001

14. 1×10
    5×1
    8×0.1
    7×0.01
    1×0.001
    5×0.0001

15. 2×10
1×1
3×0.1
8×0.01
5×0.001
1×0.0001
2×0.00001
16. 4×1
2×0.1
7×0.01
17. 4×1
0×0.1
0×0.01
6×0.001
7×0.0001
18. 2×1
0×0.1
5×0.01
19. 0×1
8×0.1
5×0.01
20. 0×1
2×0.1
0×0.01
8×0.001
7×0.0001
1×0.00001

21. $\dfrac{25}{100}$

22. $\dfrac{17}{100}$

23. $\dfrac{3}{10}$

24. $\dfrac{5}{10}$

25. $\dfrac{125}{1000}$

26. $\dfrac{371}{1000}$

27. $\dfrac{7}{1000}$

28. $\dfrac{35}{10,000}$

29. $\dfrac{2}{100,000}$

30. $\dfrac{105}{100,000,000}$

Significant figures and rounding, page 53
1. 1 is most, 2 is least
2. 5 is most, 2 is least

3. 3 is most, 1 is least
4. 8 is most, 5 is least
5. 1 is most, 5 is least
6. 5 is most, 1 is least
7. 3 is most, 8 is least
8. 9 is most, 5 is least
9. 3 is most, 1 is least
10. 9 is most, 7 is least

11. 875
12. 282
13. 35.3
14. 10.9
15. 10.0
16. 3.33
17. 0.667
18. 0.556
19. 18.2
20. 1000

21. 0.20
22. 3.00
23. 1.06
24. 2.16
25. 2.00
26. 9.00
27. 0.89
28. 1.01
29. 5.06
30. 890.10

Addition and subtraction of decimal numbers, pages 54-55
1. 15.55
2. 57.2
3. 12.119
4. 114.105
5. 12.1
6. 0.0116
7. 0.8368
8. 0.00945
9. 7.72
10. 8.9478
11. 0.0585"
12. 486.125 feet
13. $276.99
14. 7.42 meters
15. 32.65'

Multiplication and division of decimal numbers, page 58
1. 10.35
2. 1322.5
3. 0.04296
4. 0.00139335
5. 32.85
6. 42.7975

7. 3.81
8. 5.83
9. 0.141
10. 0.206
11. 0.2
12. 0.781
13. 6.59
14. 17.0
15. 22.2
16. 50.4"
17. 10.5"
18. 36 tables
19. 24 pieces
20. 0.375"

Converting decimals and fractions, pages 59-60

1. $\dfrac{3}{5}$

2. $\dfrac{4}{5}$

3. $\dfrac{1}{4}$

4. $\dfrac{3}{4}$

5. $\dfrac{7}{200}$

6. $\dfrac{1}{25}$

7. $4\dfrac{1}{250}$

8. $3\dfrac{1}{8}$

9. $5\dfrac{3}{8}$

10. $2\dfrac{1}{500}$

11. 0.5
12. 0.75
13. 0.4
14. 0.625
15. 3.33
16. 6.31
17. 5.56
18. 1.44
19. 1.67
20. 6.7

TEST YOUR SKILLS—CHAPTER 3
Answers to odd-numbered problems, pages 60-63

1. 1×1000
    3×100
    2×10
    4×1
3. 8×100
    1×10
    9×1
5. $\dfrac{3}{100}$

7. $\dfrac{79}{1,000,000}$

9. $\dfrac{5}{10}$

11. $\dfrac{49}{1000}$

13. 29.1
15. 0.0739
17. 561
19. 0.727
21. 0.005
23. 0.009
25. 0.022
27. 14.6102
29. 32.15
31. 0.009
33. 1.0005
35. 108.981
37. 61.04
39. 513.525
41. 70
43. 0.6
45. 0.22
47. 1872
49. 562.5
51. 0.6
53. 3.8
55. 1.125

57. $\dfrac{7}{10}$

59. $8\dfrac{2}{5}$

61. 71.1 ft.
63. $89.60
65. 20 oz.
67. 257 pieces
69. 0.0625 in.

PRACTICE PROBLEMS—CHAPTER 4
Relationship of percentage to fractions and decimals, page 66

	Percent	Decimal	Fraction
1.	25%*	0.25	25/100
2.	15%	0.15*	15/100
3.	12.5%	0.125	1/8*
4.	65.5%*	0.655	131/200
5.	7.8%	0.078*	39/200
6.	60%	0.6	3/5*
7.	133.3%*	1.333	1 1/3
8.	287.5%	2.875*	2 7/8
9.	31.25%	0.3125	5/16*
10.	5.25%*	0.0525	21/400

* indicates the given values

Solving basic percent problems, page 68
1. P = 126
2. R = 16.7%
3. B = 15
4. P = 25.2
5. R = 16.7%
6. B = 24
7. P = 60
8. R = 33.3%
9. B = 25
10. R = 70%
11. R = 65.2%
12. P = 62
13. P = 14
14. R = 87.5%
15. 3/4" stone: P = 6750 lb.

     1/2" stone: P = 4800 lb.

     sand: P = 3450 lb.

Percent tolerance, pages 69-70
1. max. = 367.5; min. = 332.5
2. max. = 132; min. = 108
3. max. = 57.6; min. = 38.4
4. max. = 24.72; min. = 23.28
5. max. = 0.3636; min. = 0.3564
6. max. = .918; min. = .882
7. max. = 0.08241; min. = 0.08159
8. max. = 0.016192; min. = 0.015808
9. max. = 2.9725; min. = 2.8275
10. max. = 5.6112; min. = 5.5888
11. max. = 198; min. = 162; It is within tolerance.
12. max. = 1.515; min. = 1.485; It is not within tolerance.
13. max. = 16.5; min. = 13.5
14. No opinion = 14%; max. = 14.42; min. = 13.58
15. $2100

TEST YOUR SKILLS—CHAPTER 4
Answers to odd-numbered problems, pages 70-71
1. 80%, 0.8
3. 0.45, $\frac{9}{20}$
5. 62.5%, 0.625
7. 30%
9. 32
11. 33.3%
13. $3750
15. 200 screws
17. Motor 2: most efficient
19. 82.3% of capacity
21. 176 ft., 144 ft.
23. 0.255 oz., 0.245 oz.
25. 0.125625 mm, 0.124375 mm

PRACTICE PROBLEMS—CHAPTER 5
Dollars and cents, pages 74-75
1. $1.25
2. $50.00
3. $10.06
4. $2.60
5. $7.505
6. $963.01
7. $0.11
8. $12.3575
9. $8.71
10. $0.32
11. Four dollars and thirty-two cents.
12. Fifty-three dollars and ninety cents.
13. Eighty-five cents.
14. Fifty-six dollars.
15. Three dollars and six cents.
16. Fifty-five cents.
17. Four hundred and nine dollars and eighty-five and one-half cents.
18. Two hundred and thirty-five dollars and two cents.
19. Twenty-one dollars.
20. Twenty-five and one-tenth cents.
21. 63 ¢
22. 5 ¢
23. 130 ¢
24. 81.5 ¢
25. 9 ¢
26. 85 ¢
27. 201 ¢
28. 215 ¢
29. 1239 ¢
30. 38.25 ¢
31. Thirty-three cents.
32. Eighteen cents.
33. Fifteen cents.
34. Two hundred and thirty-five cents.
35. Eight hundred and seventy-three cents.
36. Forty-eight cents.
37. Twelve and one-half cents.

38. Three hundred and twenty-five cents.
39. Nine hundred and twelve cents.
40. Twenty-five and three-quarters.
41. $60.20
42. $0.50
43. $4.66
44. $45.90
45. $13.50
46. $16.60
47. $42.50
48. $14.65
49. $364.53
50. 20 oz. is better; 4 ¢ per ounce difference

Gross income pages 76-77

	Weekly Hours	Base Hourly Rate	Regular Time Pay	Overtime pay	Combined Weekly Income	Annual Gross Income (Round answer to the nearest dollar)
1.	40*	$6.25*	$250.00	$0	$250.00	$13,000
2.	35*	$9.89	$346.15	$0	$346.15	$18,000*
3.	40*	$8.50	$340.00	$0	$340.00*	$17,680
4.	50*	$10.25*	$410.00	$153.75	$563.75	$29,315
5.	48*	$16.00*	$640.00	$192.00	$832.00	$43,264
6.	40*	$16.25	$650.00	$0	$650.00*	$33,800
7.	35*	$15.40	$539.00	$0	$539.00	$28,028*
8.	60*	$5.10*	$204.00	$153.00	$357.00	$18,564
9.	45*	$14.30*	$572.00	$107.25	$679.25	$35,321
10.	50*	$8.60*	$344.00	$129.00	$473.00	$24,596

* indicates given value

Net income and payroll deductions, page 78

1. $395.60 net
2. $14,924 net
3. $271.80 net
4. $15,632 net
5. $1067 net

Sales and property tax, pages 79-80

1. $15.75
2. $252.81
3. $48.67
4. $0.36
5. $30.54
6. $381.27
7. $6695.00
8. $13,104.00
9. $120.00
10. $40.00
11. $168.00
12. $378.00
13. $737.84
14. $3125.00
15. $4770.50
16. $268.80
17. $478.80
18. $1738.64
19. $3105.00
20. $10,824.00

Interest, pages 83-84

1. $84
2. $336
3. $735
4. $10,224
5. $13
6. $103
7. $817
8. $1440
9. $9405
10. $12
11. $210
12. $609
13. $3091
14. $8060
15. $12,302
16. $53
17. $137
18. $160
19. $99
20. $375
21. $101
22. $63
23. $81
24. $274
25. $440

Merchandise pricing using percentage, page 88

	Wholesale Cost	Retail Price	Percent Markup
1.	$32.50*	$42.25*	30%
2.	$19.00*	$28.50	50%*
3.	$22.73	$50.00*	120%*
4.	$125.00*	$175.00*	40%
5.	$250.50*	$438.38	75%*
6.	$16.94	$30.50*	80%*
7.	$272.00	$680.00*	150%*
8.	$25.00*	$75.00*	200%
9.	$12.00*	$18.00*	50%*
10.	$0.25	$0.50*	200%*

* indicates given values

	Retail Price	Sale Price	Percent Discount
11.	$50.00*	$35.00*	30%
12.	$6.50*	$4.88	25%*
13.	$77.50	$62.00*	20%*
14.	$180.00*	$90.00*	50%
15.	$360.00*	$240.00	1/3*
16.	$25.71	$18.00*	30%*
17.	$12.00	$6.00*	1/2*
18.	$36.00*	$21.60*	40%
19.	$75.00*	$56.25	25%*
20.	$200.00	$120.00*	40%*

* indicates given values

TEST YOUR SKILLS—CHAPTER 5
Answers to odd-numbered problems, pages 88-90

1. $23.02
3. $3.05
5. $6.62
7. $19,968.00 per yr.
9. $785.20
11. $493.30
13. $8.76
15. 25.15%
17. 2.58¢
19. $276.00
21. $13.86
23. $304.53
25. $2715.77
27. $161.00
29. $104.00
31. 50%
33. $69.44
35. 257%
37. $89.00
39. 47.4%

PRACTICE PROBLEMS—CHAPTER 6
Line graphs, pages 97-98

1. 30 mph
2. decrease in mpg
3. 20 mpg
4. 28.6% decrease
5.

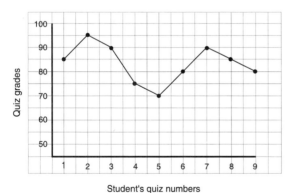

6. 83.3%
7.

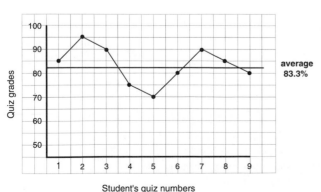

8. Constant, with slight variations.
9. 7 above; 2 below
10. 5.3%

Bar graphs, pages 98-100

1. August
2. January
3. 48.75°
4. 40°
5. 10°
6.

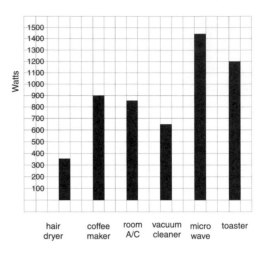

7. Most: microwave oven. Least: hair dryer.
8. 907 watts
9. microwave and toaster
10. coffee maker, hair dryer, room air conditioner, and vacuum cleaner

Circle graphs, page 101

1. engineering payroll = 36°
   management and office support payroll = 36°
   factory payroll = 108°
   supply costs = 72°
   other business expenses = 90°
   profit = 18°
2. $175,000
3. $29,167
4. $6562
5. factory payroll and supply costs
6.

Education = 126°
Road maintenance = 90°
Fire/Police = 10% = 36°
Administration = 15% = 54°
Miscellaneous = 15% = 54°

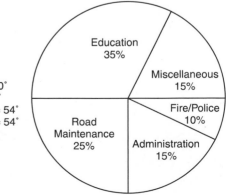
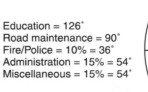

7. Miscellaneous expenses: 15%. Fire/police: 10%.
8. $2.24 million
9. $16,000
10. $80,000

TEST YOUR SKILLS—CHAPTER 6
Answers to odd-numbered problems, pages 101-104

1. $625 million
3. 48%
5. 370 million
7.

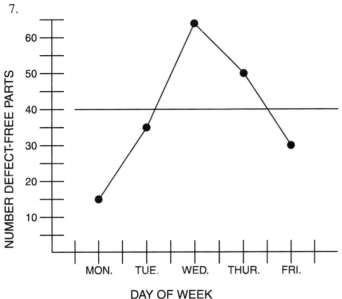

9. 39.6 parts
11. 37.9%
13. 2
15. $160,000
17. 62.5%
19. Car number is independent variable.
21. Car #2.
23. Car #4.
25. $120
27. 10%
29.

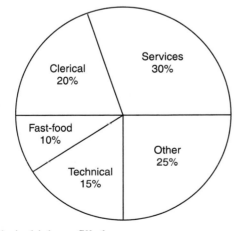

31. 80 clerical jobs unfilled
33. 25%

PRACTICE PROBLEMS—CHAPTER 7
Categories of measurement, page 108

1. acre: U.S. Conventional area
2. centimeter: metric length
3. cubic centimeter: metric volume
4. cubic inch: U.S. Conventional volume
5. cubic foot: U.S. Conventional volume
6. cubic meter: metric volume
7. cup: U.S. Conventional capacity
8. fluid ounce: U.S. Conventional capacity
9. foot: U.S. Conventional length
10. inch: U.S. Conventional length
11. gallon: U.S. Conventional capacity
12. gram: metric weight
13. kilogram: metric weight
14. kilometer: metric length
15. kiloliter: metric capacity
16. liter: metric capacity
17. meter: metric length
18. mile: U.S. Conventional length
19. milligram: metric weight
20. milliliter: metric capacity
21. ounce: U.S. Conventional weight
22. pint: U.S. Conventional capacity
23. pound: U.S. Conventional weight
24. quart: U.S. Conventional capacity
25. square centimeters: metric area
26. square foot: U.S. Conventional area
27. square inch: U.S. Conventional area
28. square kilometer: metric area
29. square meter: metric area
30. ton: U.S. Conventional weight

Converting simple units, page 112

1. 2 miles = 10,560 feet
2. 48 fluid ounces = 3 pints
3. 18 inches = 1.5 feet
4. 12 sq. in. = 0.0833 sq. ft.
5. 6 yards = 216 inches
6. 36 feet = 432 inches
7. 12 cups = 3 quarts
8. 128 cups = 8 gallons
9. 8 cu. yd. = 216 cu. ft.
10. 5 pounds = 80 ounces
11. 65 meters = 6500 cm
12. 15 grams = 150 dg
13. 350 mm = 0.35 meters
14. 3 km = 3000 meters
15. $25m^2 = 2500\ dm^2$
16. 850 cm = 8.5 meters
17. 5 liters = 5000 mL
18. $530\ dm^2 = 5.3\ m^2$
19. 10 kg = 10,000 grams
20. 5000 meters = 50 hm
21. 3 weeks = 504 hours
22. 90 hours = 5400 minutes
23. 6 days = 8640 minutes

24. January = 744 hours
25. 2 hours = 7200 seconds
26. 84 days = 12 weeks
27. 2300 min. = 38.3 hours
28. 850 hours = 35.4 days
29. 7200 seconds = 120 minutes
30. 2 years = 17520 hours

Converting combined units, pages 113-114

1. 80 miles per hour = 117 feet per second
2. 90 kilometers per hour = 90,000 meters per hour
3. 25 miles per gallon = 0.04 gallons per mile
4. 135 words per minute = 8100 words per hour
5. 80 liters per kilometer = 0.08 liters per meter
6. 35 gallons per hour = 2.33 quarts per minute
7. 10 pounds per square inch = 1440 pounds per square foot
8. 50 feet per second = 34.1 miles per hour
9. 7 pounds per gallon = 1.75 pounds per quart
10. 888 grams per minute = 53.28 kilograms per hour
11. 35 gallons
12. 13.9 minutes
13. 4800 kg
14. 488 miles
15. 8 gallons

Converting between U.S. Conventional and SI metric, pages 114-115

1. 48 miles = 77.2 km
2. 80 meters = 87.52 yd.
3. 16 cm = 6.3 in.
4. 24 ft = 7.32 m
5. 32 ft^2 = 2.976 m^2
6. 18 m^2 = 193.7 ft^2
7. 280 oz. = 7938 grams
8. 125 lb. = 56.75 kg
9. 48 kg = 105.84 lb.
10. 36 L = 9.504 gal.
11. 68.13 L
12. 68 kg = 150 lb.
13. 179.5 miles
14. 13.4 m^2
15. 560 m^3

Arithmetic with denominate numbers, page 117

1. 32 5/8 inches
2. 1173 cm
3. 334 inches
4. 16.58 m = 1658 cm = 16,580 mm
5. 16 qt
6. 30 min
7. 24 ounces
8. 5880 grams
9. 18.2 m = 59.7 ft.
10. 41.15 lb. = 18.68 kg

11. 32 ft.
12. 192 km

13. 8 in.
14. 16 grams
15. 40 gal.
16. 70,000 cm^2
17. 300 ft.^{3}
18. 48 m^3
19. 8 in.^{2}
20. 16 cm^2

Arithmetic with compound denominate numbers, page 121

1. 16 yards, 2 feet
2. 12 gallons, 1 pint
3. 18 days, 17 hours, 40 minutes
4. 12 yards, 1 foot, 8 inches
5. 8 weeks, 5 days, 9 hours, 17 minutes
6. 1 day, 11 hours, 40 minutes
7. 17 yards
8. 54 gallons, 1 quart, 1 pint
9. 1 day, 20 hours
10. 5 feet, 6 inches

Converting compound and single units, page 122

1. 210 inches; 5.833 yards
2. 76 ounces; 4.75 pounds
3. 19 pints; 9.5 quarts
4. 624.75 hours; 26.03 days
5. 13.167 feet; 158 inches
6. 2 yards, 2 feet, 4 inches
7. 2 pounds, 6 ounces
8. 14 gallons
9. 3 hours, 20 minutes
10. 19 yards, 1 foot

Time zones and travel time, pages 124-125

1. 0835
2. 0906
3. 1014
4. 1155
5. 1440
6. 1730
7. 2220
8. 2300
9. 1200
10. 2400
11. 6:30 a.m.
12. 3:20 a.m.
13. 12:40 p.m.
14. 3:45 p.m.
15. 11:10 a.m.
16. 12:00 noon
17. 9:05 a.m.
18. 1:08 a.m.
19. 10:50 p.m.
20. 6:35 p.m.
21. 10:30 a.m.
22. 5:00 a.m.
23. 12:20 a.m.

24. 8:20 a.m.
25. 4:10 p.m.
26. 4:30 a.m.
27. 6:00 a.m.
28. 1:30 p.m.
29. 0650
30. 1015
31. 8:15 a.m.
32. 6:20 a.m.

TEST YOUR SKILLS—CHAPTER 7
Answers to odd-numbered problems, pages 125-127

1. 15,840
3. 406,080
5. 8
7. 650
9. 20,000
11. 8736
13. 744
15. 3.6
17. 4800
19. 0.06
21. 96.54
23. 5.04
25. 23.76
27. 2 yd.^{2}, 122 in.^{2}
29. 1 ton, 1700 lb.
31. 2 wk., 8 days, 18 hr., 51 min.
33. 55 gal.
35. 1 day, 12 hr., 2 2/3 min.
37. 6.1 mi.
39. 96,000,000 m^3
41. 62.1 mph
43. 1034.4 g/cm^2
45. $10,200 labor cost
47. 5:08 p.m.
49. 5 hr., 20 min.

TEST YOUR SKILLS—CHAPTER 8
Answers to odd-numbered problems, pages 154-187

1. a = 1'1 5/8"
   b = 1'2 1/4"
   c = 1'3 3/8"
3. a = 11 9/32"
   b = 11 21/32"
   c = 12 7/8"
5. a = 0.92'
   b = 1.05'
   c = 1.1'
7. a = 5.5 cm
   b. = 15.9 cm
   c = 76 mm
9. a = 2'4 3/4"
   b = 3'4"
   c = 7'9"
11. a = 2'6 1/2"
   b = 3'5"

c = 1 7/16"
13. 0.141"
15. 0.003"
17. 14.81 mm
19. 28.68 mm
21. 16.98 mm
23. 0.3552"
25. 0.2426"
27. 41.70 mm
29. a = 2.4 ACV
   b = 5.4 ACV
   c = 8.8 ACV
31. range = 2.5 DCmA
   scale = 0, 0.5, 1.0, 1.5, 2.0, 2.5 DCmA
   line = 0.05 DCmA
33. range = 250 DCV
   scale = 0, 50, 100, 150, 200, 250 DCV
   line = 5 DCV

PRACTICE PROBLEMS—CHAPTER 9
Perimeter, pages 173-174

1. 84 inches
2. 97 feet
3. 142 inches
4. 46 feet
5. 74 cm
6. 100 yards
7. 200 feet
8. 61 m
9. 1727 feet
10. 151 bricks
11. 45 inches
12. 314 mm
13. 350 feet
14. 100 feet
15. 62 cm

Area, pages 177-178

1. 1728 tiles
2. 25 boxes will fit in the closet
3. 2400 ft.^{2}
4. (a) 672 ft.^{2} combined surface area
   (b) 2 gallons of paint are needed
5. 4 rolls of wallpaper are needed
6. 2500 cm^2
7. 224 ceiling tiles
8. 72 tiles
9. 165 in.^{2}
10. 32 ft.^{2}
11. 48 sheets of plywood
12. 201 in.^{2}
13. 78.5 ft.^{2}
14. 21.3 yd.^{2}
15. 1095 m^2

Volume, page 183

1. 64 in.^{3}

2. 9 ft.^{3}
3. 100.48 in.^{3}
4. 800 cm^3
5. 4992 ft.^{3}
6. 960 ft.^{3}
7. 1432.4 in.^{3}
8. 24 yd.^{3}
9. 1.04 yd.^{3}
10. 63.19 oz.

22. +
23. +
24. –
25. +
26. –
27. –
28. +
29. +
30. +

TEST YOUR SKILLS—CHAPTER 9
Answers to odd-numbered problems, pages 183-187

1. 180 ft.
3. 70 ft.
5. 52 in.
7. 60 in.
9. 20 m
11. 18.67 ft. (18 ft., 8 in.)
13. 52 in.
15. 75.40 in.
17. 56.55 in.
19. 9.55 cm
21. 360 in.^{2}
23. 70 tiles
25. 2 rolls
27. 4 gal.
29. 11.31 cm
31. 128 ft.^{2}
33. 706.86 in.^{2}
35. 1344 in.^{3}
37. 1792 ft.^{3}
39. 308,831.12 in.^{3}

PRACTICE PROBLEMS—CHAPTER 10
The number line, pages 191-192

1. +10
2. -600
3. -35
4. -3
5. +32
6. 32
7. 15
8. 1
9. 0
10. 5
11. 4 < 9
12. -5 > -8
13. 0 > -6
14. 9 > 0
15. 6 > -6
16. 6 - 3 > 4 - 9
17. 3 wrong answers out of 25 > a grade of 75%
18. 14 × 10 < 12 × 12
19. 1 gallon < 5 quarts
20. 125 pounds < 60 kilograms
21. –

Addition and subtraction of signed numbers, pages 194-195

1. -2
2. 5
3. 3
4. -6
5. -3
6. 2
7. -10
8. -12
9. -18
10. 3
11. $\dfrac{1}{6}$
12. $\dfrac{1}{8}$
13. $1\dfrac{1}{2}$
14. $\dfrac{5}{18}$
15. $-1\dfrac{1}{10}$
16. $\dfrac{3}{7}$
17. $\dfrac{1}{8}$
18. $-1\dfrac{3}{20}$
19. $1\dfrac{4}{15}$
20. $-\dfrac{2}{3}$
21. - 0.8
22. - 28.3
23. - 41.0
24. - 6.9
25. - 3.5

Multiplication and division of signed numbers, pages 195-196

1. -4

2. -12
3. 6
4. 18
5. -16
6. -24
7. -9
8. 20
9. -3.72
10. 0.1

11. $-\dfrac{1}{3}$

12. $\dfrac{3}{20}$

13. $\dfrac{7}{12}$

14. $\dfrac{2}{27}$

15. -3
16. -6
17. 6

18. $-\dfrac{1}{2} = -0.5$

19. -9
20. -8
21. 9

22. $-\dfrac{1}{8} = -0.125$

23. 2
24. -2

25. $\dfrac{1}{4} = 0.25$

26. $-\dfrac{1}{3} = -0.333$

27. -3
28. -2
29. -0.333
30. 0.25

Applications of signed numbers, page 197
1. $751.00
2. 115 total sales
3. 170 pounds
4. 4.7° average
5. 39"

Powers and roots, page 201
1. 9

2. 16
3. 25
4. -32
5. 1
6. -27
7. 1
8. -4

9. $\dfrac{1}{4}$

10. -1

11. $\dfrac{1}{3}$

12. $\dfrac{1}{16}$

13. 3
14. 9
15. 3
16. 2
17. 0.4
18. 0.5
19. -2

20. $\dfrac{1}{5}$

21. 14 feet
22. 4.2
23. 0.71
24. 2.4
25. 0.89

Order of operations, page 203
1. 12
2. 36
3. -2
4. -5
5. 8

6. $-\dfrac{2}{5}$

7. $-\dfrac{1}{3}$

8. $-3\dfrac{1}{4}$

9. $-\dfrac{1}{3}$

10. $-8\dfrac{1}{2}$

TEST YOUR SKILLS—CHAPTER 10
Answers to odd-numbered problems, pages 203-206

1. +50
3. +20
5. +20
7. 14
9. 10
11. <
13. <
15. >
17. +
19. −
21. −
23. +
25. +
27. 12
29. -6
31. 18
33. -14
35. $-\dfrac{1}{4}$
37. $\dfrac{1}{2}$
39. $-1\dfrac{5}{12}$
41. 4.01
43. -20
45. 17
47. -24
49. 65.1
51. -15.87
53. 50
55. $\dfrac{5}{18}$
57. -6
59. 2
61. 3
63. 7.5
65. -7.5
67. $370.14
69. 48 in.
71. -12°
73. 29 feet deep
75. $27,000 gross
77. 64
79. 5
81. -2
83. 6.3
85. 2
87. 4
89. -23
91. -35
93. 16
95. -3

PRACTICE PROBLEMS—CHAPTER 11
Basic terminology, pages 209-210

1. $-15x^2 + 3x - 5$

Term	Coefficient	Variables	Constant Term
$-15x^2$	-15	x	—
+3x	+3	x	—
-5	—	—	-5

2. $x^2 - 2x + 1$

Term	Coefficient	Variables	Constant Term
x^2	+1	x	—
-2x	-2	x	—
+1	—	—	+1

3. $-y + x + 3x^2 + 1$

Term	Coefficient	Variables	Constant Term
-y	-1	y	—
x	+1	x	—
$3x^2$	+3	x	—
+1	—	—	+1

4. $15 - 3y + x + 2x^2$

Term	Coefficient	Variables	Constant Term
15	—	—	+15
-3y	-3	y	—
+x	+1	x	—
$2x^2$	+2	x	—

5. $ab - 4a + 0 - a^1$

Term	Coefficient	Variables	Constant Term
ab	+1	a,b	—
-4a	-4	a	—
+0	—	—	0
$-a^1$	-1	a	—

6. $3xy + 1 - 5x^1 + y$

Term	Coefficient	Variables	Constant Term
3xy	+3	x,y	—
+1	—	—	1
$-5x^1$	-5	x	—
+y	+1	y	—

7. $-32xy + 1 - x^2$

Term	Coefficient	Variables	Constant Term
-32xy	-32	x,y	—
+1	—	—	1
x^2	-1	x	—

8. $5 - 6ab + a - 1$

Term	Coefficient	Variables	Constant Term
5	—	—	+5
-6ab	-6	a,b	—
+a	+1	a	—
-1	—	—	-1

9. $15 - -2x + 3y^2$

Term	Coefficient	Variables	Constant Term
15	—	—	+15
-2x	-1,-2	x	—
$+3y^2$	+3	y	—

10. $-4x + -2y - -x^2 + 1$

Term	Coefficient	Variables	Constant Term
-4x	-4	x	—
-2y	-2	y	—
$-x^2$	-1,-1	x	—
+1	—	—	+1

11. $6x + 3x^2 - 5xy - 1$

Term	Factors
6x	+6•x
$+3x^2$	+3•x•x
-5xy	-5•x•y
-1	-1

12. $-4ab^2 + 3a^2b - 5ab$

Term	Factors
$-4ab^2$	-4•a•b•b
$+3a^2b$	+3•a•a•b
-5ab	-5•a•b

13. $(4)(5)(-3)x + 1x^2 - x3$

Term	Factors
(4)(5)(-3)x	+4•5•-3•x
$+1x^2$	+1•x•x
$-x^3$	-1•x•x•x

14. $5a^4 - (2)(-1)(-6)ab$

Term	Factors
$5a^4$	5•a•a•a•a
-(2)(-1)(-6)ab	-2•-1•-6•a•b

15. $0 + xyz + -1x - -2y$

Term	Factors
0	0
+xyz	+1•x•y•z
+ -1x	+1•-1•x
- -2y	-1•-2•y

16. $3y - -4x + -1xyz + 8(9)$

Term	Factors
3y	+3•y
- -4x	-1•-4•x
+ -1xyz	+1•-1•x•y•z
+8(9)	+8•9

17. $4^2 + x^2 - x0 + 18(-2)$

Term	Factors
4^2	+4•4
$+x^2$	+x•x
$-x^0$	-1
+18(-2)	+18•-2

18. $a^2b^2 + 4ab - 5^2 + 12(0)$

Term	Factors
a^2b^2	+1•a•a•b•b
+4ab	+4•a•b
-5^2	-5•-5
+12(0)	+12•0

19. $x^0 + y^1 - xy^2 - 4xyz^3$

Term	Factors
x^0	+1
$+y^1$	+1•y
$-xy^2$	-1•x•y•y
$-4xyz^3$	-4•x•y•z•z•z

20. $6^1 + 5x^0 - (5)(3)xy^4 - 0$

Term	Factors
6^1	+6
$+5^0$	+5•1
$-(5)(3)xy^4$	-1•5•3•x•y•y•y•y
-0	0

Addition and subtraction of monomials, page 210
1. 8a
2. 5x
3. 7y
4. 2z
5. -3b
6. 4c
7. 9xy
8. -16ab
9. 6a
10. 3x – 6y
11. -2 – xy – 6y
12. $8x^2 - 6x - 4xy + 11$
13. $-a^2 - 8ab - 4a - 6b + 1$
14. -a + 3b
15. $7x^2 - 7x - 1$

The laws of exponents, pages 212-213
1. 2
2. x
3. -xy
4. 3ab
5. 1
6. 1
7. 1
8. 4b
9. 8192
10. $6x^8$
11. $-x^8y^5$
12. a^5b^5
13. 125
14. $-x^4$

15. $-\dfrac{3x^3y^8}{4}$

16. $\dfrac{a^2b^4}{2}$

17. $\dfrac{1}{x}$

18. $-\dfrac{b^4}{a^2}$

19. $\dfrac{1}{y^6}$

20. $-\dfrac{3y^9}{2x^5}$

21. 512
22. 10,000
23. $4x^2y^2$

24. $\dfrac{3}{a^2}$

25. 4

26. $\dfrac{x^3}{y^3}$

27. $-\dfrac{a^6}{b^6}$

28. $-\dfrac{x}{y}$

29. 6561
30. x^6
31. $64a^{15}$
32. $81x^8y^{12}$
33. 4

34. $\sqrt[3]{x}$

35. $\dfrac{1}{\sqrt{ab}}$

36. $\dfrac{4}{\sqrt{x}}$

37. 4
38. a
39. x^2

40. $\dfrac{5}{\sqrt{y^3}}$

Multiplication and division of monomials, page 214
1. $-10x^5$
2. $-6a^5$
3. $24a^{10}$
4. $-120x^{12}$
5. $44y^2$

6. $-\dfrac{60}{c^2}$

7. $-14x^4y^3$
8. $-15x^7y^4$
9. $-9b^2$

10. $\dfrac{3}{a^3b^3}$

11. $\dfrac{x}{2}$

12. $-\dfrac{3}{x^2}$

13. $-\dfrac{b^3}{10a^{10}c^3}$

14. $-\dfrac{1}{8a^4b^5c^5}$

15. x^2y^8
16. $-4x^6y^4$

17. $\dfrac{a^6}{b^7}$

18. $-\dfrac{x^7}{2y^2}$

19. $-\dfrac{15ac}{4b^{10}}$

20. $\dfrac{x^{22}}{y^{10}}$

Addition and subtraction of polynomials, page 215

1. $2x^2 + 2x + 2$
2. $a^2 - 2a - 7$
3. $x^2 - 6x - 4$
4. $-3xy^2 - 4x - 2y$
5. $-5a^2b - 10ab - 5a + 4b$
6. $-a^2 + 4ab^2 - 4a$
7. $2x + 7$
8. $2x^2 - 12x + 3$
9. $-7a^2 + 15a$
10. $-2x^2 - x - 9$

Multiplication of polynomials, pages 215-216

1. $6x^2 + 6x$
2. $12a^3 - 18a^2 + 6a$
3. $10x^2y^2 - 35x^2y + 15xy$
4. $-2a^3b + 3a^2b^2 + 5a^4b^2 - a^3b^3 - 2a^2b$
5. $3x^3y^3 - 3x^2y^4 + 9xy^3 - 12x^2 - 6x^2y^3$
6. $6x^2 + 7x + 2$
7. $-20a^2 - 22a - 6$
8. $-15x^3 - 9x^2 - 10x + 6$
9. $-20a^2b - 10ab^2 + 12a^2 + 8b^2 + 22ab + 4a + 2b$
10. $12x^4 - 28x^3y + 54x^2y^2 + 18x^3 - 8xy^2 - 48x^2y - 30x^2 + 40y$

Division of polynomials, pages 216-217

1. $2x + 1$
2. $-4x^2 - 2$
3. $3x^2 + x - 4$
4. $-10xy + 2$
5. $\dfrac{-1}{2b} + \dfrac{1}{a} - 2$
6. $\dfrac{-3a}{b} + \dfrac{4b^3c^2}{a^5} - \dfrac{b^6}{3c^4}$
7. $-12x^3 - \dfrac{6x^2}{y^5} - \dfrac{9}{y^2} - 24x - \dfrac{12}{y^5} - \dfrac{18}{x^2y^2}$
8. $\dfrac{x^7}{4y^6} + \dfrac{x^5}{2y^5}$
9. $2x + 4$
10. $3a + 3$

Factoring, page 219

1. $3a(2a + 3)$

2. $6a(2 - a^3)$
3. $3x(x - 1)(x - 2)$
4. $8x(2 - x)$
5. $(a - b)(a + b)$
6. $(2x - y)(2x + y)$
7. $9(2x - y)(2x + y)$
8. $(7a - 9b^2)(7a + 9b^2)$
9. $(a + b)^2$
10. $(x + 3)^2$
11. $(x + 4)^2$
12. $4(a + 2)^2$
13. $5x(x + 1)^2$
14. $2a^3(a + 5)^2$
15. $(a - 2)^2$
16. $(4x - 3)^2$
17. $(a + 3)(a + 5)$
18. $(a - 2)(a - 6)$

19. $\dfrac{x-3}{x + 2}$

20. $\dfrac{a + 1}{4a}$

TEST YOUR SKILLS—CHAPTER 11
Answers to odd-numbered problems, pages 219-221

1.

Term	Coefficient	Variables	Constant Term
$-5x$	-5	x	
$+6y^2$	$+6$	y	
-7			-7

3.

Term	Coefficient	Variables	Constant Term
$+9$			$+9$
$+8\mu$	$+8$	μ	
$-v$	-1	v	

5.

Term	Coefficient	Variables	Constant Term
$-3x^2y^3$	-3	x, y	
-10			-10

7.

Term	Factors
$+5$	
$-(3)(4x)$	$-1 \bullet 3 \bullet 4 \bullet x$
$+2x^2$	$+2 \bullet x \bullet x$
-10	

9.

Term	Factors
+10	
−2(x + y)	−2•(x + y)
+(3x − 2)	+1•(3x − 2)

11. −2a + 3y
13. 0
15. -d − 7
17. -7 − 4c
19. 6xy
21. -15a^2 + 1
23. 2xyz

25. $\dfrac{2}{y^2}$

27. b
29. x^{2}y^{2}z^2
31. 6a + b
33. a + 4c
35. 4x + 4y − 3z
37. 8x + y − 1
39. -c − 3d
41. 8a^{2}b^{4}c
43. 24b^{3}c
45. -8a^3 + 4a^2 + 12a^{2}b
47. -a + 5b
49. 3x^2 + 5xy − 2y^2
51. 6x + 4
53. x^3 − x
55. x − 2
57. 3(x + 2y)
59. (2x + 1)^2

61. $\dfrac{1}{7x}$

63. $1 - \dfrac{5}{x}$

65. $\dfrac{x - 2}{x + 2}$

PRACTICE PROBLEMS—CHAPTER 12
Isolating the unknown using addition/subtraction, pages 225-226

1. x = 4
2. x = 3
3. x = 6
4. y = 13
5. a = 22
6. x = -5
7. b = 8
8. a = -3
9. x = 7
10. a = -15
11. y = -6
12. b = 8
13. x = -2
14. a = -4
15. y = 17
16. x = -12
17. x = 23
18. a = -5
19. y = 33
20. a = 2

Isolating the unknown using multiplication/division, pages 227-228

1. a = 2
2. x = 2
3. b = 2
4. y = 6
5. x = -2
6. y = -4
7. y = 6
8. a = 9
9. a = -4
10. x = -6
11. b = 3
12. y = 5
13. c = -5
14. a = -11
15. y = -1

16. x = $-\dfrac{1}{7}$

17. a = $\dfrac{1}{4}$

18. b = $\dfrac{1}{10}$

19. a = $\dfrac{1}{2}$

20. x = $-\dfrac{1}{3}$

21. x = 2
22. y = 4
23. a = 12
24. b = 24

25. x = $-\dfrac{1}{2}$

26. y = $-\dfrac{1}{3}$

27. a = -1
28. b = -2
29. y = 12
30. x = 16

Isolating the unknown in combination equations, page 230

1. x = -1
2. x = 2
3. y = -4
4. a = -5

5. x = -6
6. a = 3
7. a = -2
8. x = 5
9. y = 4
10. x = 6

Isolating the unknown in fractional equations, pages 231-232
  1. x = 1

  2. $a = -\dfrac{1}{5}$

  3. y = -6
  4. b = 4
  5. a = 12
  6. y = -30
  7. b = 2
  8. b = -15
  9. x = -17.33
  10. x = -4

Isolating the unknown in decimal equations, page 233
  1. x = -1
  2. a = 3
  3. y = -3
  4. b = 6
  5. c = 1.22

Isolating the unknown in equations with powers and roots, page 234
  1. a = 3
  2. y = 8
  3. x = 5
  4. a = 6
  5. x = 2
  6. b = 3
  7. y = 49
  8. a = 81
  9. x = 25
  10. y = 24

Inequalities, page 235
  1. x > 3
  2. a < 6
  3. y > -12
  4. b < -17
  5. a ≤ -9
  6. x ≥ -6
  7. b ≤ 3
  8. y ≥ 8
  9. x > -14
  10. a < -7
  11. y < 5
  12. b > 7
  13. a > 25
  14. x < 12

  15. b ≥ 3
  16. y ≥ 4
  17. x ≥ 10
  18. x ≤ 6
  19. x > 2
  20. a < 7

Equations with two unknowns/systems of equations, addition/subtraction method, pages 239-240
  1. x = 5, y = 4
  2. x = 2, y = -3
  3. x = -6, y = -2
  4. x = 2, y = 2
  5. x = -2, y = -1
  6. x = -3, y = 5
  7. a = 0, b = 9
  8. a = -7, b = 0
  9. x = 4, y = -4
  10. x = -1, y = 1

Equations with two unknowns/systems of equations, substitution method, page 242
  1. x = 1, y = -3
  2. a = 5, b = 5
  3. x = 2, y = -8
  4. x = -6, y = 2
  5. x = -6, y = 17

  6. $x = -\dfrac{1}{3}, \; y = \dfrac{1}{3}$

  7. x = -4, y = -4
  8. x = -2, y = 6
  9. x = 7, y = -3
  10. x = 0, y = 5

TEST YOUR SKILLS—CHAPTER 12
Answers to odd-numbered problems, pages 242-243
  1. x = 5
  3. y = -10
  5. z = 12
  7. a = 2

  9. $d = \dfrac{1}{2}$

  11. x = -10
  13. x = -4

  15. $y = -2\dfrac{1}{2}$

  17. y = -1
  19. a = 7
  21. x = 10
  23. x = 5
  25. y = 100
  27. y < 4
  29. b ≤ 4
  31. x = 2, y = 4

33. x = 2, y = 1
35. x = 1, y = 1
37. x = 1, y = 5
39. x = 4, y = -1

PRACTICE PROBLEMS—CHAPTER 13
Rectangular Coordinate System, page 255

1. x intercept = 2
   slope = undefined (vertical)

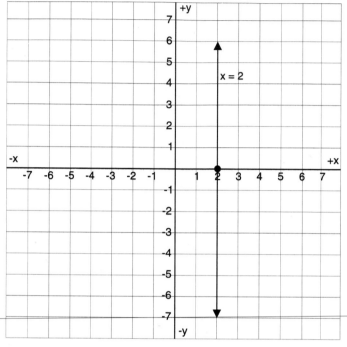

2. x intercept = -6
   slope = undefined (vertical)

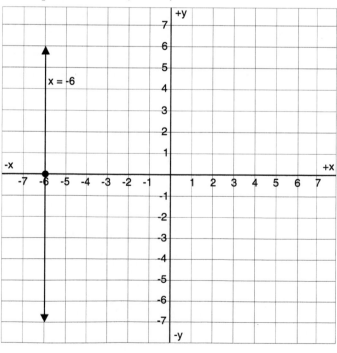

3. y intercept = -2
   slope = 0 (horizontal)

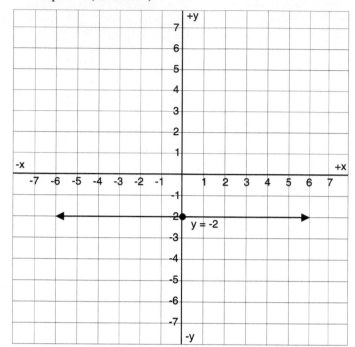

4. y intercept = 4
   slope = 0 (horizontal)

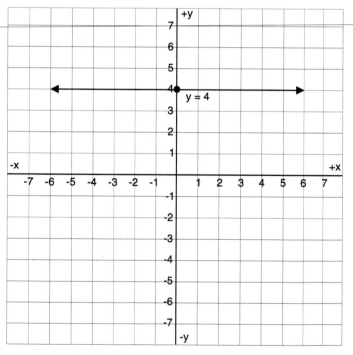

5. x intercept > 3
   slope is undefined (vertical)

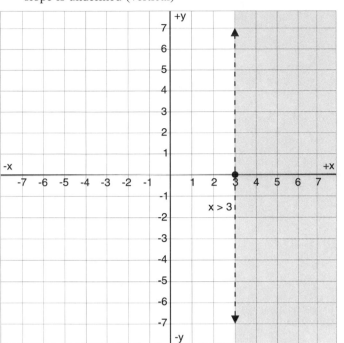

7. y intercept < -4
   slope = 0 (horizontal)

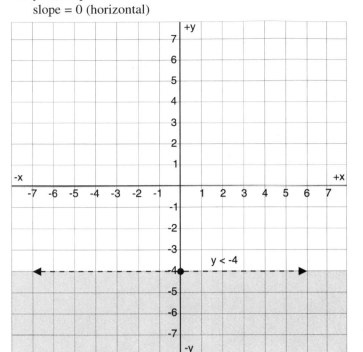

6. x intercept ≤ 2
   slope is undefined (vertical)

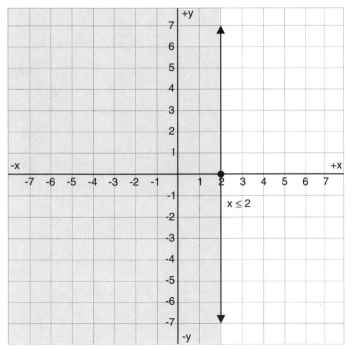

8. y intercept ≥ -1
   slope = 0 (horizontal)

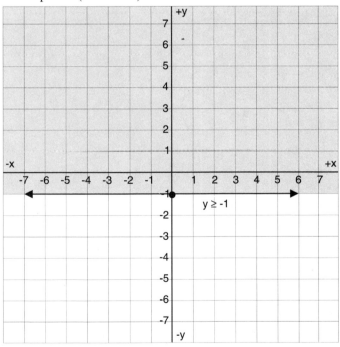

9. x intercept > 0
    slope is undefined (vertical)

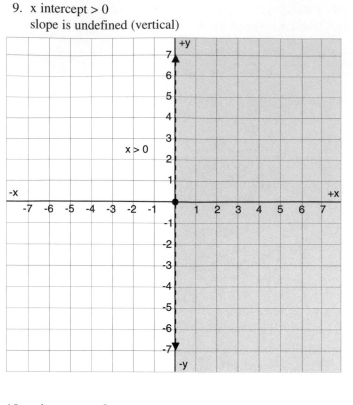

Graphs of two-variable equations, page 255

11. x = 0, y = 0
     x = 4, y = -4
     slope = -1

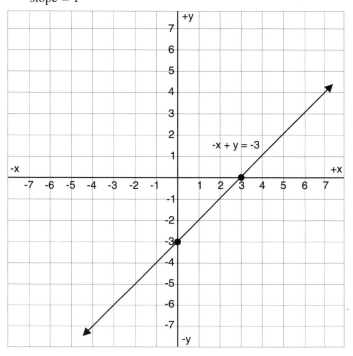

10. y intercept < 0
     slope = 0 (horizontal)

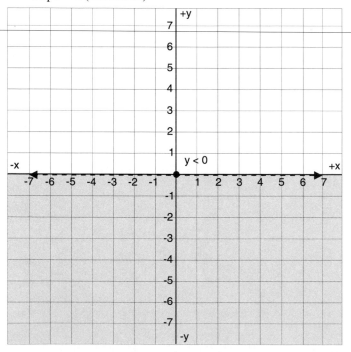

12. x = 0, y = -3
     y = 0, x = 3
     slope = 1

13. $x = 0$, $y = 4\frac{1}{3}$

   $y = 0$, $x = -6\frac{1}{2}$

   slope $= \frac{2}{3}$

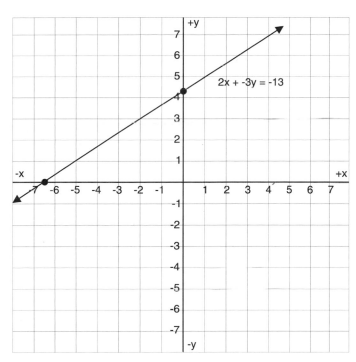

2x + -3y = -13

14. $x = 0$, $y = -6$
   $y = 0$, $x = 3$
   slope $= 2$

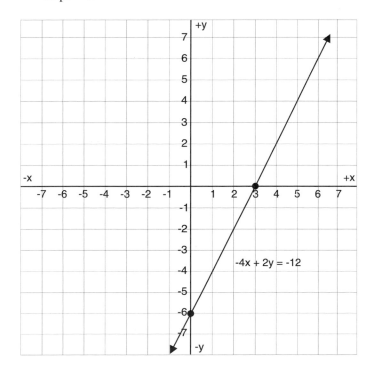

-4x + 2y = -12

15. $x = 0$, $y = 5$
   $y = 0$, $x = -10$

   slope $= \frac{1}{2}$

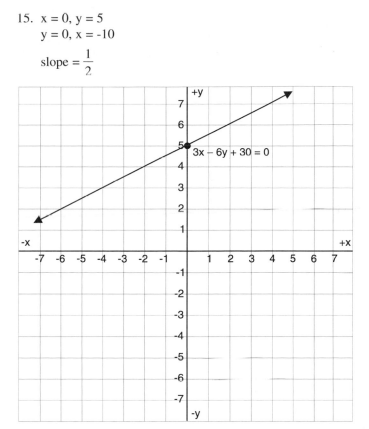

3x − 6y + 30 = 0

16. $x = 0$, $y = -1\frac{1}{2}$

   $y = 0$, $x = 3$

   slope $= \frac{1}{2}$

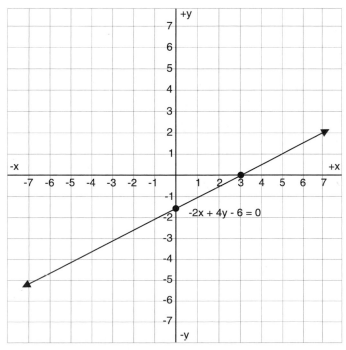

-2x + 4y - 6 = 0

17. $x = 0, y < -8$

$y = 0, x < -1\frac{3}{5}$

slope = -5

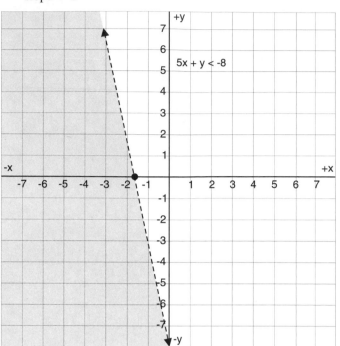

19. $x = 0, y < 5$

$y = 0, x < 2\frac{1}{2}$

slope = -2

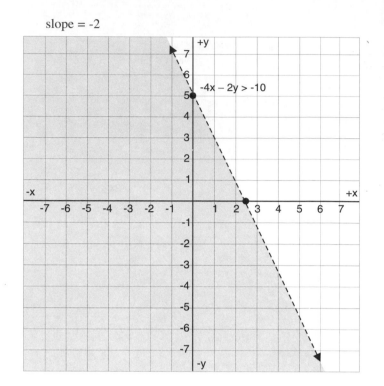

18. $x = 0, y < \frac{2}{3}$

$y = 0, x > -\frac{2}{7}$

slope = $2\frac{1}{3}$

20. $x = 0, y < -1\frac{4}{5}$

$y = 0, x > 9$

slope = $\frac{1}{5}$

Graphic solution of simultaneous linear equations, page 260

1.

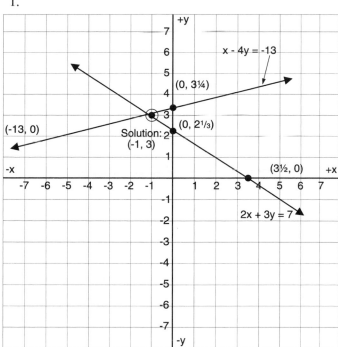

2.

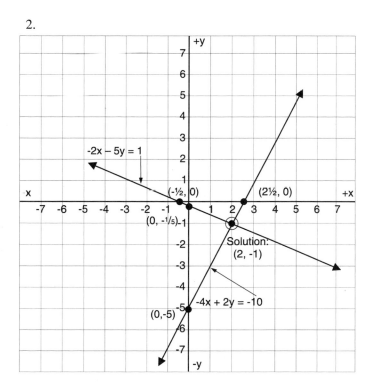

3.

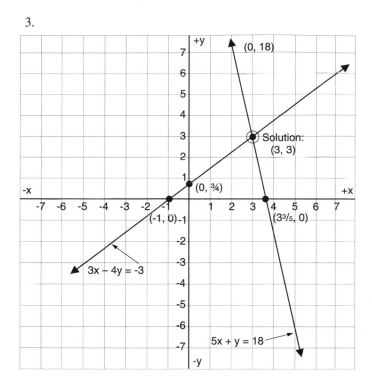

4.

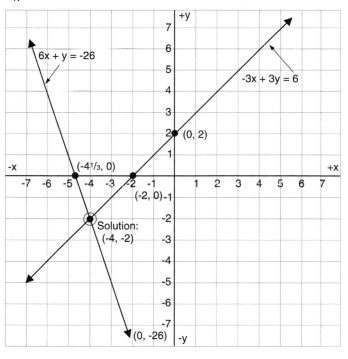

5.

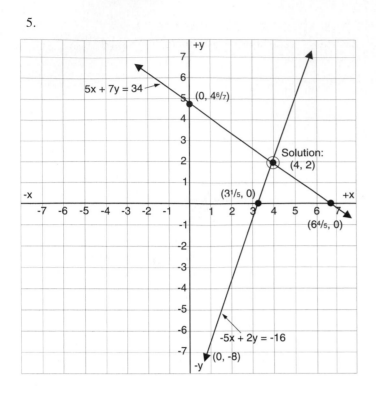

5x + 7y = 34

(0, 4⁶/₇)

Solution: (4, 2)

(3¹/₅, 0)

(6⁴/₅, 0)

-5x + 2y = -16

(0, -8)

7.

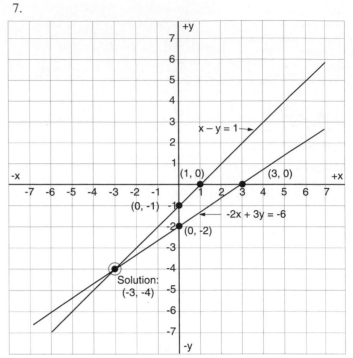

x − y = 1

(1, 0)   (3, 0)

(0, -1)

-2x + 3y = -6

(0, -2)

Solution: (-3, -4)

6.

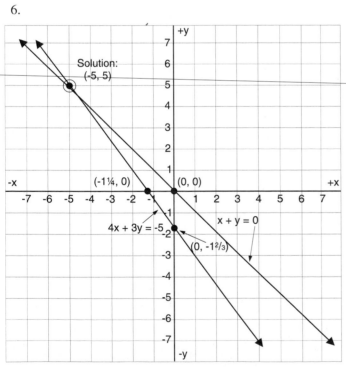

Solution: (-5, 5)

(-1¼, 0)   (0, 0)

4x + 3y = -5

x + y = 0

(0, -1²/₃)

8.

7x + 2y = 1

(0, ½)

-3 − 2y = 3

(-1, 0)

(¹/₇, 0)

(0, -1½)

Solution: (1, -3)

9.

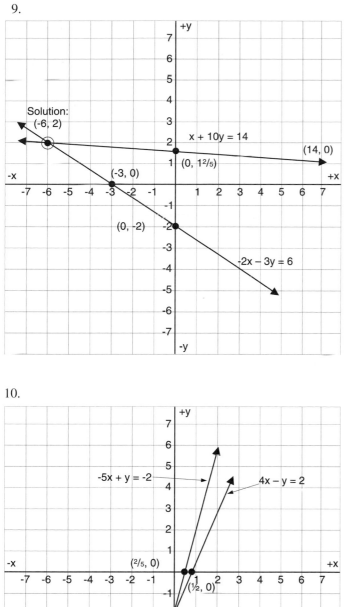

Solution: (-6, 2)

x + 10y = 14

(14, 0)

(0, 1²/₅)

(-3, 0)

(0, -2)

-2x - 3y = 6

10.

-5x + y = -2

4x - y = 2

(²/₅, 0)

(½, 0)

Solution: (0, -2)

TEST YOUR SKILLS—CHAPTER 13

Answers to odd-numbered problems, pages 260-261

1. a. (2,1) b. (3,5)
   c. (-3, 4) d. (-5, 2)
   e. (-4, -2) f. (-6, -4)
   g. (-4, -6) h. (2, -6)
   i. (3, -4) j. (6, -3)
   k. (0, -2) l. (5, 0)

3. $(1\frac{1}{2}, 4\frac{1}{2})$

Solution: (1½, 4½)

x − y = 6

y = x + 3

Proof: choose (0, 0)

4x + 2y < 10

PRACTICE PROBLEMS—CHAPTER 14
Changing word expressions into algebraic expressions, pages 267-268

Note: * indicates given number.

1.

Husband's Age (years)	Wife's Age (years)
28*	28 – 5
31 + 5	31*
43*	3 – 5
42 + 5	42*
62*	62 – 5

2.

Meal Cost (dollars and cents)	Tip Amount (dollars and cents)
15.00*	15.00 × 0.10
1.20 ÷ 0.10	1.20*
25.00*	25.00 × 0.10
2.50 ÷ 0.10	2.50*
9.00*	9.00 × 0.10

3.

Board Length (inches)	Each Piece (inches)
96*	96 ÷ 5
36 × 5	36*
168*	168 ÷ 5
24 × 5	24*
120*	120 ÷ 5

4.

Pacific Time Zone	Eastern Time Zone
9:30 a.m.*	9:30 + 3
4:00 – 3	4:00 a.m.*
11:00 p.m.*	11:00 + 3
2:15 – 3	2:15 p.m.*
5:25 a.m *	5:25 + 3

5.

Gallons	Quarts
3*	3 × 4
12 ÷ 4	12*
0.5*	0.5 × 4
18 ÷ 4	18*
6 *	6 × 4

TEST YOUR SKILLS—CHAPTER 14
Answers to odd-numbered problems, pages 279-281

1. a. 2
   b. 9
   c. $\dfrac{x}{3}$

   d. 3(y + 4)
   e. $\dfrac{4z}{3}$
3. a. 31
   b. 13
   c. 38 – x
   d. 38 – 2y
   e. 36 – z
5. side a = 17 in.
   side b = 15 in.
   side c = 5 in.
7. roofing nails at start = 12 lb.
   drywall screws at start = 6 lb.
9. # of pennies = 24
   # of nickels = 6
   # of dimes = 18
11. speed of car = 75 mph
13. number = 6
15. 1st integer = 5
    2nd integer = 7
    3rd integer = 9
17. amount of red = 14 gal.
    amount of white = 56 gal.
19. water to be added = 5 cups
21. insecticide to be added = 7 gal.
23. time working together = 9 hr.
25. height = 9 in.
27. height of flagpole = 45 ft.
29. number defective = 42

PRACTICE PROBLEMS—CHAPTER 15
Solving right triangles, pages 293-294

Note: * indicates given values.

	Angles		Sides		
	A	B	a	b	c
1.	35°*	55°	6'*	8.6'	10.5'
2.	40°*	50°	10.1"	12"*	15.7"
3.	60°*	30°	4.3 cm	2.5 cm	5"*
4.	70°	20°*	10 m*	3.6 m	10.6 m
5.	45°	45°*	15 yd.	15 yd.*	21.2 yd.
6.	15°	75°*	2.1"	7.7"	8"*
7.	37.9°	52.1°	7'*	9'*	11.4'
8.	44.4°	45.6°	14"*	14.3"	20"*
9.	45.6°	44.4°	35.7 cm	35 cm*	50 cm*
10.	45°*	45°	42.4"	42.4"	60"*

Note: * indicates given values.
Oblique Triangle A

	Angles			Sides		
	A	B	C	x	y	z
1.	25°*	105°*	50°	4'*	2.2'	5'
2.	40°*	80°	60°	21.6"	16"*	24.5"
3.	20°	135°*	25°*	15 cm*	12.1 cm	25.1 cm
4.	60°	75°*	45°*	20.4'	25'*	27.9'
5.	45°*	75°	60°*	18.4 yd.	15 yd.*	20.5 yd.

Oblique Triangle B

	Angles			Sides		
	A	B	C	x	y	z
6.	40°	35°*	105°*	21"*	12.5"	14"
7.	20°*	28°	132°	8'*	5'	3.7'
8.	30°*	35°	115°*	36.3"	22.9"	20"*
9.	35°*	45°	100°*	60 cm	43 cm*	35 cm*
10.	10°	60°*	110°*	324"	298'	60"*

Solving oblique triangles, page 301
Note: * indicates given values.

	Angles			Sides		
	A	B	C	a	b	c
1.	30°*	40°*	110°	6'*	7.7'	11.3'
2.	100°*	40°	40°	12"*	7.8'	7.8"*
3.	50°	105°*	25°*	7.9 cm	10 cm*	4.4 cm
4.	37.1°	75°*	67.9°	25'*	40'*	38.4'
5.	106.9°	28.1°	45°*	60.9 yd.	30 yd.*	45 yd.*
6.	37.7°	125.4°	16.9°	21"*	28"*	10"
7.	65°	35°	80°*	8'*	5'*	8.7'
8.	46.9°	30°*	103.1°	18"*	12.3"	24"*
9.	40°*	21.5°	118.5°	44 cm	25 cm*	60 cm*
10.	41.4°	55.8°	82.8°	40"*	50"*	60"*

TEST YOUR SKILLS—CHAPTER 15
Answers to odd-numbered problems, pages 303-311
1. equilateral
3. scalene
5. length of roof rafter = 13.4 ft.
7. length of guywire = 16 ft.
   height of antenna = 13.9 ft.
   B = 30°
9. travel distance = 424.3 yd.
   width of river = 300 yd.
   B = 45°
11. length of ladder = 21.5 ft.
    A = 68.2°
    B = 21.8°
13. length of stairway stringer = 14.4 ft.
    A = 33.7°
    B = 56.3°

15. length of cross bridging = 18.2 in.
    A = 52.8°
    B = 37.2°
17. height of wall = 34.6 ft.
    length of base = 20 ft.
    B = 30°
19. length of run = 11.3 ft.
    height of rise = 11.3 ft.
    B = 45°
21. B = 40°
    length of westerly boundary = 272.25 ft.
    length of northerly boundary = 511.67 ft.
23. B = 60°
    C = 60°
    C = 6 in.
25. length of porch roof = 10.98 ft.
    A = 35.3°
    B = 29.7°
27. A = 89.71°
    B = 89.71°
    C = 0.58°

PRACTICE PROBLEMS—CHAPTER 16
Scientific notation, pages 320-321
1. 6.8×10^8
2. 2.1×10^3
3. 1.57×10^4
4. 7.5×10^7
5. 1.25×10^{-4}
6. 1.8×10^{-2}
7. 2.5×10^{-1}
8. 3.4×10^{-5}
9. 6.0×10^{-3}
10. 2.5×10^1

11. 5.0×10^5
12. 1.2×10^8
13. 8.3×10^6
14. 2.1×10^7
15. 2.5×10^2
16. 3.04×10^{-1}
17. 7.5×10^0
18. 1.7×10^0
19. 6.7×10^{-2}
20. 5.6×10^{-3}
21. 1.4×10^2
22. 1.25×10^1
23. 1.0×10^{-5}
24. 4.6×10^{-7}
25. 2.03×10^{-8}
26. 2.5×10^{-8}

27. 35,000
28. 2,800,000
29. 0.0501
30. 0.001203
31. 120,000

32. 4.3
33. 0.000035
34. 0.00000000068
35. 1200
36. 500

37. 3.5 E05
38. 1.2 E04
39. 5.6 E03
40. 5.0 E01
41. 3.5 E-03
42. 2.9 E-01
43. 4.9 E-05
44. 1.6 E-04
45. 2.5 E08
46. 1.2 E-01

47. 5700
48. 820,000
49. 0.048
50. 0.00012
51. 6,300,000
52. 12,000
53. 250
54. 0.001
55. 4.2

Engineering notation, page 322
1. 1.5 kW
2. 250,000 mV
3. 12.5 ms
4. 5000 µA
5. 0.120 MHz
6. 20,000 kV
7. 0.025 V
8. 0.53 ms
9. 0.00005 A
10. 4.0 µW

Arithmetic with powers of 10, pages 325-326
1. 9.71×10^4
2. 5.4×10^{-3}
3. 6.087 E4
4. 1.28 E0
5. 79 kilo or 0.079 mega
6. 18 micro or 0.018 milli
7. 7.18×10^4
8. 4.4×10^{-3}
9. 1.02 E6
10. 1.5 E-4
11. 4.0 mega or 4000 kilo
12. 2.0 milli or 2000 micro
13. 1.2×10^7
14. 3.0×10^2
15. 1.4 E-7
16. 5.0 E-1
17. 300 (no exponents)
18. .06 (no exponents)
19. 3.0×10^{-2}
20. 2.0×10^{-9}
21. 8.0×10^{10}
22. 2.0 E2
23. 2.0 E0
24. 2.5 E3
25. 5.0 milli
26. 2.5 micro
27. 5.0 mega
28. 2.0×10^{-5}
29. 4.0×10^3
30. 3.0 E9

TEST YOUR SKILLS—CHAPTER 16
Answers to odd-numbered problems, pages 326-329
1. 8.9×10^5
3. 3.25×10^2
5. 1.25×10^{-2}
7. 3.56×10^1
9. $1.38 \times 10^\circ$
11. 8.9×10^7
13. 1.25×10^4
15. 6.75×10^3
17. 2.58×10^{-6}
19. 6.01×10^{-6}
21. 7600
23. 25,800,000
25. 0.0607
27. 653
29. 4.25
31. 2.5 E03
33. 1.67 E05
35. 1.0 E01
37. 1.25 E03
39. 6.7 E08
41. 3,560,000
43. 0.00000000067
45. 12.2
47. 0.000807
49. 0.0065
51. 3,500 kV, 3.5 MV
53. 0.004 kHz, 4000 mHz
55. 9 µF, 9000 nF
57. 108,000 µV, 0.108 V
59. 13 µs, 0.000013 s
61. 7.57×10^3
63. 1.93 E04
65. 1.075 milli
67. 5.64×10^0
69. 2.97 E−05
71. 87.8 mega
73. 9.36×10^7
75. 3.9 E01
77. 1.403 giga
79. 4.0×10^0
81. 3.1 E-06

83. 12
85. 1.5×10^{-4} sec.
87. 10.0011 ft. pipe
89. 20,000 gal. tank
91. 2.753×10^{-1} mi.
93. lunar: 9.324×10^7 mi.
    solar: 9.276×10^7 mi.
95. 1.248×10^{16} electrons

PRACTICE PROBLEMS—CHAPTER 17
Board measure, page 333
1. 7.6 bd. ft.
2. 14 bd. ft.
3. 21.3 bd. ft.
4. 27.8 bd. ft.
5. 16 bd. ft.
6. 2 bd. ft.
7. 4 bd. ft.
8. 33.3 bd. ft.
9. 33.3 bd. ft.
10. 40 bd. ft.
11. 60 ft.
12. 60 ft.
13. 1200 ft.
14. 1440 ft.
15. 560 ft.
16. 450 ft.
17. 120 ft.
18. 180 ft.
19. 408 ft.
20. 700 ft.

Linear measure, page 337
1. 178'2"
2. 180'
3. 23'
4. 29'2"

Framing members, page 339
1. 20
2. 24
3. 30
4. 9
5. 11
6. 14
7. 19
8. 26
9. 7
10. 12

Square measure, page 341
1. 3460 tiles
2. 865 tiles
3. 19 sheets
4. 900 tiles
5. 24,192 tiles

Cubic measure, page 344
1. 1.2 cu. yd.
2. 19.0 cu. yd.
3. 2.3 cu.yd. $115
4. 111 cu. yd.
5. 138.9 cu. yd.

TEST YOUR SKILLS—CHAPTER 17
Answers to odd-numbered problems, pages 344-347
1. a. 6.7 bd. ft.
   b. 10.7 bd. ft.
   c. 3 bd. ft.
   d. 12 bd. ft.
   e. 5 bd. ft.
   f. 10 bd. ft.
   g. 12 bd. ft.
   h. 4 bd. ft.
   i. 9.7 bd. ft.
   j. 426.7 bd. ft.
3. a. 124'2"
   b. 4-10', 3-14', 3-16'
5. 22'4"
7. a. 19
   b. 31
   c. 16
   d. 9
   e. 8
   f. 9
   g. 20
   h. 5
   i. 12
   j. 9
9. 24 tiles
11. a. 24 sheets
    b. 14 sheets
    c. 15 sheets
    d. 27 sheets
13. a. 10 yd.^{3}
    b. $600

Basic geometric formulas

plane figures:

polygon: $p = $ side $a + $ side $b + $ side $c + \ldots$
rectangle: $p = 2l + 2w$; $A = lw$
square: $p = 4s$; $A = s^2$
triangle: $p = $ side $a + $ side $b + $ side c; $A = 1/2 \, (b \times h)$
equilateral triangle: $p = 3s$; $A = 1/2 \, (b \times h)$
circle: $c = 2\pi r = \pi d$; $A = \pi r^2 = \pi d^2/4$
annulus: $A = \pi/4 \, (D^2 - d^2)$

solids:

prism: $V = A_b \times d$; $A_T = A_{\text{face 1}} + A_{\text{face 2}} + A_{\text{face 3}} + \ldots$
rectangular prism: $V = lwh$; $A_T = 2 \, (lw + wh + lh)$
cube: $V = s^3$; $A_T = 6s^2$
triangular prism: $V = 1/2 \, (b \times h \times d)$; $A_T = (p_b \times d) + 2A_b$
right circular cone: $V = 1/3 \, (\pi r^2 h)$; $A_T = \pi rs + \pi r^2$
right circular cylinder: $V = \pi r^2 h$; $A_T = 2\pi rh + 2\pi r^2$
sphere: $V = 4/3 \, (\pi r^3)$; $A_T = 4\pi r^2$

Linear equations

general form: $Ax + By + C = 0$
slope: $m = $ rise/run $= (y - y_,) / (x - x_,)$
point-slope form: $y - y_, = m \, (x - x_,)$
slope-intercept form: $y = mx + b$

Basic trigonometric formulas

Pythagorean theorem: $c^2 = a^2 + b^2$
trig functions: $\sin \theta = y/r = $ length of opposite side/length of hypotenuse
$\qquad \cos \theta = x/r = $ length of adjacent side/length of hypotenuse
$\qquad \tan \theta = y/x = $ length of opposite side/length of adjacent side
law of sines: $a/\sin A = b/\sin B = c/\sin C$
law of cosines: $a^2 = b^2 + c^2 - 2bc \, (\cos A)$
$\qquad b^2 = a^2 + c^2 - 2ac \, (\cos B)$
$\qquad c^2 = a^2 + b^2 - 2ab \, (\cos C)$

Miscellaneous formulas

percentage: portion $= $ base $\times$ rate
annual percentage rate: APR $= $ average annual interest/outstanding principal
Ohm's Law: voltage $= $ current $\times$ resistance
Basic power formula: power $= $ current $\times$ voltage
board measure: board feet $= $ nominal size $\times$ length (in inches)/144
$\qquad$ board feet $= $ nominal size $\times$ length (in feet)/12
cubic measure: volume in cubic yards $= $ length $\times$ width $\times$ depth (all in feet)/27